Teaching About Genocide:
Issues, Approaches, and
Resources

Teaching About Genocide: Issues, Approaches, and Resources

Edited by

Samuel Totten

University of Arkansas, Fayetteville

INFORMATION AGE
PUBLISHING

80 Mason Street • Greenwich, Connecticut 06830 • www.infoagepub.com

Library of Congress Cataloging-in-Publication Data

Teaching about genocide : issues, approaches, and resources / edited by
Samuel Totten.
    p. cm.
Includes bibliographical references.
  ISBN 1-59311-074-X (pbk.) — ISBN 1-59311-075-8 (hardcover)
  1. Genocide—Study and teaching. 2. Genocide—Case studies. I.
Totten, Samuel.
  HV6322.7.T43 2003
  304.6'63'071—dc22

                                                    2003024774

CONTENTS

INTRODUCTION

Samuel Totten

Teaching About Genocide: Issues, Approaches and Resources, which is comprised of essays by some of the most noted scholars working in the field of genocide studies today, addresses a host of issues, including but not limited to the following: the significance of establishing strong rationales prior to teaching about a specific aspect of genocide; a broad overview of the history of genocide; issues surrounding the definition of genocide and ways to address them in the classroom; case histories of major genocides perpetrated in the twentieth century (along with a listing of key antecedents of each); instructional strategies useful for teaching about genocide; a discussion of the human rights/genocide connection; and an examination of the complexities surrounding the issue of the intervention and prevention of genocide, along with the delineation of strategies for addressing the latter in the classroom.

Since the early 1980s, the field of genocide studies has grown immensely. Indeed, what was originally comprised of an extremely small coterie of scholars in the early to mid-1980s, the field has grown exponentially. Activity in genocide studies increased most dramatically throughout the 1990s and early 2000s as an ever-increasing number of scholars in a wide variety of fields gravitated to the field. This was/is due, at least in part, to the yeoman efforts of many of the pioneers in the field. Indeed, many of those who were active in the field in the early 1980s are the ones

Teaching About Genocide: Issues, Approaches, and Resources, vii–xii
Copyright © 2003 by Information Age Publishing
All rights of reproduction in any form reserved.

who eventually established or coestablished the first centers of genocide studies, founded and published the first newsletters on the issue, developed and coedited the first encyclopedia of genocide, founded the first major journal dedicated to genocide studies, and established the Association of Genocide Scholars. While the field has made quantum leaps over the past decade or so, there is, as one might surmise, much work still to be done. This is particularly true in the field of education.

Even though there has been a major influx of interest and activity in the field of genocide studies, it is still true that the subject of genocide (other than the Holocaust) is relatively rarely addressed—other than by a small number of scholars and teachers—in our nation's schools and universities. Granted, there are many more courses on genocide taught at the university level today than there were a decade ago, but the number is still infinitesimal compared to the number taught on the Holocaust. In that regard, the dearth of attention to genocide in our nation's schools, college and universities constitutes what Elliot Eisner, a professor of education at Stanford University, refers to as the "null curriculum." More specifically, it is his thesis that "what schools do not teach may be as important as what they do teach. I argue this position because ignorance is not simply a void, it has important effects on the kinds of options one is able to consider, the alternatives one can examine, and the perspectives with which one can view a situation or problem" (Eisner, 1979, p. 83).

This constitutes a major concern, especially in light of the fact that every decade since 1948 (which is when the United Nations (U.N.) Convention on the Prevention and Punishment of Genocide was initially passed the by the U.N. General Assembly) has witnessed at least one genocide. Equally alarming is the fact that three major genocides, alone, were perpetrated between 1988 and 1995: the Iraqi gassing of the Kurds in northern Iraq in the late 1980s, the 1994 Rwandan genocide, and the genocide perpetrated in the former Yugoslavia in the early to mid-1990s. Additionally, there were other situations in the 1990s that either threatened to be or verged on becoming genocidal, including those in the Sudan, East Timor, and Burundi, to name but three. In light of this, it is not a little disturbing that educators at the high school, college, and university levels allow genocide to continue to constitute the "null curriculum."

For there to even be a hope that potential perpetrators of genocide can be contained and prevented from carrying out their evil acts, there is a critical need for a well-informed and caring populace across the globe. Without it, politicians are likely to continue issuing sound bites about the need to prevent genocide and *then not acting when genocide rears its ugly head*. When issuing such sound bites, politicians and others (including educators and students), more often than not, revert to turning powerful admonitions into nothing more than clichés. More specifically, too many for too long have latched on to repeating such phrases as "Never Again," "Remember!" and "Those who do not remember the past are condemned to repeat it," and then leaving it at that—meaning, that they rarely, if ever, act on the admonitions. Classic examples of this phenomenon are rife. For example,

In 1979 President Jimmy Carter [U.S. President from 1976 through 1980] declared that out of the memory of the Holocaust, "we must forge an unshakable oath with all civilized people that never again will the world stand silent, never again will the world fail to act in time to prevent this terrible crime of genocide." [And yet, from 1975 through 1979, while the Khmer Rouge committed genocide against its own people in Cambodia, the Carter Administration did nothing to attempt to halt it.] Five years later, President Ronald Reagan [U.S. President from 1980 to 1988] declared. "Like you, I say in a forthright voice, 'Never Again!'" [And yet, not five years later (1987-1988), his administration did nothing as Saddam Hussein's troops destroyed thousands of Iraqi Kurdish villages and murdered almost 100,000 Iraqi Kurds, most of whom were not armed and a great number of whom were defenseless women and children.] President George Bush Sr. [U.S. President from 1988 to 1992] joined the chorus in 1991. Speaking "as a World War II veteran, as an American, and now as President of the United States," Bush said his visit to Auschwitz had left him with "the determination, not just to remember, but also to act." [And yet, in 1991 and 1992, while tens of thousands of Bosnians were murdered by Serbs in an effort to "ethnically cleanse" Bosnia, the Bush Administration was mute.] Before becoming president, candidate Bill Clinton chided Bush over Bosnia. "If the horrors of the Holocaust taught us anything," Clinton said, "It is the high cost of remaining silent and paralyzed in the in the face of genocide." Once in office and at the opening of the Holocaust Museum, Clinton [U.S. President from 1992 to 2000] faulted American's inaction during World War II. "Even as our fragmentary awareness of crimes grew into indisputable facts, far too little was done," he said. "We must not permit that to happen again." [And yet, the Clinton Administration did virtually nothing as some 800,000 Tutsis and moderate Hutus were slain in 100 days in Rwanda in 1994, and did nothing again in July 1995 to prevent the Serb slaughter of some 7,000 Muslims in Srebrenica, the largest massacre in Europe since the Holocaust.] (Power, 2002, p. xxi, the information in brackets was added by the author)

Those who care about the human rights of all people(s) across the globe—no matter their color, ethnicity, nationality, politics, gender or religion—have an important role to play in educating the young about genocide. That said, it is a given that to teach about genocide is not easy. First, genocide is a complex topic. To even begin to understand what genocide is (and is not) takes considerable study. Likewise, to even begin to grasp why and how a single case of genocide was perpetrated requires time-consuming study of multiple sources of information. Furthermore, to tackle the issues germane to the intervention and prevention of genocide is to enter the complex world of international law (where terms are carefully—and sometimes maddeningly—parsed) and the world of international politics where such concepts as realpolitik, sovereignty, and political will and how they play out across the globe are able to tax the spirit of even the most committed human rights activists. All that and more, though, must be tackled and understood if there is going to be any hope in the future of preventing genocide from being perpetrated.

If and when educators choose to teach about genocide, what is needed by them, and ultimately by their students, is a clear understanding as to how and why genocide is perpetrated. For that reason, a good amount of space in this book is dedicated to delineating the antecedents germane to key cases of genocide that were perpetrated in the twentieth century. Such insights, it is hoped, will provide students with a clearer and deeper understanding of those situations and decisions (and lack of action) that contributes to and sometimes results in genocide. This will assist students to begin to understand the "whys" behind the "whats," "wheres," "whens," and hows" of specific cases of genocide.

Some may question whether, as is argued here, education truly has an important part to play in preventing genocide. Obviously, I think it does. And in that regard, I am in complete agreement with Ben Whitaker (1985), the author the *Revised and Updated Report on the Question of the Prevention and Punishment of the Crime of Genocide* , when he asserts that

> The results of research [on the causes of genocide] could help form one part of a wide educational programme throughout the world against such aberrations, starting an early age in schools. Without a strong basis of international public support, even the most perfectly redrafted U.N. Convention on Genocide will be of little value. Conventions and good governments can give a lead, but the mobilization of public awareness and vigilance is essential to guard against any recurrence of genocide and other crimes against humanity and human rights.... As a further safeguard, public awareness should be developed internationally to reinforce the individual's responsibility, based on the knowledge that it is illegal to obey a superior order or law that violates human rights. (p. 42)

So, yes, education can make a difference. Individual courses taught by individual instructors are, of course, just a start. But they are a start in the right direction. Ultimately, what is needed, as Whitaker suggests, is a concerted and systematic worldwide effort. Every movement, though, must start somewhere, and it is my hope that each instructor who teaches his/

her students about genocide is, ultimately, paving the way for a larger and more powerful effort along these lines.

It is also my hope as the editor of this book that the essays herein are found to be thought-provoking and useful for those who have the courage and care to tackle such a complex, horrific, and vitally significant issue in the classroom.

REFERENCES

Eisner, E. (1979). *The educational imagination: On the design and evaluation of school programs*. New York: Macmillan.

Power, S. (2002). *"A problem from hell": America and the age of genocide*. New York: Basic Books.

Whitaker, B. (1985). *Revised and updated report on the question of the prevention and punishment of the crime of genocide*. (E/CN.4/Sub.2/1985/6, 2 July 1985)

CHAPTER 1

EDUCATING ABOUT GENOCIDE, YES

But What Kind of Education?

Carol Rittner

"Genocide has shown me the darkest part of humanity,
but I shall strive to be the light that drives back the darkness."

—Student Comment on a Final Exam
The Richard Stockton College of New Jersey, December 2001

When I look back over the last 35 years or so (from 1968 to the present)—
years when colleges, universities, and high schools across the United
States began teaching about the Holocaust; years when states began man-
dating study of the Holocaust and other genocides in their schools; years
when communities across the United States were building museums and
erecting monuments commemorating the victims of the Holocaust; years
when politicians and religious leaders, Jews and Christians alike, teachers,
news commentators, and ordinary people alike were proclaiming "Never
Again!" as a testament to America's determination never to let something
like the Holocaust happen ever again to the Jews, or to anyone else—I
have to ask how it was possible for so many people to find so many rea-

Teaching About Genocide: Issues, Approaches, and Resources, 1–5
Copyright © 2004 by Information Age Publishing
All rights of reproduction in any form reserved.

sons to justify American inaction in the face of genocide committed in the post-Holocaust. As a survivor of the 1994 genocide in Rwanda asked, "Is 'Never Again!' an empty phrase, meant for some people, but not for others?" (quote from the film *Wasted Lives*, Smith, 2000).

United States responses to genocide during the twentieth century (the Armenian genocide, the Holocaust, the genocide in Cambodia following the Vietnam War, Saddam Hussein's gassing of the Kurds in Northern Iraq in the late 1980s, and the Hutu slaughter of the Tutsis during the 1994 genocide in Rwanda) were astonishingly similar across time, geography, ideology, and geopolitical balance, and those responses can be summed up in one word: noninterference. Governments in Europe, Asia, Africa, and the Middle East were killing their own people, but the United States refused either to condemn what was happening or to intervene to stop what was happening. As Samantha Power, author of the Pulitzer prize-winning *"A Problem from Hell": America in the Age of Genocide*, said in February 2003 during a panel presentation at the Kennedy Library in Boston, "for diplomats, maintaining a position of neutrality is paramount in their work." If one is a diplomat, then one is to "be diplomatic." Gentlemen are not to denounce but to negotiate. "The problem with negotiation," she noted, "is that it can be a stalling device for murder, for mass murder. And in many cases ... that's exactly what it became." The question is, when do we cease to rely on negotiation and take more robust measures?

In her book, Power (2002) also writes about how U. S. political leaders have interpreted, and continue to interpret, public silence as public indifference; how the U.S. government usually "takes very few steps along the continuum of intervention to deter genocide" (Power, 2000, p. xviii) despite its rhetoric deploring human rights abuses; and how officials do everything they can to avoid using the word "genocide."

Still, if we are honest, we have to admit that in a democratic society we cannot just blame our leaders, elected or appointed. We cannot just blame "those people in Washington who don't know what they're doing!" "To understand why the United States did not do more to stem genocide, it is not enough ... to focus on the actions of presidents or their foreign policy teams. In a democracy, even an administration disinclined to act can be pressured into doing so. This pressure can come from inside or outside [government]" (Power, 2002, p. 508).

That said, in my view, when we teach about genocide we should not teach only about facts and figures, or about historical context and recent, or age-old, ethnic, religious, or national conflicts, important as all of that may be. Rather, we also must raise ethical questions, explore the human capacity for selfishness as well as the human capacity for compassion. In teaching about genocide, it is imperative to raise questions of a more personal nature, such as, for example, about the human capacity to resist evil. In doing so, we should help students to confront some of the most difficult questions any of us can confront: How do ordinary human beings—people presumably raised to distinguish right from wrong, people with families of their own, people who say they love their parents, spouses, children, friends and relatives—come to participate in the vicious slaughter of powerless men, women, and children? Why does the teaching of good and evil in organized societal and religious institutions fail to prevent genocide? Why do the structures of civil society—education, law, religion, diplomacy—not stop such evil? Where did, where does moral and religious education flounder? Why do moral and religious education fail to create more resistance to evil and not encourage more doing of good?

What kind of teaching, what kind of education can help students grapple with these kinds of questions? What kind of teaching, what kind of education can prepare people in a democratic society to do more than simply know about genocide, but, if and when faced with genocide, empower people to act to stop genocide? Of course I cannot definitively answer these questions, but I can offer some opinions about what I think we need.

What we need is an education that does more than provide students with the skills to work in a postindustrial, postmodern world, important as that may be. We need what the late Bart Giamatti (1988), president of Yale University, called, an education for freedom, one in which there is constant emphasis on the development of questions and on what one might call a "critical imagination"; that is, an emphasis on developing in students the ability to question assumptions, to challenge what is taken for granted, and to approach knowledge and truth as the stuff of human invention. Without such an education, people will simply "follow the herd," do whatever they are told, even murder men, women, and children who are perceived as "different" from themselves. We need an education that addresses why we human beings make wars, destroy lives, brutalize and devalue others, and put national interest above humaneness. We need an education that teaches the common humanness of "the other," an education that stresses the values of caring and that emphasizes personal responsibility and compassion, an education that prepares the individual for doing acts of good.

A society that teaches the common humanness of the other, that stresses the values of caring, and that emphasizes compassion and responsibility prepares the individual for doing acts of good. The ability of legitimate authority to appeal to values of caring, justice and inclusiveness facilitates the doing of good. A society that is not afraid to put the interests of those who are suffering above national self-interest will be the best teacher of the educational values that we need to teach our students so that when U.S. citizens are faced with genocide, we do not retreat into realpolitik, do not say it is "none of our business," do not look away, turn our backs, or "pass by on the other side."

I know that is a lot to ask of any education, but it is vital to draw attention to the importance of teaching about why "ordinary people" become *genocidaires*, as well as to the importance of teaching about why other "ordinary people" risk their careers and social status, in some instances, even their lives, to help their fellow human beings during a genocide. Indeed, it is vital also to ask such questions as "Why do some people help and others harm?" and "Why do some people sit on the sidelines, and others jump into the fray?" Of course we must dissect evil, study its many faces, discover how it can camouflage itself as national interest, prudence, job security, "It's none of my business" or "What can I do? I'm only one person," but we also have to try to understand why other people refuse to find refuge in such excuses in order to justify inaction when others are suffering.

During the past 20 years or more, scholars such as Nechama Tec, Sam and Pearl Oliner, Eva Fogelman, Phillip Hallie, Ervin Staub, David Gushee, and Kristen Monroe have tried to fathom why some non-Jews in Europe risked their lives to help the Jews during Hitler's Third Reich (1933-1945). We need to incorporate their studies and books about prosocial behavior—behavior that makes things better, that contributes to the common good—into our teaching about genocide, not just when we teach about the Holocaust, but when we teach about any genocide. We have to search out and find such moral exemplars, people Yad Vashem (the national memorial and research center in Israel) calls *hasidei ummot ha-olam* (the righteous ones among the nations of the world), and that Samantha Power (2003) calls "upstanders"—people who did not lose sight of right and wrong in a time of genocide, those who refused to believe they could not influence other people, foreign policy, even events in their own or other countries. Indeed, we must incorporate into our study of genocide stories about people who refused to remain silent in a time of genocide, people who stand out because they stood up for what is right.

Of course teaching about genocide must involve establishing rationales for such teaching, must include historical overviews of specific genocides, wrestling with various definitions of genocide, and developing and using effective teaching strategies. Without a doubt teaching about genocide should incorporate comparative studies of genocide and encourage students to grapple with issues of human rights and social responsibility. All of that is important and necessary, but, we also have to confront the tough questions, the questions about ethics, resistance, and responsibility, the questions about personal and political choices, about religious teachings and civil society. We need to teach our students about people who tried to make a difference, who did not give up trying to save the lives of people threatened by genocide. It is imperative that we do so if we want to nurture the kind of people who are prepared to stand up rather than to stand back.

And of course we must show students "the darkest part of humanity," but we also should empower them to "strive to be the light that drives back the darkness." To paraphrase something Holocaust scholar John Roth (1999) wrote about the Holocaust, again and again genocide illustrates how difficult it is for good people to move from knowledge of what the killers are doing to comprehension of the significance of their horrendous atrocities, then to action aimed at thwarting genocidal success. If, however, we can learn and teach something about the nature of people like Raoul Wallenberg, Oskar Schindler, and the people of *Zegota*, the Polish Council for Aid to Jews during World War II and the Holocaust, if we can learn and teach something about the nature of "upstanders" like George Kenny, Marshall Harris, Jon Western, and Steven Walker, American foreign service officers who quit in disgust rather than support American foreign policy that sanctioned nonintervention in the face of genocide in the former Yugoslavia, if we can learn and teach something about the nature of people like Phillipe Gaillard and his Red Cross colleagues who stayed in Rwanda and tried to save lives during the 1994 genocide, rather than fleeing, and if we can learn and teach about what can be done to nurture such moral qualities in the lives of our students— and ourselves—perhaps then teaching about genocide will be well worth the effort we expend.

REFERENCES

Giamatti, A. B. (1988). *A free and ordered space: The real world of the university.* New York: W. W. Norton.

Power, S. (2002). *"A problem from hell": America and the age of genocide.* New York: Basic Books.

Power, S. (2003, February 3). *Bystanders to genocide.* Talk at the Kennedy Library [Transcript], Boston, MA.

Roth, J. K. (Ed.) (1999). *Ethics after the Holocaust: Perspectives, critiques, and responses.* St. Paul, MN: Paragon House.

Smith, S. (Producer). (2000). Wasted lives [Film]. England: Aegis Trust Foundation.

CHAPTER 2

ISSUES OF RATIONALE

Teaching About Genocide

Samuel Totten

INTRODUCTION

When teaching about genocide—whether it is at the university, college, or secondary level of schooling—it is essential to establish a solid set of rationales. Lessons and units of study bereft of controlling principles often lack clearly delineated goals and objectives—and potentially, a sound conceptual and historical focus.

In this chapter, numerous concerns regarding rationale statements are addressed, including but not limited to the following: their purpose, their use in developing goals and objectives, their value in guiding content selection, their influence in selecting pedagogical strategies, their use in helping one avoid pitfalls common to many studies of genocide, and why there is a critical need to revisit rationale statements throughout the study.

THE PURPOSE OF RATIONALE STATEMENTS

Though speaking specifically of the Holocaust, the following statement by the authors of the U.S. Holocaust Memorial Museum's *Guidelines for*

Teaching About Genocide: Issues, Approaches, and Resources, 7–22
Copyright © 2004 by Information Age Publishing

7

Teaching About the Holocaust is germane to those who plan to teach about any aspect of genocide:

> Because the objective of teaching any subject is to engage the intellectual curiosity of the student in order to inspire critical thought and personal growth, it is helpful to structure lesson plans on the Holocaust by considering throughout questions of rationale. Before addressing what and how to teach, one should contemplate the following: Why should students learn this history? What are the most significant lessons students can learn about the Holocaust? Why is a particular reading, image, document, or film an appropriate medium for conveying the lessons about the Holocaust which you wish to teach?" (Parsons & Totten, 1993, p. 1)

Regardless of whether the instructor is experienced or inexperienced in teaching about genocide, questions of rationale should always be considered. Among other questions, one should ask him/herself the following: "Why am I teaching about genocide in the first place?" "Why am I teaching about this particular aspect or case of genocide?" "What are the most essential topics, issues, and questions that need to be addressed when teaching about genocide, and why is that so?"

Expanding on the above, noted genocide scholar Helen Fein (2002) asserts that

> In planning courses, teachers must decide initially not only on definition and boundaries but on which genocides to probe, whether they focus on a disciplinary or interdisciplinary perspective. Whether one selects the Holocaust or another genocide to concentrate on or starts with a comparative approach seems to me a question not only of the intrinsic interest of the case (often enhanced by who one is and what one has gone through [and significantly, the make-up of one's students]) but on approaches to knowledge. One may be a "hedgehog," immersed in knowing all one can about one thing [or wanting students to experience such an approach], or a "fox," wishing to ferret out the essential about many things [or again, desiring one's students to experience such an approach], a distinction first made by the Greek philosopher Archilocus. It is of course, also possible to combine both strategies. (p. 3)

All of the above, of course, directly impacts the development and use of rationale statements.

DEVELOPING RATIONALE STATEMENTS

When developing rationale statements, instructors often find it helpful to ask themselves a series of questions such as the following:

- Why is genocide important to study?
- What do I want my students to walk away with, and why?
- How deeply do we need to go into the complexities of and debates over the definition of genocide, and why?
- Should we focus on a single case of genocide versus a comparative study of two or more genocides—or vice versa?
- Should we focus on distant cases versus more recent cases of genocide?
- How important is it to include information and analyses of issues germane to the intervention and prevention of genocide, including the international community's record of carrying out preventive measures?
- If I can only plant one seed in my students' minds about genocide for them to ponder over the long haul of their lives, what would it be and why?

In regard to conducting a study about *a specific case of genocide*, instructors might posit such questions as:

- Why study this genocide versus another?
- What are the essential antecedents, preconditions, issues, and facts that are key to understanding this particular genocidal event?
- What do I want my students to glean from this study?

On a different but related note, when developing rationale statements one should develop them on both cognitive and affective levels—the mind, knowledge, and thinking as well as the heart, emotions, and feelings. To address one level and not the other is likely to result in an incomplete curriculum, one that is bereft of the essential components that make us human.

If, after designing a series of rationale statements, one discovers that the focus is on the "whats, wheres and whens" of the historical case and is bereft of the "whys," the instructor must reconsider the goals, objectives and content for the unit of study. Instructors who focus primarily on the whats, whens, and hows of a genocide but ignore the whys leave students bereft of essential knowledge about the historical trends that ultimately culminated in the genocide they have studied.

Finally, if an instructor discovers that the study is comprised solely of a series of concepts and facts but neglects to address the importance the study has for both the individual and society, he/she should consider adding components that address the latter.

INVOLVING STUDENTS IN THE
DEVELOPMENT OF RATIONALES

It needs to be noted that a good place to begin with the development of rationale statements for a study of genocide is with the students themselves. This immediately encourages them to begin thinking about why one would want or need to study such a topic; and in doing so, it begins to personalize the study for them. It may also motivate them to become more engaged in the study and to begin to see the relevance it has for their own lives as well as the society, country and world in which they live.

Initially, the instructor can ask the students to think of reasons why people should study genocide. After the students record their responses in journals or on a sheet of paper, the teacher can record a variety of the responses on the board or overhead projector screen. Later, the responses can be transferred to a bulletin board where they could remain during the course of the study in order to draw attention to them and/or comment on them.

An even more engaging way to get the students involved in the development of rationale statements is to have them write down those questions they have about the specific case(s) of genocide under study and/or what they wish to have answered during the course of the study. Such questions, in conjunction with the teacher's rationales, could set the stage for the study.

EXAMPLES OF RATIONALE STATEMENTS FOR
TEACHING GENOCIDE

The rationales that various instructors have generated in regard to the question "Why teach about genocide?" are numerous and eclectic. Among some of the more thought-provoking examples are the following:

- to attempt to ascertain why genocide is perpetrated;
- to teach students why, how, what, when and where a specific genocide took place (e.g., the Ottoman Turk genocide of the Armenians, the German genocide of the Jews, the Khmer Rogue genocide of its perceived class enemies and others, the Hutu genocide of the Tutis and moderate Hutus in Rwanda 1994), including the key historical trends/antecedents and preconditions that led up to and culminated in the genocide;
- to examine the incremental nature of prejudice, stereotyping, racism, discrimination and how such can, in certain situations, lead up to and result in genocide;

- to compare and contrast various genocides in order to ascertain the similarities and differences between and amongst genocides (including various types of genocide, various types of preconditions, and the actions and reactions or lack thereof by the perpetrators, victims, collaborators, bystanders , etc);
- to become "cognizant that 'little' prejudices can easily be transformed into far more serious ones" (Lipstadt, 1995, p. 29);
- "to illustrate the effects of peer pressure, individual responsibility, and the process of decision making under the most extreme conditions" (Schwartz, 1990, p. 101);
- to "make students more sensitive to ethnic and religious hatred" (Lipstadt, 1995, p. 29);
- to develop in students an awareness of the value of pluralism and diversity in a pluralistic society;
- "to develop a deeper appreciation of the relationship of rights and duties, and to realize that human rights and the corresponding duties they entail are not the birthright of the few but the birthright of all—every man, woman, and child in the world today" (Branson & Torney-Purta, 1982, p. 5);
- to "become sensitized to inhumanity and suffering whenever they occur" (Fleischner as cited in Strom & Parsons, 1982, p. 6);
- to provide a context for exploring the dangers of remaining silent, apathetic, and indifferent in the face of others' oppression;
- "to teach civic virtue … [which is related to] the importance of responsible citizenship and mature iconoclasm" (Friedlander, 1979, pp. 532-533);
- to understand that genocide is *not* an accident in history; that it is not inevitable (a paraphrase from Parsons & Totten, 1993, p. 1);
- to come to understand the complexities involved in attempting to intervene in or prevent genocide;
- to gain insights into how international law has evolved over the past century—and particularly the last half century—in regard to the crime of genocide; and
- to gain an understanding as to why the United Nations (U.N.) and individual states, more often than not, have chosen not to intervene and prevent genocide.

Instructors—in conjunction with, if they wish, their students—need to decide what the focus of the study will be, and then develop appropriate rationale statements. Just as each class of students is unique, each study of genocide, by necessity, will be unique. To a certain extent this is likely to be dictated by the instructor's background, the instructor's knowledge base, what the instructor perceives as being of the utmost significance vis-

à-vis the history, the levels and the abilities of the students, the time allotted for the study, the type and amount of resources available, and/or a
combination of these concerns.

Prior to proceeding to the next section, a caveat is in order. Some educators (particularly curriculum developers) have a tendency to borrow
rationales from others and then use them in place of developing their
own rationale statements or goals. To use other educators' rationale statements in this fashion is the same as never designing a rationale statement
in the first place.

THE NEED FOR CAREFUL USE OF
LANGUAGE IN RATIONALES

The language used in creating the rationale(s) is critical. Of the utmost
importance is that rationales not constitute "comparisons of pain." That
is, one should not assert that "This [specific case of genocide] is the most
horrific example of genocide in the history of humanity." To make such
an assertion minimizes the horror and suffering experienced by humans
in other genocides. Terms such as *unimaginable* or *unbelievable* are often
used when speaking about various genocides, but these terms may send a
message to students that the genocide perpetrated was so "unreal" that it
is pointless to try and learn about what happened. The simple but profound fact is that many genocides have been perpetrated, and all have
been systematically planned and implemented by human beings. So, they
were imaginable. Furthermore, the facts about them are real and thus
believable. As Lawrence Langer (1978) more eloquently notes:

> What does it mean to say that an event is beyond the imagination? It was not
> beyond the imagination of the men who authorized it; or those who exe-
> cuted it; or those who suffered it. Once an event occurs, can it any longer be
> said to be "beyond the imagination?" Inaccessible, yes; … contrary to "all
> those human values on which art is traditionally based," of course. What we
> confront is not the unimaginable, but the intolerable, a condition of exist-
> ence that so diminishes our own humanity that we prefer to assign it to an
> alien realm. (p. 5)

One should also avoid the use of clichés in rationale statements. Too
often teachers and curriculum developers have mindlessly latched on to
such phrases as "never again," "always remember," "never forget," and
"those who do not remember history are condemned to repeat it," without giving ample thought and consideration as to their meaning. Spoken
by survivors, these admonitions *are* powerful *and* meaningful. That is not
true, though, when they are spoken by politicians, after-dinner speakers,

conference participants, and teachers *who neglect to acknowledge the fact that genocide has been perpetrated time and again since 1945*. To use such phrases simply because they sound good is problematic at best. (For a more detailed discussion of this issue, see Totten, 2002, "Teaching the Holocaust: The Imperative to Avoid Clichés.")

FACTORS INFLUENCING THE FOCUS OF RATIONALE STATEMENTS

Among the most important factors influencing the focus of rationale statements are: (1) one's aims in teaching the concept and/or history of genocide; (2) one's knowledge of the history of a specific genocide; (3) the particular course one is teaching; (4) the levels and abilities of one's students; (5) the available time for study; and (6) the instructional resources available.

Obviously, the major factor influencing the focus of one's rationale statements is one's aim in teaching this subject. Is it to provide a deep understanding as to why genocide is perpetrated? An overview of the history of genocide? An examination of a specific genocide? A comparative study of two or more genocides? An examination of the complexities of the intervention and prevention of genocide? To focus on both the incremental nature of genocide and how and why, in various situations, the latter results in genocide? To focus on the complexity and danger of the bystander syndrome? To focus on the literature (fiction, poetry, drama) of a specific genocidal act? To teach lessons about living in a world where genocide is perpetrated on a regular basis? The focus will, more or less, dictate the readings and films, the class activities and assignments, and the final assessment one uses during the course of the study—and if they do not, they should.

In an article entitled "Reflections on Studying Genocide for Three Decades," Fein (2002) delineates potential "bases" to consider when developing a study of genocide—all of which, ideally and ultimately—should impact the development of rationale statements: "1. Historical perspective on genocide from ancient times to modern...; 2. Totalitarianism: genocide, terror and famine in Fascist and communist States...; 3. Total genocide in the 20th century (e.g., a. that of the Hereros of South-West Africa under German rule (1904-1917); b. the Armenians in the Ottoman Empire (1915-1918); c. Jews and Roma by Nazi Germany (1941-1945); and d. the Tutsi in Rwanda in 1994 ...); 4. Genocide of indigenous peoples in the 18th and 19th century and in contemporary (developed and underdeveloped) states; 5. Differentiation by time (e.g., contemporary genocides since 1945 ...); and 6. Differentiation by region

(e.g., genocide and ethnic cleansing in Asia or Africa since decolonization—such as Uganda, Rwanda and Sudan)" (pp. 3-4).

A host of other possibilities also exist in regard to the potential focus of
a study. Among some of the many, for example, are:

1. The genesis and evolution of the development and ratification of
   the U.N. Convention on the Prevention and Punishment of Genocide;
2. Typologies of genocide;
3. The complexity of intervention and prevention of genocide;
4. International law and the intervention and prevention of genocide;
5. The perpetration of genocide and the issue of impunity;
6. The impact, thus far, of the International Criminal Tribunal of the
   Former Yugoslavia and the International Criminal Tribunal for
   Rwanda;
7. The battle over the establishment of the International Court (ICC);
   and
8. The potential impact of the ICC.

Those teaching advanced students need to focus, at least some time, on
the issue of typologies. That said, as Fein (1990) notes: "General explanations are difficult because of the diversity of situations and motives for
genocide yet the quest for general explanations persist" (pp. 30-31).
While it is true that there is not a total consensus among scholars on
which label (or type) best describes certain genocides, there is, as Fein
(1990) points out, enough agreement between scholars to merit the use of
such typologies and to introduce students to them. Among some of the
many typologies that scholars have created to explain specific types of
genocide over the years are as follows: *ideological* (e.g., the Ottoman genocide of the Armenians (1915-1918); the Holocaust (1933-1945); the Indonesian genocide of communists and suspected communists (1965-1966);
and the Khmer Rouge-perpetrated genocide of its fellow Cambodians
(1975-1978)); *retributive* (e.g., Burundi, 1972; East Timor, mid-1970s;
Rwanda, 1994); *despotic* (e.g., those committed by the Soviet Union such
as the manmade famine in Ukraine in 1932-1933 and the genocides of
various minority groups in the Soviet Union in the 1940s); and *developmental* (e.g., the extermination of the Ache in Paraguay) (Fein, 1990, pp.
86, 88-89). Some genocides may fall under dual categories/typologies. For
example, Fein (1990) argues that "The genocide in Burundi was not only
a retributive reaction to the threat of a Hutu coup in 1965 and 1972, it
could [also] be viewed as a pre-emptive slaughter" (p. 89). More specifically, "the selective massacre of educated Hutus in 1972-73 served to
reduce Hutus' ability to challenge the Tutsi" (p. 89).

Even those who have a working knowledge of the theories of genocide, typologies of genocide, and/or the specifics regarding a particular genocidal act (including the antecedents and preconditions) need to seriously consider what their goals are for such a study. Doing so will generally result in a more focused and stronger study.

The particular course one teaches will also dictate the focus of one's rationale statements. For example, the rationale statement in a history course will differ from one in a literature course because the focus of each is bound to be fundamentally different. That said, there is bound to be overlap in the content of the rationale statements. For example, the study of human behavior could very well be the focus of both a historical and literary rationale statement.

In developing a course on genocide, it is vital to take into consideration the levels, abilities, and backgrounds of one's students. The factors inherent in these concerns will dictate the sophistication and depth of the study, the resources used, and the pedagogical strategies and learning activities employed. To neglect these issues is pedagogically unsound. Further, it is an invitation to a study that will likely be too sophisticated or, conversely, simplistic.

Time is always a critical factor in the classroom. Most instructors are limited in how much time they can allot to such a subject. This results in the need to make some difficult decisions in regard to what one can and cannot include in such a study. An in-depth study of a topic is generally preferable to superficial coverage of a plethora of topics. As Newmann (1988) states, "less in this context does not mean less knowledge or information, for depth can be achieved only through the mastery of considerable information. Rather, less refers to less mastery of information that provides only a superficial acquaintance with a topic" (p. 347). Continuing, Newmann (1988) argues that

> we usually try to teach too much.... We are addicted to coverage. This addiction seems endemic in high school, especially in history.... Beyond simply wasting time or failing to impart knowledge of lasting value, superficial coverage has a more insidious consequence: it reinforces habits of mindlessness. Classrooms become places in which material must be learned—even though students find it nonsensical because their teachers have no time to explain. Students are denied opportunities to explore related areas that arouse their curiosity for fear of straying too far from the official list of topics to be covered.... The alternative to coverage, though difficult to achieve, is depth: the sustained study of a given topic that leads students beyond superficial exposure to rich, complex understanding. (p. 346)

Given that the range of available instructional resources will influence classroom practices, the development of statements of rationale should coincide with a determined effort to expand available resources to encompass the goals of the course. While this is time-consuming and costly, over time a comprehensive set of instructional resources, including books, first-person accounts, primary documents, on-line resources, computer software, audio-visual materials, fiction and poetry, and a list of guest speakers, can be developed to support curricular rationales and programs.

THE USE OF RATIONALE STATEMENTS TO DEVELOP GOALS AND OBJECTIVES FOR THE STUDY

Limited value exists in developing rationale statements if they do not influence the actual goals and objectives of the course. Together, rationale statements, goals and objectives will assist one in developing the content, the pedagogical strategies, the resources, and the learning activities to be employed.

Again, learning objectives should take into consideration the developmental level of prospective students in the course, address all levels of Bloom's taxonomy or a comparable taxonomy of thinking or cognitive operations, include both content and process concerns, and provide opportunities for extension of learning beyond the classroom so that other "communities" experience the benefits of what students have learned in the course.

RATIONALES SHOULD DIRECT THE CONTENT USED IN THE STUDY

In addition to the question, "Why teach about genocide (or this specific aspect or case of genocide)?", there are several other key questions that teachers must ask themselves prior to engaging their students in such a study: "What are the most important lessons we want our students to learn from this study?"; and, as the Coalition of Essential School's personnel (1989) continually ask in the course of their pedagogical efforts, "So what? What does it [the information and new-found knowledge] matter? What does it all mean?" (p. 2). This process should guide instructors to develop and implement a study that is relevant and meaningful for their students as well as assist them (instructors) in shaping an unwieldy and massive amount of information into something more manageable.

It is almost a given that it is easier to teach about the whos, whats, wheres, whens and hows of a historical event versus the why(s).[1] *But, again, to neglect to address the why(s) behind the whats, wheres, whos, whens and hows of a genocidal event leaves students with a superficial understanding of the event and with a skewed view of history and the way in which genocide can unfold.*

By developing strong rationale statements that focus on the whys and hows of a genocidal event as well those pertaining to the whats, wheres and whens, teachers are more likely to not only include absolutely critical information about the genocide but also a more well-rounded perspective of the events leading up to it. Concomitantly, they are more likely to include a balanced and accurate view of the perpetrators, collaborators, bystanders, rescuers, and victims. To skip over this initial work (e.g., developing rationale statements that provide an accurate and comprehensive view of the history) is to enter a study blind.

It is also essential to place the study of any genocidal event within a historical context in order to allow students to see the relationship of the political, social, and economic factors that impacted the times and events that resulted in that history; and this should be clearly stated in a rationale statement—*if for no other reason than a reminder of the critical nature of such a component for the study.* Thus, if one is teaching, for example, about the Holocaust and an emphasis is placed almost exclusively on the political factors leading to the rise of the Nazis, while scant attention is paid to a social phenomenon such as antisemitism, the ultimate organization of the unit and its historical context will be directly affected. Similarly, creating a context that suggests the economic misery brought on by the depression was the *main* "cause" of the Nazis' assumption of power to the exclusion of the social and political factors that motivated large and significant segments of the German public will, likewise, impact *and skew* the final organization of the unit.

It is also incumbent on instructors to help students understand that behind the statistics of genocide (e.g., the numbers killed) are real people, comprised of families of grandparents, parents, and children—and this, too, should be reflected in statements of rationale. Such a rationale statement may influence the instructor to incorporate first-person accounts into the study for such resources have the means to assist in moving the study "from a welter of statistics, remote places and events, to one that is immersed in the 'personal' and 'particular'" (Totten, 1987, p. 63).

Finally, instructors need to carefully examine their set of rationale statements in order to determine whether a course of study is comprehensive, thorough, and historically accurate versus one that is likely to be unduly limited in one form or another.

USING RATIONALES TO GUIDE THE SELECTION OF PEDAGOGICAL STRATEGIES

Strong rationales should also guide the selection of effective pedagogical strategies. The complexity and emotionally-charged nature of the study of genocide dictates that not just any instructional strategy or learning activity will do. Each must be thought-provoking, engaging, and not be comprised of those that, even inadvertently, minimize, romanticize, or simplify the history.

Some instructors at both the secondary and college levels engage their students in simulations where they, the students, are to act out "actual situations" experienced by the perpetrators, victims, and others. As the authors of the U.S. Holocaust Memorial Museum's *Guidelines for Teaching About the Holocaust* state, this is a dubious practice: "Even when teachers take great care to prepare a class for such an activity, simulating experiences from the Holocaust remains pedagogically unsound. The activity may engage students, but they often forget the purpose of the lesson, and even worse, they are left with the impression at the conclusion of the activity that they now know what it was like during the Holocaust...The problem with trying to simulate situations from the Holocaust is that complex events and actions are over-simplified, and students are left with a skewed view of history" (Parsons & Totten, 1993, pp. 7, 8). This concern is germane to the study of any and all cases of genocide. There is absolutely no way that students will *ever* be able to experience what it was like for the victims of genocide to be forced from their homes, herded into desolate deserts or ghettos (or burned alive in their homes or churches) or tortured and murdered at the hands of their assailants. *Simulations that purport to provide students with such experiences trivialize and mock the experiences of those who were killed as well as those who survived.*

Instructors prone to using simulations as part of a unit on genocide should ask themselves the question, "Why should simulations be used when there already exists a wealth of actual case studies on many genocides?" It is one thing for a teacher to broaden student perspectives by asking "What do you *think* you *might* have done?" (which is more appropriate than asking, "What would you have done?"), and quite another to launch a class into a simulation that attempts to recreate choices and human behavior that were made and carried out under conditions of horrific duress.

Strong rationales can also assist with the opening and closing of lessons and units on genocide. Opening and closing lessons in a study of genocide are important because they set the tone and context for the entire course of study. For example, if one of the rationales for the study is "to help students think about the use and abuse of power, and the roles and responsibilities of individuals, organizations, and nations when confronted with civil rights violations and/or policies of genocide," an opening can easily be tailored to begin to explore such issues. Additionally, strong openings are able to establish a reflective tone whereby students come to appreciate the need to make careful distinctions when weighing various ideas, motives, and behaviors; indicate to the students that their ideas and opinions about this history are important; tie the history to the students' lives; and/or establish that the history has multiple interpretations and ramifications. On the other hand, a strong closing can encourage students to connect this history to the world they live in and encourage them to continue to examine this history. (For a detailed discussion of various ways to begin and close lessons or units on genocide, see Chapter 6: "Instructional Strategies and Learning Activities: Teaching About Genocide" by Samuel Totten.)

Ultimately, the pedagogy used in such a study should be one that is student centered; that is, one in which the students are not passive but rather actively engaged in the study. It should be a study that, in the best sense of the word, complicates the students' thinking, engages students in critical and creative thought, and involves in-depth versus superficial coverage of information. Again, this should be reflected in the statements of rationale.

USING RATIONALE STATEMENTS TO AVOID PITFALLS DURING THE COURSE OF STUDY

Another use of rationale statements is to safeguard against some of the many pitfalls that have plagued some lessons and units on genocide (e.g., assaulting students with one horrific image after another, skewing the history by minimizing or overstating various aspects of the genocide, providing simple answers to complex questions/situations).

Instructors need to be judicious in their approach to such a study and be sure that in teaching about genocide they do not constantly bombard their students with one horrific image after another to the point where the students are overwhelmed with the horror. Students are essentially a "captive audience." *Assaulting* them with horrific images is antithetical to good teaching. The assumption that students will seek to "understand" human behavior after being exposed to horrible images is a fallacy. Instead of becoming engaged with the history, students may have a tendency to ignore it or "shut down" their attention in order to protect themselves from the ghastly images. *At one and the same time, educators need to avoid denying the reality of the horror of genocide by minimizing the fact that the perpetrators committed ghastly crimes.* What is required is a very fine balancing act. To attempt to teach about genocide without presenting the hard and brutal facts is miseducational. That said, when the horror is explored, it should be done only to the extent necessary to achieve the objective(s) of the lesson.

On a different note, it is important to keep in mind that generalizations without modifying and qualifying words (e.g., sometimes, usually, etc.) tend to stereotype group behavior and historical reality (Parsons & Totten, 1993, p. 4).

When teaching any piece of history, it is imperative to avoid allowing simple or simplistic answers or notions to serve as an explanation for complex behavior or situations. *Common knowledge of a historical event does not constitute accurate knowledge.* Rather, accurate knowledge of a historical event is based on the collection of accurate data. Its effective interpretation is predicated on using organizing principles and concepts drawn from legitimate scholarship. It is this type of information and data that should be used in the classroom. A good method to assess the accuracy of information is to do ample reading of the major scholars of the particular genocide under study and *to check and cross-check sources.*

Yet again, all of the above concerns should be at the forefront of one's mind when developing rationales for the course of study.

REVISITING RATIONALES THROUGHOUT THE STUDY

Since teaching about complex human behavior often results in examining multiple aspects of events and deeds, students need to continuously think about why they are studying this history. By repeatedly highlighting questions of rationale throughout a course, a signal is sent to students that *this is not simply another piece of history to wade through*, but that it has important lessons for both contemporary and future generations. What those lessons are will have to be extrapolated, discussed, and wrestled with by the teacher and the students.

CONCLUSION

Well constructed rationales for a study of genocide represent the foundation for successful curriculum design, instructional planning, teaching, and evaluation of student progress. The ongoing refinement of rationales is a critical dimension of reflective practice, because it fosters a critical approach to improving and strengthening genocide education.

NOTE

1.   The Holocaust serves as an example of this situation. More specifically, historian Lucy Dawidowicz (1992) correctly asserted that most Holocaust curricula are "better at describing what happened during the Holocaust than explaining why it happened" (p. 69). In a survey of 25 curricula Dawidowicz (1992) found that "most curricula plunge right into the story of Hitler's Germany; a few provide some background on the Weimar Republic, presumably to explain Hitler's rise to power. Though all curricula discuss Nazi anti-Semitism, preferring generic terms like 'racism' and 'prejudice' instead of the specific 'anti-Semitism,' 15 of the 25 never even suggest that anti-Semitism had a history *before* Hitler. Of those that do, barely a handful present coherent historical accounts, however brief.... A small number of curricula include lessons which survey the pre-Nazi history of Jews in Europe, presumably to humanize the image of the Jews depicted in Nazi propaganda." (p. 69)

     In regard to the issue of antisemitism—which is often the issue that most curricula developed for use by secondary level teachers neglect to address—Dawidowicz (1992) persuasively argues that avoidance of that topic, "and especially its roots in Christian doctrine" (p. 71), skews history and provides a distorted picture of the cause of the Holocaust. "To be sure, Christianity cannot be held responsible for Hitler, but the Nazis would not have succeeded in disseminating their brand of racist anti-Semitism had they not been confident of the pervasiveness, firmness, and durability of Christian hatred of Jews" (Dawidowicz, 1992, p. 71). "Omitting all references to Christian anti-Semitism is one way some curricula avoid the sensitivities of the subject. The more acceptable and common pedagogic strategy is to generalize the highly particular nature and history of anti-Semitism by subsuming (and camouflaging) it under general rubrics like scapegoating prejudice, and bigotry.... These abstract words suggest that hatred of the Jews is not a thing in itself, but a symptom of 'larger' troubles, though no explanation is given as to why the Jews, rather than dervishes, for instance, are consistently chosen as the scapegoat" (Dawidowicz, 1992, p. 73).

REFERENCES

Branson, M. S., & Torney-Purta, J. (Eds.). (1982). Introduction. In *International human rights society and the schools* (pp. 1-5). Washington, DC: National Council for the Social Studies.

The Coalition of Essential Schools. (1989, June). Asking the essential questions: Curriculum development. *Horace, 5*(5), 1-6.

Dawidowicz, L. (Ed.). (1992). How they teach the Holocaust. In *What is the use of Jewish history?* (pp. 65-83). New York: Schocken Books

Fein, H. (2002). Reflections on studying genocide for three decades. In J. Apsel & H. Fein (Eds.), *Teaching about genocide: An interdisciplinary guidebook with syllabi for college and university teachers* (pp. 1-5). New York: American Sociological Association.

Friedlander, H. (1979). Toward a methodology of teaching about the Holocaust. *Teachers College Record, 80*(3), 519-542.

Langer, L. L. (1978). *The age of atrocity: Death in modern literature.* Boston: Beacon Press.

Lipstadt, D. (1995, March 6). Not facing the history. *The New Republic*, pp. 26-27, 29.

Newmann, F. (1988, January). Can depth replace coverage in the high school curriculum? *Phi Delta Kappan*, pp. 345-348.

Parsons, W. S., & Totten, S. (1993). *Guidelines for teaching about the Holocaust.* Washington, DC: United States Holocaust Memorial Museum.

Schwartz, D. (1990, February). Who will tell them after we're gone?: Reflections on teaching the Holocaust. *The History Teacher, 23*(2), 95-110.

Strom, M. S., & Parsons, W. S. (1982). *Facing history and ourselves: Holocaust and human behavior.* Watertown, MA: Intentional Educations.

Totten, S. (1987). The personal face of genocide: Words of witnesses. *Social Science Record, 24*(2, Special Issue—Genocide: Issues, approaches, resources), 63-67.

Totten, S. (Ed.). (2002). The Imperative to Avoid Clichés. In *Holocaust education: Issues and approaches* (pp. 139-150). Boston: Allyn & Bacon.

CHAPTER 3

THE HISTORY OF GENOCIDE

An Overview

Paul R. Bartrop and Samuel Totten

INTRODUCTION

Genocide is a crime that has been committed throughout the ages. Indeed, every century of recorded history has been marred by genocidal acts. It was not until the twentieth century, though, that this particular act of mass murder was given the name "genocide." The term itself was originally coined in 1944 by Raphael Lemkin, a Polish-Jewish jurist who lost most of his family members in the maw of the Holocaust. It was a term appropriated by the fledging United Nations to describe the planned, systematic mass murder, in whole or part, of a particular group of people in its landmark convention, the United Nations Convention on the Prevention and Punishment of the Crime of Genocide (UNCG).

In this chapter, we provide an overview of the history of genocide. In doing so, we discuss, albeit briefly, the following: genocide through the ages, genocide in the twentieth century, Raphael Lemkin's efforts to staunch genocide, the development and ratification of the U.N Convention on the Prevention and Punishment of the Crime of Genocide (1948), definitional issues, genocide during the Cold War years and the reaction

Teaching About Genocide: Issues, Approaches, and Resources, 23–55
Copyright © 2004 by Information Age Publishing

of the international community, genocide in the post-Cold War years (late 1980s-2003), and the possibility and likelihood of intervening and preventing future cases of genocide.

GENOCIDE THROUGH THE AGES

As Chalk and Jonassohn (1988) have noted, "The first genocidal killing is lost in antiquity" (p. 41). That said, as previously noted, genocide is a new word for an ancient practice, and it has taken many forms in the past. The Hebrew Bible contains a number of important passages that refer to mass destruction of a kind which we would, today, identify as genocide. The Greeks engaged in the practice, as chronicled by Thucydides in the famous case of the island inhabitants of Melos, as did the Romans—most notably in the fate that befell the inhabitants of Rome's arch-enemy Carthage, where both the people were destroyed and the land on which they lived was despoiled. (For a detailed discussion of the history of genocide, including the earliest genocides, see Chalk & Jonassohn, 1991.)

In the aftermath of the Roman victories over the Jews of Palestine (Judaea) during the first century CE, at which time the Temple was destroyed (70 CE) and the last remnants of Jewish opposition to Roman rule under Simeon Bar Kochba were snuffed out at Betar (135 CE), the Jews were a devastated people. Over half a million had been killed in the aftermath of the wars, their cities had been laid waste (Katz, 1994, pp. 80-81), and the survivors were dispersed through slave markets across the known world (Safrai, 1976, pp. 330-335). In what was a clear case of genocide, the Jewish state was extinguished, and would not appear again for over 1,800 years.

Other cases of mass killing in the premodern period are similarly prominent, embracing a variety of locations and situations. As the Mongols under Genghis Khan swept through central Asia and eastern Europe during the thirteenth century, they wrought havoc and destruction on a massive scale. There is little doubt that this was done as a matter of deliberate policy; the more brutal the Mongols were, the more their reputation for violence spread. This, in turn, made it easier to conquer new territories and cities. The death and destruction wrought by the Mongols was a direct outcome of the need to find a practical method of ruling a large empire using limited resources.

Another example of what may be termed a premodern genocide was the persecution and eradication in the early thirteenth century of the Cathars (or Albigensians) of France, who were accused by the church of heresy. In its drive to wipe out all traces of dissent, the French church fell on the free-thinking people of the Languedoc region, destroying them utterly (O'Shea, 2001). Their example introduces the issue of doctrine (or ideology) to our understanding of genocide. By this stage, Europe had moved far down the road toward becoming a persecuting society established on notions of religious intolerance, a frightening portent of things to come at the dawn of the modern age. The development of such attitudes, exemplified in the European perspective of the undesirability of "the Other," ultimately became internalized as part of European civilization, and it was this development which dominated European society just at the time Europe began to expand overseas into non-European cultures. This ethos of intolerance and persecution, coupled with an advanced military technology, was to have a devastating effect of the societies being invaded—though, it must be recalled, in numerous ways the Europeans were only applying to populations overseas the kind of actions they had already been doing to themselves for centuries.

The destruction of the indigenous peoples of the Americas represents one of the most extensive human catastrophes in history. The pace and magnitude of the devastation varied from region to region over the years, but it can safely be concluded that in the two-and-a-half centuries following Christopher Columbus's "discovery" of the Americas in 1492, probably 95% of the pre-Columbian population had been eradicated—by disease as well as by deliberate policy on the part of the Spanish, the French, the English, and, ultimately, by the American-born heirs of those colonizing nations (Stannard, 1992). It was a massive case of population collapse, sometimes intentional, sometimes not, but nonetheless effective in clearing the way for the European expropriation of two continents. This was a horrific case of mass human destruction, in which tens of millions of people lost their lives.

In the case of the indigenous peoples within the United States, the destruction did not stop once the people had died or been killed. Policies of population removal, dispossession of lands, forced assimilation and confinement to "reservations" meant that even the survivors were denied the opportunity to retain a sense of peoplehood.

In order to reinforce the "European-ness" of the genocidal impulse at this time, we need look no further than the fate of the Jews in eastern Poland and the Ukraine in the mid-seventeenth century. A series of massacres perpetrated by the Ukrainian Cossacks under the leadership of Bogdan Chmielnicki saw the death of up to 100,000 Jews and the destruction of perhaps 700 communities between 1648 and 1654 (though instances of anti-Jewish pogroms kept appearing for several years after this).

Religious persecution resulting in mass death was a constant throughout the early modern period, some of the most terrible examples taking place before and during the Thirty Years' War. Some areas in the lands that were later to comprise Germany were depopulated by as much as 90 percent, while in France the struggle between Catholics and Protestants (in their French variant, Huguenots) was played out after the revocation of the Edict of Nantes. On St. Bartholomew's Day (August 24, 1572) massacres of the Huguenot leadership took place all over Paris, and in the days that followed scores of thousands more were killed all over France. Up to 100,000 men, women and children were murdered as a result of these persecutions.

The pattern of European-inspired brutality that had hit the Americas from 1492, and been refined in Europe itself in the years that followed, was repeated in other lands of European colonial settlement later, particularly in Australia—after the continent had been claimed for Britain and settled as a convict colony (or rather, series of colonies) from 1788 onwards. For well over a century, for example, it was generally accepted in Australian popular wisdom that a woman named Truggernanna (sometimes spelled Truganini), of the Bruny Island people, was "the last Tasmanian," and that after her death in 1876 an entire people had been exterminated—the first "total" genocide in history. The nonsense of this myth has since been demonstrated many times over, but the story of "the last Tasmanian" has very deep roots which are still being fed today (Reynolds, 1995). This does not alter the fact, however, that the indigenous Tasmanians suffered a population collapse (largely through warfare with the encroaching settlers, through diseases, and, after the concentration and removal of the remaining tribal members to Tasmania's Flinders Island, through a pitiful longing for their lost homeland and way of life in which they simply pined away) as complete as any that had ever taken place elsewhere, and probably more than most.

The situation of the Aborigines in mainland Australia was very different, and serves as yet another dimension to the study of genocide before the modern era. The question that must be asked is straightforward: did the destruction of Aboriginal society in the century following the arrival of the First Fleet in 1788 constitute an act of genocide? (Reynolds, 2001). For some, unequivocally so; for others, the answer is nowhere near as obvious. Here, we find ourselves studying a situation in which there was no definite state-initiated plan of mass extermination; indeed, it was frequently the case that colonial governments went to great lengths to maintain Aboriginal security in the face of settler and pastoralist encroachments, and of punishments (even hangings) of those found guilty of the murder of Aborigines. Despite this, there were immense and very intensive periods of killing in the bush, accompanied by enormous population losses commensurate with those in the Americas as a result of disease and starvation.

There is no doubt that the nineteenth century saw the effective destruction of Aboriginal society by European settlement, but where genocide is concerned this must be understood against two essential facts. In the first place, there was no unified stance on the Aborigines throughout the nineteenth century, as the Australian continent was divided into six separate British colonies, mostly self-governing from the middle of the century, until federation in 1901. Secondly, no government at any time displayed the necessary intent, in word or in deed, to prove the existence of a genocidal policy. This in no way mitigates the catastrophe which destroyed the Aborigines, but neither does the history show that that tragedy was the result of what might be termed genocide.

If there is debate among some scholars concerning the question of genocide in nineteenth century Australia, there can be no doubt that one series of actions in the twentieth century fits the United Nations Genocide Convention perfectly. This relates to the forcible removal of children of part-Aboriginal descent from their parents and subsequent placement in a non-Aboriginal environment for the purpose of "breeding out the color." The policy, which was set in place by state and federal governments, was to last in various forms until the 1970s. It decimated at least two generations of Aborigines of mixed descent, and in a major federal government inquiry in the late 1990s, the allegation of genocide of these "Stolen Generations" was for the first time raised in an official capacity (Bartrop, 2001; Reynolds, 2001, pp. 162-165).

Returning to the nineteenth century, colonial expansion across the globe was the most likely scenario in which genocidal episodes took place at this point in time, and it is here that we must be very careful to distinguish between instances of death and destruction that took place in wartime—during which innocent people were killed as the by-product of conflict (what in the twentieth century would come to be called "collateral damage")—and massacres and other deliberate violations of human rights in which wanton and wholesale annihilation took place. These latter events clearly fell under a different heading from the sort of destruction that usually occurred in wartime. This must be understood clearly. There is a relationship between war and genocide, of course, but war does not have to be present for genocide to be perpetrated (Bartrop, 2002).

Genocide, though while occurring often throughout history, reached its zenith during the twentieth century, and it is to this period we now turn.

GENOCIDE IN THE TWENTIETH CENTURY

The twentieth century has been deemed by some to be "the century of genocide" (Smith, 1987; Totten, Parsons, & Charny, 1997). That is due to the fact that the twentieth century was a century plagued by genocide. Indeed, genocidal acts were perpetrated in every decade of the century, and, over the course of the century, on at least four different continents (Africa, Asia, Europe, and South America)—and arguably, Australia (i.e., the "Stolen Generations"). Among such genocides were: the German-perpetrated genocide of the Herero in southwest Africa in 1904; the Ottoman Turk genocide of the Armenians between 1915 and 1918; the Soviet man-made famine in Ukraine in 1932-1933; the Holocaust (1933-1945); the Bangladesh Genocide (1971); the Khmer Rouge-perpetrated genocide in Kampuchea (Cambodia) 1975-1979; the Iraqi genocide of its Kurdish population in the late 1980s; the 1994 Rwandan genocide of the Tutsis and moderate Hutus by the extremist Hutus; and the genocide committed in the former Yugoslavia in the early to mid-1990s). There were also the almost totally unnoticed genocides of many indigenous groups across the globe. (Space precludes providing detailed discussions of individual genocides, but Chapter 3, "Genocide in the Twentieth Century," provides overviews of nine specific genocides perpetrated in the twentieth century. The overviews are written by some of the most noted scholars in the field of genocide studies, each of whom is a specialist on the genocide he/she writes about. Another excellent source for accurate and relatively detailed information is Totten, Parsons, & Charny, 2004.)

Three major genocides were perpetrated over the course of the first half of the century: the Ottoman Turk genocide of the Armenians, the Soviet manmade famine in Ukraine, and the Holocaust.

While the history of the genocide of the Armenians by the Ottoman Turks from 1915 onward is now well known, that of the enormous massacres of Armenians between 1894 and 1896, and again in 1909, is less so. Yet it was these massacres that, in numerous ways, led the Turks to adopt a mind-set conducive to their later genocidal actions. The worst of the massacres took place in 1895, when at least 100,000 (and certainly more, possibly as high as 200,000 to 300,000) Armenian civilians were killed by Turkish mobs acting under instruction from Sultan Abdul Hamid II (a direction which gave the slaughter its name as the "Hamidian Massacres"). While these statistics were horrible enough, they were but a prelude to the even greater number of killed—at least one million, according to most accounts, though on the balance of probabilities probably closer to 1.5 million—in the genocide that was perpetrated by the Young Turks after April 1915. At that time, well after the outbreak of World War I, the Turkish nationalist revolutionary government known as the Young Turks (formally, the Committee of Union and Progress) launched a wholesale assault against the Armenian population of the empire. It was far more extensive than the earlier massacres, and saw all the relevant agencies of government directed toward the singular aim of totally eradicating the

Armenian presence in Turkey. That the genocide took place under the cover of war was more than just a matter of interest; the war was in reality a crucial part of the genocide's success. By conducting deportations of Armenians in places far off the beaten track, forcing many victims (primarily women and young children, including babies) into underpopulated desert regions of the empire, the Turks were able to exploit the war situation for the purpose of achieving their genocidal aims. The eventual result was a loss of life in a very short space of time.

In Ukraine, over the course of 1932-1933, between five and seven million peasants—most of them ethnic Ukrainians—starved to death due to a Soviet manmade famine as a result of Josef Stalin's government having seized all crops and methods of food production. In this case, the huge number of deaths was the product of ideologically-driven social destruction. Purportedly, the goal was the destruction of the so-called "kulak" class: the independent peasantry who were sacrificed in Stalin's urgent desire to collectivize the agricultural sector along communist lines. In fact, anyone who supported Ukrainization (Ukrainian nationalism, freedom, and the expression of its culture) was targeted, including the poorest of the poor. Accompanying this was the forced integration of a diverse array of religious and national groups into the existing Soviet political structure, a move which took place in an environment of forced Russification. This had a catastrophic effect on Ukrainian national aspirations, hopes which had been initially dashed in the immediate aftermath of the Great War and the effective reconquest of the country by the nascent Red Army in the early 1920s.

As the Soviets swept through Ukraine, Stalin's agents pointed to the need to wage a class war for the good of the entire Soviet Union and the communist revolution. The genocide, a product of an extreme form of social engineering, transformed traditional forms of land holding and land use, and literally stripped the countryside of everything that could be consumed by the peasantry.

As mentioned above, while Stalin was committed to the destruction of the kulak class, people from a variety of classes, religions, and ethnic/national groups were victims of the deliberately-inflicted starvation policies—although, again, Ukrainians formed the majority of those who lost their lives. Despite this, by even the strictest of definitional interpretations the massive destruction wrought by the Soviet regime can be classified as genocide.

Casting a giant shadow over the myriad positive achievements of the twentieth century is the Nazi Holocaust of the Jewish people, a genocidal event that resulted in a horrific break with all of the humanitarian traditions that had been developing in Europe over the previous thousand years. The relationship between mass death and the industrial state as manifested in the Holocaust was both intimate and interdependent.

The perpetrators of the Nazi Holocaust had but one aim in mind: the physical removal of all Jews falling under their rule and beyond. A change of status through religious conversion or naturalization was not acceptable to the Nazis. What was planned and what was intended was a total annihilation, from which none would be allowed to escape. The Nazis intended nothing less than the physical destruction, through murder, of every Jew who fell into their net.

The means the Nazis employed in order to achieve their murderous aims, especially from early 1942 onwards, was the death (or extermination) camp (*Vernichtungslager*), though earlier murderous measures such as mass shootings by mobile killing squads known as Einsatzgruppen, which accompanied the Nazi invasion of the Soviet Union in mid-1941, also played a huge role in the extermination program. So, too, did brutal treatment and deliberately-induced starvation policies in ghettos and other segregated areas, and the early use of gas vans.

The six death camps established in Poland for the mass murder of the Jews—Auschwitz-Birkenau, Belzec, Chelmno, Majdanek, Sobibor, and Treblinka—were a departure from anything previously visualized in both their design and character. With the exception of Auschwitz (which must always fit into a category of its own in any discussion of the Holocaust), these camps were institutions designed to methodically and efficiently murder millions of people. The mass murders took place in specially-designed gas chambers. The death camps became the most lucid and unequivocal statement National Socialism made about itself, demonstrating beyond doubt that it was an antihuman ideology in which respect for life and moral goodness counted for nothing.

The Nazi death camps add to our knowledge of genocide in two important ways. In the first place, it shows us how a regime dedicated to mass murder mobilized all its resources for the purpose of feeding the demands of an industry that had been deliberately assigned the tasks of incarceration, degradation, and annihilation. Second, the history and fate of the camps demonstrates that genocidal regimes can be aware that their activities are of a criminal nature; just as the Turks did before them, the Nazis chose to carry out their murderous assignments in places far removed from key population centers, accompanied by an exhaustive effort to destroy as much evidence of the killing as possible prior to being overrun by the advancing Allied forces.

A number of the aforementioned genocides (meaning the long list of genocides perpetrated over the course of the twentieth century) either went undetected until the mass killing was well under way and tens of thousands or more had already been murdered; or, more commonly, they were detected by outside governments and/or the United Nations (in the latter case, after 1945), but the officials of such entities were, for various reasons, hesitant or simply unwilling to deem the actions genocide. Primarily, though, officials of governments (other than those where the genocide was being perpetrated) were cognizant of the fact that if they deemed a situation to be "genocide," not only would there be a moral imperative to attempt to prevent it, but—if their nation was a signatory to the UN Convention on Genocide—a legal imperative as well. In many cases, realpolitik also played a role in "hearing no evil, seeing no evil." And just like the officials of individual nations, the officials of the United Nations acted similarly.

The governments of individual states and officials of the United Nations, though, are not the only entities that have been tentative about deeming a series of mass murders "genocidal" or outright genocide. Indeed, many journalists, human rights activists, humanitarian personnel, and even genocide scholars have often been hesitant—most often early on, but sometimes even well into the killing period—to designate a situation "genocidal." It is important to note that some of these individuals and groups, particularly scholars, were not being circumspect for cowardly or evasive reasons—nor due to any other ulterior motive—but rather because they were (and are) intent on using the term only when it is truly applicable. Such individuals are cognizant and appreciative of the fact that the terms "genocide" and genocidal" are highly charged terms, and thus look askance at the fact that the terms are often applied to situations that are not technically genocidal (Totten, 1999, pp. 35-36). Furthermore, they appreciate the fact that a loose use of the term/concept only serves to water down its import.

Conversely, there are others—more often than not these individuals and/or groups include certain journalists, politicians, and others who are not well-versed in the field of genocide studies—who tend to assert hastily that a situation is genocidal even when there is no hard evidence to support such an assertion. Again, a fast and loose use of the term is counterproductive in that it degenerates, over time, into a situation where it loses its real value and purpose.

Unfortunately, one cannot be sanguine about ending genocidal acts in the near future. Political, racial, religious, and social conflicts are rife across the globe and seemingly will be for a long time to come. Furthermore, there is as yet no single, central, and highly effective genocide early warning mechanism in place. Concomitantly, both the United Nations and individual states are not a little tentative about providing the needed resources and personnel to halt massive human rights abuses at their outset, when a situation may be slouching toward genocide. And yet, certain progress—some of it significant—has been made in addressing the problem of genocide, its prevention, and its punishment.

RAPHAEL LEMKIN'S EFFORTS TO PREVENT GENOCIDE

No history of genocide would be complete without acknowledgment of the significant efforts and accomplishments of Raphael Lemkin (1900-1959). Lemkin was born in what was historically known as White Russia. As he grew up, he heard stories about vicious pogroms that were being perpetrated in the region where he lived. Such stories influenced him greatly. As he states in his autobiography, *Totally Unofficial Man*, "I could not define history with my childish mind, but I saw it vividly and strongly with my eyes, as a huge torture place of the innocent" (quoted in Lemkin, 2002, p. 370). Early in life, he also became aware of the fact that Ottoman Turkey perpetrated a genocide against its Armenian population, murdering more than a million women, children, and men. He was incensed that following World War I the 150 Turkish war criminals who had been arrested and interned by the British government on the island of Malta had been released. This shocking turn of events caused him to ponder the following:

> A nation that killed and the guilty persons were set free. Why is a man punished when he kills another man? Why is the killing of a million a lesser crime than the killing of a single individual? I didn't know all the answers, but I felt that a law against this type of racial or religious murder must be adopted by the world. (Lemkin, 2002, p. 371)

Eventually, Lemkin studied international criminal law, during which he became obsessed with the then current debate about the sanctity of state sovereignty. As he relates in his autobiography:

> [My professors] evoked the argument about sovereignty of states. "But sovereignty of states," I answered, "implies conducting an independent foreign and internal policy, building of schools, construction of roads, in brief, all types of activity directed towards the welfare of people. Sovereignty," I argued, "cannot be conceived as the right to kill millions of innocent people." (Lemkin, 2002, p. 371)

In October 1933, at an international conference for the unification of Penal Law in Madrid, he put forth the notion that there was a dire need "to outlaw the destruction of national, racial, and religious groups, [and that] the crime was so big that nothing less than declaring it an international offense would be adequate. This should be done by international treaty or convention" (Lemkin, 2002, p. 372).

The draft international law he proposed along such lines was not met with any enthusiasm; and indeed, he was told that the formulation of such laws was "not necessary, because they apply to crime(s) which occur seldom in history" (Lemkin, 2002, p. 367).

In 1939, with the Nazis on the move and following the initial bombing of Warsaw where he was living, Lemkin fled Poland, first to Sweden and then the United States (where he moved to Durham, North Carolina, having taken an appointment at Duke University). At the time, he sensed that the Nazis' intentions included more than simply conquering other nations for their land and wealth: "I knew that this was more than war, that this was the beginning of genocide, on a large scale" (Lemkin, 2002, p. 367).

Throughout World War II, Lemkin continued to work at a torrid pace to convince world leaders, including U.S. President Franklin Delano Roosevelt, of the critical need to pass an international law against the perpetration of genocide, but his Herculean efforts came to nil. Throughout this period, he was obsessed with the issue of genocide, writing about it, talking to anyone and everyone about the need to establish an international convention against genocide, and attempting to figure out ways to convince government officials and others to support his efforts. During this period he wrote what is considered his magnum opus, *Axis Rule in Occupied Europe: Laws of Occupation, Analysis of Government, Proposals for Redress* (1944). It was also at this time that he coined the term "genocide."

Following the war, Lemkin served as an advisor to United States Justice Robert Jackson at the International Military Tribunal at Nuremberg, Germany, hoping the Nazi war criminals would be found guilty of genocide. It was not to be. Of the Nuremberg judgment, Lemkin asserted that

> it relieved only in part the moral tensions in the world. [T]he purely juridical consequences of the trials were wholly insufficient, [i.e.] the refusal of the Nuremberg Tribunal to establish for the future a precedent against this type of international crime. In brief, the Allies decided in Nuremberg a case against a past Hitler, but refused to envisage future Hitlers, or like situations. They did not want to or could not establish a rule of international law which would prevent and punish future crimes of the same type....
>
> The Tribunal declared that it is bound by the Statute of the International Military Tribunal which did not contain the charge of genocide. In brief, the Germans were punished only for crimes committed during or in connection with the war of aggression (Lemkin, 2002, pp. 384, 385).

Lemkin eventually turned to the newly formed United Nations. Within a couple of days of returning to the United States from Europe, he had convinced the governments of Cuba, India, and Panama to sponsor his resolution that genocide be a crime under international law, without any limitations to war or peace. The resolution he prepared also "called for the preparation of an international treaty against genocide" (Lemkin, 2002, p. 368).

Initially, the Soviet Union argued against the resolution, but in the 1946 Assembly it changed its position and argued in favor of the Convention. Initially, the Soviet Union looked askance as such legislation, arguing that it was not needed. Many scholars have argued that the latter position was likely due to the heinous acts that the U.S.S.R. committed against both its own people and those of others, and that it feared being held accountable for them. Eventually, though, diplomatic efforts by other nations convinced the U.S.S.R. to support the nascent effort to prevent and punish genocide. Ultimately, the Assembly adopted the resolution to include the issue of genocide on its agenda for discussion. Subsequently, a special subcommittee of the Legal Committee was established to prepare the text of the resolution on genocide. Lemkin wrote that "The subcommittee decided to include, in the resolution, a declaration that genocide is a crime under international law, a condemnation of this crime by all civilized nations and the decision to prepare a convention on the prevention and punishment of the crime of genocide and to present it to the next Assembly" (quoted in Lemkin, 2002, pp. 386-387). Ultimately, there was an unanimous vote by the Legal Committee to support the resolution, and several days later the resolution was, likewise, adopted unanimously by the General Assembly.

In July 1948, a discussion of the resolution was undertaken by the U.N.'s Economic and Social Council in Geneva. Lemkin traveled to Geneva to act as the spokesman for the establishment of a U.N. Convention on Genocide. Of this period, Lemkin (2002) wrote that

> There were still three weeks before the Economic and Social Council would start discussing the Genocide Convention. I felt that the issue must be made important in the eyes of the delegates. Major John A. F. Ennals (of the World Association) organized two lectures on genocide; one in the building of the United Nations to which the delegates were invited, the other in the summer school which the World Federation of the U.N. Associations was running for foreign students. After both lectures a discussion followed. Interest was aroused by my historical examples dating from antiquity through the middle ages to modern times. Then questions were asked. I did not refrain from reading aloud from my historical files in considerable detail. In such a way I conveyed the impression that genocide is not the result of a casual mood of a ruler but that it has its place in history.

What can you do to prevent such a thing from happening?

You do exactly the same as to prevent other crimes. We have to deal with this matter on two levels: national and international. Nationally, we must make it a crime in our criminal codes and punish through national courts in the same way as we punish for larceny and arson.

On the international level, I continued, we must make every nation responsible to the world community, either by bringing up cases of genocide in the World Court of Justice in the Hague to which all civilized nations belong or in all organs of the U.N. The main thing is to make the nations of the world feel that minorities and weaker nations are not chickens in the hands of a farmer, to be slaughtered, but they are groups of people of great value to themselves and to world civilization (p. 388).

On September 15, 1948, in Paris, the Legal Committee met to discuss those acts which constituted genocide. Speaking of the discussion that ensued, Lemkin (2002) reported that

The delegates seem[ed] to be lost in an endless discussion on the motives of the crime of genocide. Erling Wikborg of the Norwegian delegation proposed to include in the definition also the destruction of a group "in part." He argued that when intellectual leaders, who provide the forces of cohesion to the group, are destroyed, then the group is destroyed as such, as a group.

Finally, the following formula emerged from the discussion: "deliberately inflicting on the group conditions of life calculated to bring about its physical destruction in whole or in part." In order to avoid misapplication of this article in cases when people objectively suffer from bad conditions such as extreme poverty, unsanitary conditions and the like, which generally prevail in a country or locality, the Committee required that the intent to destroy the groups should be strengthened by the additional expression of intent such as "deliberately inflicting conditions of life calculated to bring about..."

Two days later the representatives of twenty two nations signed the Convention. [A s]ignature meant the intention to ratify by the parliament. It is an act of government, which must be followed up later by an act of parliament, if a nation decides to ratify a treaty (pp. 392, 395).

In large part because of Lemkin's indefatigable work, the United Nations Convention on the Prevention and Punishment of the Crime of Genocide was adopted unanimously and without abstentions by the United Nations General Assembly on December 9, 1948.

Next, Lemkin began work on obtaining at least 20 ratifications, which would enable the General Assembly of 1950 to draw up the protocol for bringing the Convention into force. To do so, he focused on the small nations of Latin America, who, he considered, needed the protection of international law more than the bigger, more powerful nations. As a result, Ecuador became the first Latin American nation to ratify the Genocide Convention. Ultimately, the Convention was ratified by enough countries to go into effect on January 12, 1951 (LeBlanc, 1991, p. 1). It would take the United States almost four decades after Lemkin's death, to ratify—during the Reagan Administration—the Convention.

In the last half of the twentieth century, the Convention was applied sporadically and selectively, and remains a source of continuing debate and discussion among legal and genocide scholars the world over. To this day, however, it remains the cornerstone of international legal and moral-ethical reasoning regarding genocide and its prevention. (Lemkin's auto-biography, *Totally Unofficial Man*, is now, for the first time, available in print in Totten & Jacobs, 2002.)

UNITED NATIONS CONVENTION ON THE PREVENTION AND PUNISHMENT OF GENOCIDE (1948)

In the aftermath, then, of World War II and the Nazi extermination of six million Jews and of approximately five million other people (includ-ing but not limited to the Gypsies, mentally and physically handi-capped, and Slavic peoples), the United Nations adopted a resolution on December 9, 1946, recommending that international attention and cooperation be focused on the prevention and punishment of genocide. It was, in fact, this horrific slaughter by the Nazis, along with Lemkin's continuing efforts and prodding, that prompted the member states of the United Nations to formally recognize genocide as a crime in inter-national law.

As alluded to in the above section, from the outset the development of the U.N Genocide Convention was plagued by great difficulty and contro-versy. As might be expected, nations with diverse philosophies, cultures, "historical experience[,] and sensitivities to human suffering" (Kuper, 1985, p. 10) presented different interpretations as to what constituted genocide, and argued in favor of a definition and wording in the Conven-tion that fit their particular perspective(s).

On December 11, 1946, the United Nations General Assembly passed this initial resolution (96-I):

> Genocide is a denial of the right of existence of entire human groups, as homicide is the denial of the right to life of individual human beings. Many instances of such crimes of genocide have occurred, when racial, religious, political, and other groups have been destroyed entirely or in part. The General Assembly, Therefore, Affirms that genocide is a crime under inter-national law which the civilized world condemns, for the commission of which principals and accomplices—whether private individuals, public offi-

cials or statesmen, and whether the crime is committed on religious, racial, political or any other grounds—are punishable. (United Nations, 1978, pp. 6-7)

Of the utmost significance here is that the initial resolution included "*political or any other grounds*" (italics added). However, a brouhaha erupted when the Soviet Union, Poland, and other nations argued against the inclusion of political groups. The Russians argued that the inclusion of political groups would not conform "with the scientific definition of genocide and would, in practice, distort the perspective in which the crime should be viewed and impair the efficacy of the Convention" (U.N. Economic and Social Council, 1948, p. 721). Similarly, the Poles added that "the inclusion of provisions relating to political groups, which because of their mutability and lack of distinguishing characteristics did not lend themselves to definition, would weaken and blur the whole Convention" (U.N. Economic and Social Council, 1948, p. 712).

Yet another argument against the inclusion of political groups was that unlike national, racial, or religious groups, membership in political groups was voluntary. However, in a later session, the French argued that "whereas in the past crimes of genocide had been committed on racial or religious grounds, it was clear that in the future they would be committed mainly on political grounds" (U.N. Economic and Social Council, 1948, p. 723). Representatives of many other nations concurred with this position and offered strong support in favor of its recognition.

A particularly moving argument put forth *against the exclusion* of political groups was that "those who committed the crime of genocide might use the pretext of the political opinions of a racial or religious group to persecute and destroy it, without becoming liable to international sanctions" (U.N. Legal Committee, October 1948, p. 100).

Ultimately,

Political groups survived for many sessions and seemed securely ensconced in the Convention, but on 29 November 1948, the issue was reopened in the Legal Committee on a motion by the delegations of Iran, Egypt, and Uruguay. A compromise seems to have been reached behind the scenes. The United States delegation, though still committed to the principle of extending protection to political groups, was conciliatory. It feared non-ratification of the Convention, and rejection of the proposal for an international tribunal, if political groups were included. On a vote, political groups were expunged (Kuper, 1981, p. 29).

The upshot was that the United Nations Convention on the Prevention and Punishment of Genocide came to define genocide in the following manner:

In the present Convention, genocide means any of the following acts committed with the intent to destroy, in whole or in part, a national, ethnical, racial or religious group, as such:

a.   Killing members of the group;
b.   Causing serious bodily or mental harm to members of the group;
c.   Deliberately inflicting on the group conditions of life calculated to bring about its physical destruction in whole or in part;
d.   Imposing measures intended to prevent births within the group;
e.   Forcibly transferring children of the group to another group.

The aforementioned arguments and counter-arguments resulted in what can best be described as a "compromise definition."

The exclusion of political and social groups from the U.N. Convention has been roundly criticized by many scholars (Chalk, 1989; Charny, 1988a, 1988b; Drost, 1959; Hawk, 1987; Kuper, 1981, 1985; Totten, 2002; Whitaker, 1985). It is also worth noting that in 1985, Ben Whitaker, the U.N. rapporteur on genocide, made a recommendation, in his much heralded report entitled *Revised and Updated Report on the Question of the Prevention and Punishment of the Crime of Genocide*, that political groups be protected under the UNCG.

Hawk (1987), for one, asserts that "The absence of political groups from the coverage of the Genocide Convention has unfortunately had the effect of diverting discussion from what to do to deter or remedy a concrete situation of mass killings into a debilitating, confusing debate over the question of whether a situation is, legally genocide" (p. 6). In a similar but somewhat different vein, Whitaker (1988) argues that

> There are strong arguments for the addition of at least political groups to the United Nations definition. In some cases of mass crimes it is not easy to determine which of overlapping political, economic, racial, social, or religious factors was the determinant, motivating one. For example, were comparatively recent selective genocides in Burundi and Kampuchea intrinsically political or ethnic in their intent? Most genocides (including that by Nazis) have at least some political tinge, and it can be argued that to leave political and other groups beyond the protection of the [U.N. Convention on Genocide] offers a dangerous loophole that would permit any designated group to be exterminated, ostensibly under the excuse that this was for political reasons. (p. 53)

As for the controversy over the issue of cultural genocide, Kuper (1981) notes the following:

> In this controversy, the roles of the national delegations were somewhat reversed. The Soviet Bloc pressed for inclusion of cultural genocide in the Convention, the Western European democracies opposed. The issue was not

whether groups should be protected against attempts to destroy their culture.... The issue was rather whether the protection of culture should be
extended through the Convention on Genocide or in conventions on
human rights and rights of minorities. This conflict was not sharply ideological, but presumably the representatives of the colonial powers would have
been somewhat on the defensive, sensitive to criticism of their policies in
non-self governing territories.

In the result, cultural genocide was excluded from the Convention,
though vestiges remain. The Convention makes special reference to the
forcible transfer of children from one group to another, and the word ethnical has been added to the list of groups covered by the Convention. This
would have the effect of extending protection to groups with a distinctive
culture or language. (p.31).

Regarding the need for the inclusion of social groups in the definition
of genocide, Chalk (1989) argues that

Within living memory the governments of the Soviet Union, Germany and
Kampuchea had defined class, mental and physical defects, and sexual preference as primary classifications in their societies. In the hands of rulers
who claimed a monopoly on truth and of a bureaucracy which did their bid-
ding, membership of these social categories had proven lethal to millions of
human beings. It seem[s] obvious that researchers of genocide must investigate the destruction of such social groups or surrender any hope of explaining the modern world in all its complexity. (p. 151)

It is instructive to note that in 1959, Pieter N. Drost, a Dutch law professor, argued ardently that the United Nations needed to redefine genocide as "the deliberate destruction of physical life of individual human
beings by reason of their membership of any human collectivity as such"
(p. 125). By doing so, the definition would be more inclusive—and thus
not exclude political and social groups that were targeted for annihilation.
Totten (2002) has argued that

one needs to realize and appreciate the fact that the very wording of the
Convention includes the following: "in the *present* Convention" (italics
added). The use of "present" denotes that the Convention, as it stands, is
not set in stone for eternity. Indeed, it is possible to revise it so that it is clear
in its statements and more efficacious in its reach and intent to protect people from the insidious scourge of genocide. It seems that if that the world
community truly cares about preventing genocide then it is reasonable to
radically overhaul this remarkable yet flawed Convention in order to tighten
it up, delete the ambiguity, and make it inclusive enough to ensure that all
groups of people are placed under its protective umbrella. (p. 566)

All that said, the Genocide Convention, along with the United Nations Declaration on Human Rights, constitutes a major milestone in the history of international law vis-à-vis the attempt to provide protection for hundreds of millions across the globe from horrific abuse.

The hope among many was that the Convention on the Prevention and Punishment of the Crime of Genocide would put "teeth" into the cry of "Never Again." The aforementioned hope, though, was one that was not to be realized. As R. J. Rummel (2002), a political scientist and a scholar of genocide and democide studies, has trenchantly noted, "instead of 'Never Again,' the fact of the matter is, genocide reappeared in the last half of the twentieth century again, and again, and again" (p. 173).

DEFINITIONAL ISSUES OF GENOCIDE:
THE DEBATE CONTINUES

The on going debate (and one must add, confusion) over the most accurate and "workable" definition of genocide has played into the hands of those who do not wish to admit to the rest of the world (e.g., the perpetrators themselves) or wish to recognize (e.g., intergovernmental organizations such as the United Nations and/or individual nations) that a genocide is in the making. In certain instances, it has also hamstrung government analysts and even scholars and activists in ascertaining whether a situation is moving toward genocide or has actually exploded into genocide.

While the United Nations and most individual governments accede to the definition used in the U.N. Convention on the Prevention and Punishment of Genocide, scholars and activists have, as noted above, debated the usefulness of the latter. Over and above the debate regarding the exclusion of social and political groups from the Convention's definition of genocide, heated debates have centered around the wording and the actual meaning of key words in the UNCG. Indeed, at the heart of the often contentious, and seemingly endless, debate over the efficacy of the UNCG's definition is the issue of "intent." (Again, in part, the definition used in the UNCG reads as follows: "In the present Convention, genocide means any of the following acts committed with *intent* to destroy, in whole or in part, a national, ethnical, racial or religious groups, as such...." (italics added)). In *Genocide: Its Political Use in the Twentieth Century*, Kuper (1981) asserted that "The inclusion of intent in the definition of genocide introduces a subjective element, which would often prove difficult to establish.... In contemporary extra-judicial discussions of allegations of genocide, the question of intent has become a controversial issue, providing a ready basis for denial of guilt" (p. 33). The inclusion of the word

"intent," then, seems to suggest that a situation can only be deemed genocide if, in fact, there is hard evidence that genocide was the actual intent
of the perpetrator. But, as Kuper ponders, what if a perpetrator of genocide asserts that he/she never had any intent to commit genocide, and
what if there is no documentation or any corroborative evidence whatsoever that there was an actual plan to commit genocide?

Tellingly, when confronted with the accusation that Paraguay committed genocide against the Achè people, the Paraguayan defense minister
argued as follows: "Although there are victims and victimizer, there is not
the third element necessary to establish the crime of genocide—that is
'intent.' Therefore, as there is no 'intent,' one cannot speak of 'genocide'"
(cited in Alvarez, 2001, p. 52). The latter case provides a counterweight to
those who claim, as some scholars do, that the debate over the definition
of genocide is little more than an academic exercise. Indeed, in the case
of Paraguay, not only does the Convention's definition play into the hands
of deniers, but it could possibly be used (though, hopefully without success) as a defense in court against the charge of genocide.

In *The Prevention of Genocide*, Kuper (1985) further argued that "There
are also problems in determining the conditions under which intention
can be imputed" (p. 12). Continuing, he stated that "I will assume that
intent is established if the foreseeable consequences of an act are, or seem
likely to be, the destruction of a group. But this may be controversial....
[Ultimately,] it would be for a court to decide the issue in the circumstances of the particular case" (Kuper, 1985, pp. 12, 13).

Some scholars, such as Barta (1987) and Huttenbach (1988), have suggested that intent should be eliminated as a criterion, thus avoiding, altogether, the ambiguity inherent in it.

Helen Fein (1990), a sociologist and executive director of the Institute
for the Study of Genocide in New York City, though, makes the following
argument vis-à-vis the meaning of "intent":

> As sociologists, immersed in the distinctions between "manifest" and
> "latent" function as a paradigm of intended and unintended action, we have
> needlessly confused the meaning of intent. Intent or purposeful action—or
> inaction—is not the same in law or every-day language as either motive or
> function. An actor performs an act, we say, with intent if there are foresee-
> able ends or consequences...
>
> As some attempt to make a case under the Convention (not, so far, insti-
> gating the UN to act) we can see that the Convention has greater flexibility
> than understood by some:
>
>> The "intent" required by the Convention as a necessary constituent
>> element of the crime of genocide cannot be confused with, or inter-
>> preted to mean, "motive".... The "intent" clause of Article II of the
>> Genocide Convention requires only that the various destructive acts—

killings, causing mental and physical harm, deliberately inflicted conditions of life, etc.—have a purposeful or deliberate character as opposed to an accidental or unintentional character. (Hannum & Hawk, 1986, pp. 140-146, as cited in Fein, 1990, pp. 19-20)

Unfortunately, many, both inside and outside of government and the academy, do not seem to be cognizant of, or, at the least, appreciative of, the above points made by Fein. And that is a major problem for when the very wording in such a significant definition used by the international community to ascertain whether genocide is being committed or not is confusing, it constitutes a major stumbling block. Likewise, it sends the wrong message to past, current and future perpetrators of genocide who may think that they can use semantics to avoid prosecution.

Other wording in the UNCG that has caused consternation is the phrase, "in whole or in part." As Kuper (1985), again, notes: "The ambiguity lies in determining what number or what proportion would constitute a part for purpose of the definition.... Presumably, the Convention is intended to deal with acts against large numbers, relative to the size of the persecuted groups, and it would rest with the courts to adjudicate on this issue" (p. 12).

In light of the ambiguities in the wording of the Convention, it is plain to see why many scholars have committed great amounts of time and effort to analyzing, interpreting, and rephrasing the wording of the definition in the Convention in an attempt to strengthen it. For the same reason, many scholars have developed their own definitions of genocide. While some have been successful in developing stronger definitions, others have muddied the waters with definitions that are as ambiguous, if not more so, than that used in the UNCG. Presently, at least, such definitions are irrelevant vis-à-vis international law, for only the definition in the United Nations Convention on the Prevention and Punishment of the Crime of Genocide is used to prosecute the perpetrators of genocide. (For an informative, instructive, and detailed discussion of definitions of genocide, including their strengths and weaknesses, see Fein, 1990.)

The purpose of broaching the above issues is not to make excuses for or to "rationalize away" the fact that the international community has been agonizingly slow in recognizing actual genocides for what they are. Nor is it to excuse the totally inadequate and negligent way in which the international community has reacted to a genocide once one has been detected. Rather, it is to illustrate the complexity of the situation, even for scholars of genocide, in ascertaining that which constitutes genocide versus that which constitutes a series of sporadic massacres, war crimes or crimes against humanity in a region. (For a solid discussion as to how to address the issue of "definitions of genocide" in the classroom, see Chapter 3, "Wrestling with the Definition of 'Genocide': A Complex But Critical Task" by Totten, and Chapter 4, "Defining Genocide: Issues and Resolutions" by Huttenbach in this book.)

THE COLD WAR YEARS AND GENOCIDE

We move now to an examination of genocide in the post-World War II period, to a time when the world was confronted by an enormous number of bewildering stresses and strains: economic boom and bust, decolonization and wars of liberation, social protest and wide-sweeping calls for change, and the biggest threat of them all—mutually-assured total nuclear destruction (MAD). Characterizing many of these stresses were the numerous genocidal outbreaks which took place in the former colonial lands vacated (sometimes amicably, but often violently) by departing European powers in the years following 1945. Nowhere were these played out with such devastating ferocity as in Africa and Asia.

The objectives of those who shaped the post-1945 agenda increasingly became diluted as the twentieth century wore on, until the postwar cry of "Never Again!" became more and more muted; indeed, until the second half of the century began to appear as nothing other than a continual period of killing—in large wars, small wars, civil wars, and sometimes when there was no war at all. A number of these stand out as "models" of what the world became during the time that became known as the Cold War, a period approximating 1945 to 1989 (with localized variations).

The first concerns a place now often forgotten in the popular consciousness—and almost completely overlooked by genocide scholars. This is the case of a small, short-lived country in west Africa called Biafra. For many people born before the mid-1970s, the very name instantly conjures up images of babies with large staring eyes and bloated bodies, tiny stick-like limbs and a helplessness preceding death which only starvation can bring. Biafra was formed in 1967, when the Eastern State of Nigeria broke away to establish itself as an independent country. The Nigerian Civil War of 1967-1970, which followed, was the first occasion in which scenes of mass starvation were brought home to a television-dominated West, and millions throughout Europe, North America, and elsewhere were horrified by what they saw. Less apparent was the reality which lay behind this otherwise simple case of a brutal and bloody secessionist conflict, for in the Nigerian determination to defeat the Biafran breakaway state a deliberately designed genocidal policy of enforced famine was perpetrated against the population of the newly-formed country (Kuper, 1981, pp. 75-76).

The Biafran conflict led to an eventual death toll of up to a million people, mostly of the largely Christian Ibo ethnic group. The Nigerian Federal Army, and the government which supported it, was a perpetrator of genocide through a premeditated and strictly-enforced policy of starvation, as well as the military targeting of civilians. There was little doubt as to the Nigerians' genocidal intent; Biafra existed only two-and-a-half years, until its final collapse in January 1970.

The year 1971 saw an independence struggle in which East Pakistan sought to secede from West Pakistan, a move which was resisted with staggering violence. The subsequent emergence of the independent nation of Bangladesh was accompanied by some three million dead and a quarter of a million women and girls raped, the result of a calculated policy of genocide initiated by the government of West Pakistan for the purpose of terrorizing the population into accepting a continuance of Pakistani rule over the region. Ultimately the strategy did not work, as Bangladesh achieved its independence after the involvement of India in the conflict and the consequent defeat of the Pakistani forces.

The following year, 1972, another outbreak of genocidal violence took place in the tiny central African nation of Burundi, where a Hutu-instigated uprising against Tutsi domination resulted in the army subjugation and massacre of scores of thousands of Hutu civilians over a five-month period. The final total numbered up to 150,000 people, men, women, and children, and ushered in a period of Tutsi dominance which was to last for several decades—and during which time the Hutu majority population was reduced to a position of entrenched second-class subservience (Lemarchand, 1997).

It was characteristic of the era that Burundi became a location for Cold War rivalries. The Western world, and in particular the United States, saw the catastrophe that befell the Hutu as an irrelevancy so far as the bigger picture of defeating communism was concerned. The French government saw the conflict as an opportunity to reinforce its preferred Francophone client-state, while communist countries such as China and North Korea took the opportunity to assist the Tutsi junta with arms and infrastructural support as a means to woo the regime away from the West.

A little over two years after the worst of the violence ceased in Burundi, a communist tyrant, Pol Pot, and his cronies, the Khmer Rouge, won a bloody civil war in Cambodia, and began one of the most radical attempts at remodeling an existing society the world had ever seen. In taking the Cambodian people back to the Year Zero, as Pol Pot put it, at least 1.5 million people lost their lives (Kiernan, 1995, p. 334), though the figure was almost certainly higher. This would appear to be a clear-cut case of genocidal mass murder, though some strenuously question whether or not the massive human destruction that took place can truly be deemed a case of genocide when examined under the lens of the United Nations definition. The difficulty lies in the fact that the majority of the deaths that took place under Pol Pot's Khmer Rouge were perpetrated against their own Khmer (Cambodian) people (thus the appellation, "autogenocide"). Most of the victims were not targeted for reasons related to their membership in any of the groups identified under the United Nations definition, but were rather killed for social or political reasons. That said, there were specific groups that were singled out for "'eradication' by the leadership of the ruling Communist Party of Kampuchea because of their membership to a particular group; and these included Cambodia's preeminent religious group, the Buddhist monkhood, Cambodia's ethnic minorities, and the large portion of the Cambodian 'national group' deemed to [have been] tainted by 'feudal,' 'bourgeois' or 'foreign' influences'" (Hawk, 1988, p. 138).

Ultimately, though, scholars and others are confronted with a primary definitional issue: under what circumstances may they depart from the U.N. definition in order to apply the classification of genocide to an event of massive human destruction, and what are the implications of our doing so? Under international law, of course, no charge of genocide could stick in a case where those killed came from a political or social group. The thorny issue of " intent" could also come into play. Pol Pot intended to create a new type of communist utopia, to be sure, but did he intend to annihilate millions of his own people in order to do so—and were his victims targeted for the sole reason of their existence? The answer in regard to what took place in Cambodia is, sometimes yes, at other times no. The issues of "target group" and of "intent" are key sticking points militating against a satisfactory blanket application of the U.N. Convention in this case.

While the cataclysm of Cambodia was being played out, yet another Cold War genocide was taking place elsewhere in Asia, this time in the former Portuguese overseas territory of East Timor. In 1975, one of the political factions jockeying for power in the aftermath of Portuguese decolonization, Fretilin, declared the territory's independence; within weeks, Indonesian military forces invaded, declared East Timor to be that country's twenty-seventh state, and began a systematic campaign of human rights abuses which resulted in the mass murder, starvation, and death by torture of up to 200,000 people—about a third of the preinvasion East Timorese population (Dunn, 1997).

For many years, the international response to what was happening in East Timor was one of indifference. Indonesia's neighbor, Australia, was especially keen not to antagonize the populous nation to its north, and was the first (and for a long time, only) country to recognize the de jure incorporation of East Timor into the body of Indonesia (Aubrey, 1998).

United Nations resolutions calling on Indonesia to withdraw were ignored, and the United States, anxious lest a hard-line approach toward the annexation be seen by the Indonesians as a reason to look elsewhere for support—for example, to nonaligned nations—trod very softly on the whole issue (Gunn, 1997). Only in 1999, after a long period of Indonesian oppression and the threat of another outbreak of genocidal violence (this time committed by Indonesian-backed militias and units of the Indonesian army), was East Timor freed. In 2002, the first parliament, elected by universal suffrage and guaranteed by the United Nations, took its place in the community of nations.

In sum, the Cold War had a devastating effect on post-1945 hopes that a new, nongenocidal regime could be created throughout the world. Not only were peoples and groups in conflict left to fight out their differences unimpeded; all too often, as capitalist and communist states saw the possibility of achieving an advantage through either action or inaction, those committing genocidal acts were frequently aided and abetted for the most blatant of realpolitik motives. Britain, for example, refused to assist Biafra in alleviating its distress, for to do so would further undermine Nigeria at a time when oil exploration was starting to bear fruit, and a strong Nigeria was needed to keep out Soviet influence in sub-Saharan Africa, as well as to block Francophone designs.

As long as the Cold War raged, there was little chance that the kind of pressures likely to lead to a genocidal situation would find a "release valve." The great powers played a leading role in manipulating local conflicts so as to suit their own needs, after which each side was able to serve as a proxy in the greater ideological conflicts of the time. The Cold War showed with great clarity that the world's major players paid only lip service to their postwar commitment to "never again" stand by while genocide took place.

POST-COLD WAR YEARS AND GENOCIDE: A MOVE TOWARD THE INTERVENTION AND PREVENTION OF GENOCIDE?

Following the end of the Cold War, it appeared that a sea change was taking place in international relations regarding the willingness of nations to intervene to prevent genocide. More specifically, as Geoffrey Robertson (2000) notes in his book, *Crimes Against Humanity: The Struggle for Global Justice,*

> It was not until 1993, after the Cold War was over and as the spectre of "ethnic cleansing" returned to Europe, that there was sufficient superpower

resolve to apply the *proviso* to Article 2(7) [of the United Nations Charter, which reads, in part, "Nothing contained in the present Charter shall authorize the United Nations to intervene in matters which are essentially within the domestic jurisdiction of any State ..."], namely that it could be overridden by Chapter VII. This is the chapter of the Charter which permits the Security Council to order armed intervention against any state once it has determined that such a response is necessary to restore international peace and security. Since Article 55 expressly makes the observance of human rights a condition necessary for peaceful relations, the appalling crimes against humanity [and genocide] which occurred after 1945 could have been forcibly combated by the UN under its Chapter VII power, but until the Balkan atrocities in the 1990s the Security Council never sought or even thought to invoke military action upon human rights grounds. (p. 25)

But, in fact, did a sea change occur? In actuality, the verdict is still out.

Certainly, the issue of "sovereignty" in the post-Cold War period came under closer scrutiny, if not under attack, and was no longer seen, in certain cases at least, as sacrosanct. That is, a nation committing genocide or other egregious human rights violations against its own people was no longer seen as "untouchable." Furthermore, such situations were no longer automatically deemed a matter of "internal affairs."

Be that as it may, if one were to ask if a "sea change" had actually taken place in regard to the realities on the ground every time a genocide appeared on the horizon, the answer would be an unequivocal "No!" Why? Because the international community failed, time and again, to halt various genocides in a timely and effective manner.

Still, there was a change in the air for, in certain notable cases, the international community did act in concert to stave off certain potential genocides.

The disaster of Bosnia-Herzegovina, in the former Yugoslavia, dominated international news for much of the 1990s. It was the first genocide committed within a military conflict in Europe since the Holocaust, and involved the killing and displacement of Bosnia's Muslims by both local Serbs and Serbian forces from the Yugoslav National Army. These actions were justified by the perpetrators on the grounds of ideology and the desire to acquire (or retain) territory seen as sacred by the Serbs (Cigar, 1995). The questions thrown up by the genocide were many, but of equal, if not greater, concern was the position of the bystanders. Efforts to stop the genocide while it was happening were neither quick to emerge nor effective when attempted (Gow, 1997; Power, 2002, pp. 247-327; Simms, 2001).

The genocide in Bosnia—though probably the most closely reported genocide in history (Sadkovich, 1998)—was yet another case of international inaction in the face of massive human rights violations. In Bosnia, the Western powers, led by the U.N., the European Community and the North Atlantic Treaty Organization (NATO) (and preeminently, the United States and Britain), failed consistently both to resolve the war and to stop the killing. Diplomatic efforts were subjected to ridicule by the perpetrators, and military efforts reached their lowest ebb with what was effectively a surrender by Dutch peacekeepers—while acting as part of the United Nations Protection Force (UNPROFOR)—of the so-called "safe haven" of Srebrenica. Following this surrender, up to 7,000 Muslim men and boys were hunted down and killed; it was the greatest massacre on European soil since the Second World War (Honig & Both, 1996).

Also of note is the fact that the Serbians used rape as "a weapon of war" (Rieff, 2002, p. 67) and as a means, in part, to destroy the Bosnian Muslim population.

In 1999, almost as an acknowledgment of a guilty conscience concerning their failure to act in Bosnia, the combined air forces of the United States, Great Britain, France, Germany, Italy, and the Netherlands, operating together as part of NATO, attacked Serbia with the intention of forcing the Serbs to stop their persecution of the ethnic Albanian population living in the Serbian province of Kosovo (Weymouth & Hening, 2001). It was the first occasion in which a war was fought for the express purpose of stopping a genocide before its worst horrors took place. Under international law, however, the attack was illegal; it was neither called for nor approved by the United Nations. Nonetheless, after a lengthy and intensive bombing campaign, the Serbian regime of Slobodan Milosevic pulled its troops out of Kosovo. U.N. peacekeepers moved into the province, allowing the one million persecuted Kosovars, who had been expelled from the country in a huge outbreak of so-called "ethnic cleansing," to return home.

The NATO intervention in Kosovo is an example of how a potential genocide can be addressed early on if the international will to do so is present. That said, the action was not without controversy; not only was there the issue of acting without the imprimatur of the U.N. but many innocent people were also killed as a direct result of the bombing missions. Many also assert that the use of ground troops would have likely avoided a good number of the innocent deaths that resulted from the bombings.

The international community's guilty conscience, if indeed one existed, did not help the Tutsi population of the tiny African country of Rwanda earlier in the 1990s. There, in 1994, in the space of 100 days, perhaps as many as one million people were killed in a genocide. The United Nations did practically nothing to stop the killing, and indeed the killing continued right up to the point where the extremist Hutu killers, attacking innocent Tutsi and moderate (which is to say, democratic) Hutu victims, had almost virtually exhausted their killing machine. The killings were, for the most part, done by hand, with the murderers using machetes (*pangus*) or nail-studded clubs (*masus*). This was an outbreak of savagery on a grand scale. And yet, as we now know—and as the U.N. Security Council had been warned nearly four months beforehand by a defector from those who would become the killers—the Rwanda genocide was carefully planned in advance, to such a degree that death lists of names marked for murder had been prepared long before the killing actually began.

The Rwanda genocide tells us much about the nature and priorities of the international system in the 1990s, and offers lessons of which all citizens of democratic nations should take note. To a large degree, the main players in both world and regional politics did nothing to stop the killing or intervene to rescue the survivors. The United Nations Security Council scaled back the size and scope of the peacekeeping forces already present on the ground, and worked hard to ensure that no aid of a practical nature was sent to Rwanda (Barnett, 2002). Belgium, the former mandatory power, withdrew from the Rwanda altogether, along with most other Western countries. France, after the worst of the killing had taken place, established a so-called "safe zone" in the south, but its ultimate effect was to protect the tens of thousands of Hutu killers who had poured into the area escaping the advancing Tutsi rebel army (Melvern, 2000, pp. 210-226).

In view of cases such as Bosnia and Rwanda, how can the final years of the twentieth century be interpreted? If the trend over the course of the past century has been toward greater killing, greater targeting of civilians, and a greater likelihood than ever before that groups are being singled out for destruction, what hope does this offer those with a commitment to the sanctity of life?

There is, unfortunately, no ready-made answer to this question. Genocide still exists, despite the legal processes that have been instituted against it. Genocide, one of the legacies of the twentieth century, has been described as "the crime of crimes"; and if it is true that the twentieth century was an "age of genocide," we need to ask not only why this was so, but what its impact has been on world civilization, as a whole. Furthermore, it seems natural that we should try to build the means of ensuring that the twenty-first century will not go the same way, but of course, that is no easy task.

In 1993 and 1994 two ad hoc international courts were established by the United Nations Security Council for the express purpose of trying those indicted for genocide, crimes against humanity, and war crimes, as they pertained to the former Yugoslavia and Rwanda. The International Criminal Tribunal on Yugoslavia (ICTY) and the International Criminal

Tribunal on Rwanda (ICTR) have both been moderately successful (if excruciatingly slow) in bringing prosecutions, and in September 1998 the ICTR made history when it found Jean-Paul Akayesu, the former mayor of the Rwandan town of Taba, guilty of the crime of genocide. This was the first time any international court had issued such a verdict for this specific crime. Other prosecutions have followed, and case-law precedents in the law of genocide prosecutions are now growing. Quite clearly, genocide is now recognized throughout the world as a crime that is abhorrent.

It is noteworthy that during the trial of Jean-Paul Akayesu, Pierre Prosper, an African American serving as a prosecutor at the ICTR, argued ardently to convince the court that "sexual violence against women could be carried out with an intent that amounted to genocide" (Power, 2002, p. 485). As Power (2002) notes, Prosper argued that "a group could physically exist, or escape extermination, but be left so marginalized or so irrelevant to society that it was, in effect, destroyed" (p. 486). Ultimately, the ICTR found that the "systematic rape of Tutsi women in Rwanda's Taba commune was found to constitute the genocide act of 'causing serious bodily or mental harm to members of the group'" (Power, 2002, p. 486) and thus Akayesu was found guilty of genocide.

As if to demonstrate the firmness of the international community's resolve to do something about genocide—and to prove that impunity is no longer an option for those who commit it—the Rome Statute of the International Criminal Court (ICC) was adopted on July 17, 1998. The court was established by the United Nations under the aegis of the U.N. Security Council. The statute gives the court jurisdiction for the crime of genocide, crimes against humanity, and war crimes. It is both an extension of the ICTY and the ICTR, and the fulfillment of the promises first articulated in the aftermath of 1945 (Schabas, 2001, pp. 1-21). The Court became operative on July 1, 2002, after a minimum of 60 U.N. countries had ratified it. Notable among those refusing to ratify was the United States. Cognizant of the United States' intransigence, most European countries had earlier decided that the leadership of the U.S. would not be required for the purpose of establishing what was seen as a highly moral body, the purpose of which would be to assist in safeguarding the peace of the world and the lives of its citizens. The U.S. promptly sought, and received, an agreement within the U.N. that would place Americans serving in foreign postings outside the court's jurisdiction.

It is perhaps instructive to note that the ICC, after considerable debate, decided to absorb the U.N. Convention on Genocide, including its definition, directly into its charter.

INTERVENTION AND PREVENTION OF GENOCIDE:
IS IT POSSIBLE?

Since the late 1970s, an ever-increasing number of scholars—and in the mid- to late-1980s, more and more policy makers—focused attention on the development of effective methods to intervene and/or prevent genocide. Based on the fact that genocide continues to plague humanity, such efforts are, obviously, still at their most incipient stage.

The aforementioned efforts have involved a wide array of theoretical and practical components that will, hopefully, lend themselves to detecting potential situations that may lead to genocide and/or assisting in the effort to staunch genocides already under way. Among such efforts are the following: developing categories (or classifications) of genocide; defining the "nature of societies in which genocides take place" (Charny, 1988b, p. 5); developing potential predictors of genocide; developing "specified conditions of communal conflicts and the analysis of "accelerators" (Harff, 1994, pp. 25-30); developing of theoretical risk assessments (Gurr, 1994, pp. 20-24); developing various types of early warning systems by nongovernmental organizations, individual nations, and the United Nations; and, establishing of the ICC, which many hope will serve as a deterrent of sorts to those contemplating the perpetration of genocidal acts. (For a more detailed and technical discussion of many of the above-mentioned issues, see Gurr, 1994, and Harff, 1994. For a more general discussion of the above issues, see, Rittner, Roth, & Smith, 2002.)

Possibly the thorniest issue of all is that of political will. That is, the political will of the United Nations Security Council and individual states (especially, the strongest nations, including but not limited to the United States, Germany, France, Great Britain, Russia, China) to act in a timely and effective fashion to intervene and/or prevent genocide. Driven to a large extent by realpolitik and, though to a lesser extent today, the sense that sovereignty is all but sacrosanct, many, if not most, nations are still tentative about intervening in another nation's so-called "business." Indeed, all one needs to do is look at either the record of the United Nations or the United States in regard to their efforts at intervention and prevention (Power, 2002)

Totten (2002) asserts that there are a host of additional issues and concerns that genocide scholars need to take into account as they continue to work with nongovernmental personnel, policy makers, and others in the development of effective mechanisms for the intervention and prevention of genocide, and these include, but are not limited to the following: information-gathering and analysis (including satellite observation methods, systematic gathering of information, accurate analysis of information, timely dissemination of findings to key bodies); intelligence sharing; confidence building measures in states enmeshed in conflict; preventive diplomacy; Track I diplomacy; Track II diplomacy; conflict prevention; conflict management; conflict resolution; information peacekeeping; peacekeeping; diplomatic peacekeeping; peacemaking; peace enforcement; sanctions (various types, and the purposes and efficacy of); failed states (how to deal with them in an effective manner); partitioning (the efficacy of such, and to make it more effective); temporary protection measures for refugees fleeing internal and other types of conflict; the plight of displaced persons within a state; the conceptualization of various types of policing efforts; institution building; the role of the United Nations vis-à-vis intervention and prevention efforts; and, possibly most significantly, systemic issues (e.g., endemic poverty; the shocking divide between the "haves" and "have nots"; greedy consumption by the few—and the United States of America is a primary culprit of this phenomena—to the detriment of the many; and unjust governmental systems) that cause dissension and often result in violent conflict, including genocide. To neglect to address the aforementioned issues is tantamount to groping blindly at intervention and prevention.

Ultimately, as discussed above, with the establishment of the ICC (and the success, thus far, of the ICTY and ICTR), movement, even if glacial, is being made in the direction of recognizing that individual nations have no right, morally, legally, or otherwise, to commit genocide. (For a more in-depth discussion of the issues of intervention and prevention of genocide, see Chapter 9, "The Intervention and Prevention of Genocide," by Totten in this volume.)

THE FUTURE?

Knowing the innate atavism of humanity and the thirst for power and revenge by some, one can safely, if sorrowfully, predict that genocide will be perpetrated again in the future—if not, again and again. On the other side of the coin, never in the history of humanity has so much been done in such a short span of time to attempt to quell, if not completely stop, genocide from being perpetrated.

The development and ratification of the U.N. Convention on the Prevention and Punishment of Genocide, the trials of the Nazi war criminals, the revitalized and vigorous scholarly and activist focus in the 1980s, 1990s, and early 2000s on the intervention and prevention of genocide, the trials conducted by the ICTR and ICTY, and the development of the ICC all bode well for the future. That said, what is now needed is a central, highly sophisticated genocide early warning system and the international resolve (i.e., political will) to quickly and effectively act early on in an attempt to prevent genocide from becoming a reality. To accomplish the latter is going to be the most difficult component, thus far, in the attempt to prevent genocide from becoming a reality, for nations are often apt to revert to realpolitik when positions, decisions, and actions are not perceived to be in their own, often selfish, interest.

REFERENCES

Alvarez, A. (2001). *Governments, citizens, and genocide: A comparative and interdisciplinary approach.* Bloomington: Indiana University Press.

Aubrey, J. (Ed.) (1998). *Free East Timor: Australia's culpability in East Timor's genocide.* Sydney: Vintage/Random House Australia.

Barnett, M. (2002). *Eyewitness to a genocide: The United Nations and Rwanda.* Ithaca, NY: Cornell University Press.

Barta, T. (1987). Relations of genocide: Land and lives in the colonization of Australia. In I. Wallimann & M. N. Dobkowski (Eds.), *Genocide and the modern age: Etiology and case studies of mass death* (pp. 237-252). New York: Greenwood.

Bartrop, P. R. (2001). The Holocaust, the Aborigines, and the Bureaucracy of destruction: An Australian dimension of genocide." *Journal of Genocide Research, 3*(1), 75-87.

Bartrop, P. R. (2002). The relationship between war and genocide in the twentieth century: A consideration. *Journal of Genocide Research, 4*(4), 519-532.

Chalk, F. (1989). Definitions of genocide and their implications for prediction and prevention. *Holocaust and Genocide Studies, 4*(2), 149-160.

Chalk, F., & Jonassohn, C. (1988). The history and sociology of genocide. In I. W. Charny (Ed.), *Genocide: A critical bibliographic review* (pp. 39-58). New York: Facts on File.

Chalk, F., & Jonassohn, C. (1991). *The history and sociology of genocide.* New Haven, CT: Yale University Press.

Charny, I. W. (Ed.). (1988a). Intervention and prevention of genocide. In *Genocide: A critical bibliographic review* (pp. 20-38). New York: Facts on File.

Charny, I. W. (Ed.). (1988b). The study of genocide. In *Genocide: A critical bibliographic review* (pp. 1-19). New York: Facts on File.

Cigar, N. (1995). *Genocide in Bosnia: The policy of "ethnic cleansing."* College Station: Texas A & M University Press.

Drost, P. (1959). *The crimes of state* (Vol. 2). Leyden, the Netherlands: A.W. Sythoff.

Dunn, J. (1997). Genocide in East Timor. In S. Totten, W. S. Parsons, & I. W. Charny (Eds.), *Century of genocide: Eyewitness accounts and critical views* (pp. 264-290). New York: Garland.

Fein, H. (1990). Genocide: A sociological perspective. *Current Sociology, 38*(1), 1-126.

Gow, J. (1997). *Triumph of the lack of will: International diplomacy and the Yugoslav War.* New York: Columbia University Press.

Gunn, G. C. (1997). *East Timor and the United Nations: The case for intervention.* Lawrenceville, NJ: Red Sea Press.

Gurr, T. R. (1994). Testing and using a model of communal conflict for early warning. *The Journal of Ethno-Development, 4*(1), 20-24.

Harff, B. (1994). A theoretical model of genocides and politicides. *The Journal of Ethno-Development, 4*(1), 25-30.

Hawk, D. (1987, January). Quoted in the Institute of the International Conference on the Holocaust and Genocide's *Internet on the Holocaust and Genocide* (Jerusalem), Issue Eight, n.p.

Hawk, D. (1988). The Cambodian genocide. In I. W. Charny (Ed.), *Genocide: A critical bibliographic review* (pp. 137-154). New York: Facts on File.

Honig, J. W., & Both, N. (1996). *Srebrenica: Record of a war crime.* New York: Penguin Books.

Huttenbach, H. (1988). Locating the Holocaust on the genocide spectrum. *Holocaust and Genocide Studies, 3*(3), 289-304.

Katz, S. T. (1994). *The Holocaust in historical context.* New York: Oxford University Press.

Kiernan, B. (1995). The Cambodian Genocide, 1975-1979. pp. 334-371. In S. Totten, W. S. Parsons & I. W. Charny (Eds.), *Century of genocide: Eyewitness accounts and critical views.* New York: Garland.

Kuper, L. (1981). *Genocide: Its political use in the twentieth century.* New Haven, CT: Yale University Press.

Kuper, L. (1985). *The prevention of genocide.* New Haven, CT: Yale University Press.

LeBlanc, L. (1991). *The United States and the genocide convention.* Durham, NC: Duke University Press.

Lemarchand, R. (1997). The Burundi Genocide. In S. Totten, W. S. Parsons, & I. W. Charny (Eds.) *Century of genocide. Eyewitness accounts and critical views* (pp. 317-333). New York: Garland.

Lemkin, R. (1944). *Axis rule in occupied Europe: Laws of occupation, analysis of government, proposals for redress.* Washington, DC: Carnegie Endowment for International Peace.

Lemkin, R. (2002). Totally unofficial man. In S. Totten & S. Jacobs (Eds.), *Pioneers of genocide studies* (pp. 365-399). New Brunswick, NJ: Transaction.

Melvern, L. (2000). *A people betrayed: The role of the West in Rwanda's genocide.* London: Zed Books.

O'Shea, S. (2001). *The perfect heresy: The revolutionary life and death of the medieval Cathars.* London: Profile Books.

Power, S. (2002). *"A problem from hell": America and the age of genocide.* New York: Basic Books.

Reynolds, H. (1995). *Fate of a free people.* Ringwood, Victoria, Australia: Penguin Books.

Reynolds, H. (2001). *An indelible stain? The question of genocide in Australia's history.* Ringwood, Victoria, Australia: Viking/Penguin Books.

Rieff, D. (2002). Murder in the neighborhood. In N. Mills & K. Brunner (Eds.), *The new killing fields: Massacre and the politics of intervention* (pp. 55-69). New York: Basic Books.

Rittner, C., Roth, J. K., & Smith, J. M. (Eds.). (2002). *Will genocide ever end?* St. Paul, MN: Paragon House.

Robertson, G. (2000). *Crimes against humanity: The struggle for global justice*. New York: The New Press.

Rohde, D. (1997). *Endgame: The betrayal and fall of Srebrenica: Europe's worst massacre since World War II*. New York: Farrar, Straus and Giroux.

Rummel, R. J. (2002). From the study of war and revolution to democide. In S. Totten & S. L. Jacobs (Eds.), Pioneers of genocide studies (pp. 153-177). New Brunswick, NJ: Transaction.

Sadkovich, J. J. (1998). *The U.S. Media and Yugoslavia, 1991-1995*. Westport, CT: Praeger.

Safrai, S. (1976). The era of the Mishnah and Talmud (70-640). In H. H. Ben-Sasson (Ed.), *A history of the Jewish people* (pp. 307-382). London: Weiderfeld & Nicolson.

Schabas, W. A. (2001). *An Introduction to the International Criminal Court*. Cambridge: Cambridge University Press.

Simms, B. (2001). *Unfinest hour: Britain and the destruction of Bosnia*. London: Allen Lane/The Penguin Press.

Smith, R. (1987). Human destructiveness and politics: The twentieth century as an age of genocide. In I. Wallimann & M. Dobkowski (Eds.), *Genocide and the modern age: Etiology and case studies of mass death* (pp. 21-39). Westport, CT: Greenwood Press.

Stannard, D. E. (1992). *American holocaust: The conquest of the new world*. New York: Oxford University Press.

Totten, S. (1999). Technology and genocide. In I. W. Charny (Ed.), *Encyclopedia of genocide* (pp. 533-534). Santa Barbara, CA: ABC CLIO Press.

Totten, S. (2002). A matter of conscience. In S. Totten & S. Jacobs (Eds.), *Pioneers of genocide studies* (pp. 545-580). New Brunswick, NJ: Transaction.

Totten, S., Parsons, W. S., & Charny, I. W. (Eds.). (1997). *Century of genocide: Eyewitness accounts and critical views*. New York: Garland Publishers.

United Nations. (1978). *Study of the question of the prevention and punishment of the crime of genocide*. (July 4, E/CN. 4/Sub. 2/416)

United Nations Economic and Social Council. (1948). *Official records, August 26, Session 7*.

Weymouth, T., & Henig, S. (Eds.). (2001). *The Kosovo crisis: The last American war in Europe?* London: Reuters.

Whitaker, B. (1985, April). *Revised and updated report on the question of the prevention and punishment of the crime of genocide*. (E/CN.4/Sub.2/1985/6, 2 July 1985.) Submitted to the United Nations Subcommission on Prevention of Discrimination and Protection of Minorities of the Commission on Human Rights of the United Nations Economic and Social Council in Geneva.

Whitaker, B. (1988). Genocide: The ultimate crime. In P. Davies (Ed.), *Human rights* (pp. 51-56.). London and New York: Routledge.

CHAPTER 4

WRESTLING WITH THE DEFINITION OF "GENOCIDE"

A Critical Task

Samuel Totten

INTRODUCTION

What is genocide? To many—especially the "lay person"—that is relatively simple to answer. Indeed, a common answer might be that it is the effort to annihilate any group of people. It would be an excellent guess. But *technically* and *legally* speaking, it is not correct.

A major concern of many scholars and human rights activists is that there is not a consensus in regard to what constitutes the tightest, most logical, and workable definition of genocide. While the international community abides by the definition found in the United Nations Convention on the Prevention and Punishment of Genocide (UNCG), many scholars find the definition unsatisfactory and often refer to it as a "compromise" definition—and that is true for various reasons that shall be examined later in this chapter. Some scholars also have noted that the wording in the definition is ambiguous and that certain wording is often misconstrued by both scholars and policymakers. As one might imagine,

Teaching About Genocide: Issues, Approaches, and Resources, 57–74
Copyright © 2004 by Information Age Publishing
All rights of reproduction in any form reserved.

the lack of a consensus regarding the definition complicates both the study of genocide, *and, more importantly,* the efforts to prevent and/or prosecute cases of genocide.

There now exists a score or more definitions of genocide—running the gamut from *extremely* restrictive to *extremely* inclusive. While many of the definitions share certain commonalties, they also differ in significant ways from one another. A major problem surrounding such disagreement and lack of consensus over the definition of genocide is that it frequently results in acrimonious debate and disagreement among intergovernmental officials, governmental officials, scholars, and activists in regard to whether a situation is approaching genocidal proportions or not and/or is an outright case of genocide or not. The ramifications of such confusion can be, and often are, immense. The most significant problem is that it often results in a tentativeness to act—early on and in an effective manner, or *even at all*—in order to prevent or stanch genocide.

Taking all of the above into consideration, the following shall be explored herein: the coining of the term "genocide"; the United Nations' "compromise definition"; thorny issues surrounding the UN definition; other definitions of genocide; avoiding the use of the term genocide and/ or using euphemisms in the place of the term "genocide"; and methods and strategies for helping students gain a solid sense of the complexities surrounding the effort to define genocide (including a clear understanding as to what does and does not constitute genocide).

COINING OF THE TERM "GENOCIDE"

Raphael Lemkin (1900-1959), a Polish Jewish émigré and a law professor at Yale and Duke Universities who waged a one-man crusade for the development and ratification of an international convention against the perpetration of genocide, coined the term "genocide" by combining the Greek "genos" (race, tribe) and the Latin "cide" (killing). Following the Nazis' rise to power in Germany, Lemkin fled Europe and sought refuge, first, in Sweden, and then in the United States. His family, however, was murdered by the Nazis.

In *Axis Rule in Occupied Europe,* Lemkin (1944/1973) defined genocide in the following manner:

> Generally speaking, genocide does not necessarily mean the immediate destruction of a nation, except when accomplished by mass killings of all members of a nation. It is intended rather to signify a coordinated plan of different actions aiming at the destruction of essential foundations of the

life of national groups with the aim of annihilating the groups themselves. The objectives of such a plan would be the disintegration of the political and social institutions of culture, language, national feelings, religion, economic existence of national groups and the destruction of the personal security, liberty, health, dignity, and even the lives of the individuals belonging to such groups. Genocide is directed against the national group as an entity, and the actions involved are directed at individuals, not in their individual capacity, but as members of the national groups.... Genocide has two phases: one, destruction of the national pattern of the oppressed group; the other, the imposition of the national pattern of the oppressor. (p. 79)

Speaking of Lemkin's definition, Chalk and Jonassohn (1990) noted, critically, that, "Even nonlethal acts that undermined the liberty, dignity, and personal security of members of a group constituted genocide if they contributed to weakening the viability of the group. Under Lemkin's definition, acts of ethnocide—a term coined by the French [after World War II] to cover the destruction of a culture without the killing of its bearers—also qualified as genocide" (p. 9). Those who are against including ethnocide under the rubric of genocide argue that there is a qualitative and significant difference between those situations in which people are outright slain and when certain aspects of a peoples' culture are destroyed and thus such actions should not be commingled under a single definition.

Lemkin dedicated the latter part of his life working tirelessly—to the point of driving himself so hard his health literally deteriorated—to prod the international community to develop and ratify the United Nations Convention on the Punishment of the Crime of Genocide (UNCG). His crowning achievement was the ratification of the UNCG in 1948. (For a much more detailed discussion of Lemkin's background and effort to establish a convention against genocide, see his autobiography, "Totally Unofficial Man," Lemkin, 2002.)

THE UNITED NATIONS' "COMPROMISE DEFINITION"

From the outset, the development of the U.N. Genocide Convention was enmeshed in controversy. As genocide scholar Leo Kuper (1985) observed, nations with vastly different philosophies, cultures, and "historical experiences and sensitivities to human suffering" (p. 10) presented various interpretations as to what constituted genocide, and argued in favor of a definition and wording in the Convention that fit their particular perspective(s). As mentioned previously, the arguments and counterarguments resulted in what can best be described as a "compromise definition."

On December 11, 1946, the United Nations General Assembly passed the following *initial* resolution regarding the definition of genocide:

> Genocide is a denial of the right of existence of entire human groups, as homicide is the denial of the right to lives of individual human beings.... Many instances of such crimes of genocide have occurred, when racial, religious, *political*, and other groups have been destroyed entirely or in part ... [italics added].
>
> The General Assembly Therefore, Affirms that genocide is a crime under international law which the civilized world condemns, for the commission of which principals and accomplices—whether private individuals, public officials or statesmen, and whether the crime is committed on religious, racial, political or any other grounds—are punishable (quoted in Kuper, 1981, p. 23).

Of the utmost significance is that while this resolution "significantly narrowed Lemkin's definition of genocide by down playing ethnocide as one of its components, ... at the same time, it broadened the definition by adding a new category of victims—'political and other groups'—to Lemkin's list" (Chalk & Jonassohn, 1990, p. 10). That, however, would change.

The Soviet Union, Poland, and various other nations argued against the inclusion of political groups claiming that such an inclusion would not conform "with the scientific definition of genocide and would, in practice, distort the perspective in which the crime should be viewed and impair the efficacy of the Convention" (Kuper, 1981, p. 25). The Poles further asserted that "the inclusion of provisions relating to political groups, which because of their mutability and lack of distinguishing characteristics did not lend themselves to definition, would weaken and blur the whole Convention" (Kuper, 1981, p. 26).

The upshot is that political and social groups were excluded from the Convention. The sagacity of excluding such groups has been questioned, and in some cases outright criticized, by numerous scholars. Others, however, believe that the exclusion of political groups from the UNCG constituted a sound decision. For example, LeBlanc (1988) supports the exclusion of political groups because of the "difficulty inherent in selecting criteria for determining what constitutes a 'political group,' their instability over time, the right of the state to protect itself, and the potential misuses of genocide-labeling of antagonists in war and political conflict" (pp. 292-294). (For a cogent discussion of the debate surrounding the U.N. Convention on Genocide, see the chapter entitled "The Genocide Convention" in Kuper's (1981, pp. 19-39) *Genocide: Its Political Use in the Twentieth Century.*)

After a great deal of debate and compromise between and among the delegates and representatives of various nations, on December 9, 1948, the following Convention on the Prevention and Punishment of Genocide was approved by the General Assembly of the United Nations.

In the present Convention, genocide means any of the following acts committed with the intent to destroy, in whole or in part, a national, ethnical, racial or religious group, as such:

a. Killing members of the group;
b. Causing serious bodily or mental harm to members of the group;
c. Deliberately inflicting on the group conditions of life calculated to bring about its physical destruction in whole or in part;
d. Imposing measures intended to prevent births within the group;
e. Forcibly transferring children of the group to another group.

Significantly, "political groups" were no longer included, while the specific mention of national and ethnical groups had been added.

As commented on above, over the past 50 years, great debate among legal scholars, specialists in genocide studies, and others has taken place in regard to the focus and wording of the UNCG. What is addressed next is a succinct overview of some of the more pressing issues that have been the focus of such discussion.

THORNY ISSUES INHERENT IN THE UNCG

Four of the most contested issues that scholars have discussed and debated, time and again, vis-à-vis the definition of genocide used in the UNCG are the following: (1 & 2) that it, at one and the same time, is too inclusive in certain respects and too exclusive in other respects; (3) the phrase "committed with the intent" is often misunderstood and interpreted incorrectly; and (4) the wording "in whole or in part" often leads to confusion.

"Too Inclusive"?

As for being too inclusive, the Convention's definition "makes no distinction between violence intended to annihilate a group and nonlethal attacks on members of a group. 'Killing members of the group' and 'deliberately inflicting ... conditions of life calculated to bring about its physical destruction in whole or in part' are commingled in the definition with causing 'mental harm to members of the group' and 'forcibly transferring children of the group to another group'" (Chalk & Jonassohn, 1990, p. 11).

"Too Exclusive"?

As for being too exclusive, many scholars have been critical of the fact that both political and social groups are not specifically cited as coming under the protection of the UNCG. Furthermore, many scholars disagree with LeBlanc's (1988) assertion that there is great "difficulty inherent in selecting criteria for determining what constitutes a political group" (p. 292).

The debate over the exclusion of political groups is far from being simply an "academic or intellectual argument," and the Khmer Rouge-perpetrated genocide of its own people (those perceived as political enemies) is a case in point. Knowing full well that the UNCG does not include the mention of political groups, many individuals (including U.N. officials, governmental bureaucrats, and even scholars) have, for various reasons, "parsed" the victim groups murdered by the Khmer Rouge into those who come under the UNCG and those who do not. More specifically,

> Ignoring all the evidence available in Cambodia and their commitments to punish genocide, UN member states continued to refuse to invoke the genocide convention to file genocide charges at the International Court of Justice against the Cambodian government. Indeed, official UN bodies refrained even from condemning the genocide. Only in 1985 were bureaucratic inertia and political divides briefly overcome so that a UN investigation could finally be conducted. By then, because it had emerged that the Khmer Rouge had killed huge percentages of Muslim Chams, Buddhist monks, and Vietnamese *as such*, it proved relatively easy to show that the regime was guilty of genocide against distinct ethnic, national, and religious groups.... The subcommission noted that the horrors were carried out against political enemies as well as ethnic and religious minorities but found that this did not disqualify the use of the term "genocide." Indeed, in the words of Ben Whitaker, the UN special rapporteur on genocide, the [Khmer Rouge] had carried out genocide "even under the most restricted definition" (Power, 2002, pp. 153-154).

That said, specialists in international law note and insist that the debate over the exclusion of political groups from the UNCG is misguided. For example, in her article "The Crime of Political Genocide: Repairing the Genocide Convention's Blind Spot" in *The Yale Law Journal*, Beth Van Schaack (1997) argues that the Genocide Convention is not the sole authority on the crime of genocide. Rather, a higher law exists: the prohibition of genocide represents the paradigmatic *jus cogens* norm, a customary and preemptory norm of international law from which no derogation is permitted. The *jus cogens* prohibition of genocide, as expressed in a variety of sources, is broader than the Convention's prohibition as has been demonstrated with respect to the jurisdictional principle applied to acts of genocide. Notwithstanding that the framers of the Genocide Convention attempted to limit the prohibition of genocide by deliberately excluding political groups from Article II, the provision's without legal force to the extent that it is inconsistent with the *jus cogens* prohibition of genocide. Therefore, when faced with mass killings evidencing the intent to eradicate political groups in whole or in part, domestic and international adjudicatory bodies should apply the *jus cogens* prohibition of genocide and invoke the Genocide Convention (pp. 2259-2291).

The same, then, should be true regarding the exclusion of social groups from the UNCG. Likewise, it should apply to groups of specific gender (e.g., women or homosexuals, for example) should they be targeted for annihilation in whole or in part.

"Intent"

Some scholars, most notably Barta (1987) and Huttenbach (1988), have suggested that the term and concept of "intent" be eliminated as a criterion, thus avoiding altogether the ambiguity inherent in it.

In his book *Genocide: Its Political Use in the Twentieth Century*, Kuper (1981) readily acknowledges the fact that "the inclusion of intent in the definition of genocide introduces a subjective element, which would often prove difficult to establish.... In contemporary extra-judicial discussions of allegations of genocide, the questions of intent has become a controversial issue, providing a ready basis for denial of guilt" (p. 33). However, in *The Prevention of Genocide*, Kuper (1985) further comments that while "there are ... problems in determining the conditions under which intention can be imputed..., I will assume that intent is established if the foreseeable consequences of an act are, or seem likely to be, the destruction of a group. But this may be controversial ... [Ultimately,] it would be for a court to decide the issue in the circumstances of the particular case" (Kuper, 1985, pp. 12, 13).

Five years later, in 1990, Helen Fein, a sociologist and executive director of the Institute for the Study of Genocide in New York City, argued as follows in regard to the meaning of "intent":

As sociologists, immersed in the distinctions between "manifest" and "latent" function as a paradigm of intended and unintended action, we have needlessly confused the meaning of intent. Intent or purposeful action—or inaction—is not the same in law or every-day language as either motive or function. An actor performs an act, we say, with intent if there are foreseeable ends or consequences: for what purpose is different from why or for what motive is the act designed....

As some attempt to make a case under the Convention (not, so far, instigating the UN to act) we can see that the Convention has greater flexibility than understood by some: The "intent" required by the Convention as a necessary constituent element of the crime of genocide cannot be confused with, or interpreted to mean, "motive".... The "intent" clause of article II of the Genocide Convention requires only that the various destructive acts— killings, causing mental and physical harm, deliberately inflicted conditions of life, etc.—have a purposeful or deliberate character as opposed to an accidental or unintentional character (Hannum & Hawk, 1986, pp. 140-146 as cited in Fein, 1990, pp. 19-20).

In an article ("Rethinking Genocidal Intent: The Case for a Knowledge-Based Interpretation") in the *Columbia Law Review*, Alexander K. A. Greenawalt (1999) presents an argument similar to Kuper's and Fein's: "[R]elying on both the history of the Genocide Convention and on a substantive critique of the specific intent interpretation, in defined situations, principal culpability for genocide should extend to those who may personally lack a specific genocidal purpose, but who commit genocidal acts while understanding the destructive consequences of their actions" (Greenawalt, 1999, p. 2259).

Unfortunately, various intergovernmental and government officials do not seem to be cognizant of, or, at least not appreciative of, the above points. That, in and of itself, speaks to the critical need for an intense and broad-based educational effort vis-à-vis the issue of genocide.

"In Whole or in Part"

In regard to the wording "in whole or in part," Kuper (1985) makes the following observation: "The ambiguity lies in determining what number or what proportion would constitute a *part* for purpose of the definition.... Presumably the Convention is intended to deal with acts against large numbers, relative to the size of the persecuted groups, and it would rest with the courts to adjudicate on this issue" (p. 12).

The debate over numbers continues to be a nettlesome problem. In order to differentiate between various situations (e.g., a full-blown genocide and a small number of sporadic massacres), some scholars have used such terms as "genocidal process," "genocidal massacres," and "selective genocide."

While it is often difficult to differentiate between a series of massacres and genocide, one factor in attempting to make such an assessment should be to ascertain whether the massacres are systematic and ongoing. Put another way, an effort should be undertaken to ascertain whether the attack is "sustained ... [or constitutes a] continuity of attacks by the perpetrator to physically destroy" [members of a group. In order to do this, it is necessary to] "trace the time span, repetition of similar or related actions and the number of victims" (Fein, 1990, p. 25).

Early on, it is also significant to attempt to ascertain what the perpetrators' potential or possible aim is and what proportion of the population has been engulfed in the killing.

Even more significant, though, is the critical need to halt the massacres early on! Too often scholars sound like contestants in a game to see who can come up with the best estimate or guess as to whether something constitutes genocide or not, when, in reality, they should be more focused and concerned with how to stop the killing—any killing—in the first place.

The purpose of broaching these issues is *not* to make excuses for or provide a rationalization for the international community's agonizingly slow recognition of situations leading up to or constituting genocide or its repeated "nonattempts" to prevent various genocides once they were detected; rather, *it is* to illustrate the complexities and significance in defining genocide—and to highlight the fact that the latter has profound ramifications for the potential and actual victims, as well as the bystanders.

OTHER DEFINITIONS

As previously mentioned, over the past 45 years or so various scholars have recast the definition of genocide in an attempt to either make it more workable, manageable, "analytically rigorous" (Chalk & Jonassohn, 1990, p. 15), and/or to fit within their conceptual framework or typology of genocide. Also, as previously mentioned, other terms have been coined in an effort to make key distinctions between the intent, scope, and type of various crimes. Among such terms are the following: "ethnocide," "cultural genocide," "selective genocide," "genocidal process," and "genocidal massacres." Such efforts to hone the definition of genocide and to make distinctions between various acts of violence continue to this day as scholars attempt to develop a theoretically sound *and,* at the same time, practical working definition of genocide.

While some scholars have developed definitions that are ostensibly stronger than that found in the UNCG, others have muddied the waters with definitions that are as ambiguous, if not more so, than that used in the Convention. (For a detailed, informative and instructive discussion of definitions of genocide, including their strengths and weaknesses, see Fein, 1990.)

Pieter N. Drost (1959), a Dutch law professor who was extremely critical of the fact that political and other groups were excluded from the U.N. definition of genocide, created the following definition: "the deliberate destruction of physical life of individual human beings by reason of their membership of any human collectivity as such" (1959, p. 125).

In 1974, Vahakn Dadrian, a sociologist and scholar of the Armenian genocide, also developed his own definition: "Genocide is the successful attempt by a dominant group, vested with formal authority and/or with preponderant access to the overall resources of power, to reduce by coercion or lethal violence the number of a minority group whose ultimate extermination is held desirable and useful and whose respective vulnerability is a major factor contributing to the decision for genocide" (Dadrian, 1974, p. 123). As for Dadrian's "definition," Fein (1990) commented that "Here explanation has usurped definition; furthermore, it is not clear what is to be observed and classed as genocide except that the perpetrator is a representative of the dominant group and the victims are a minority group. This elementary distinction was later outmoded by the Khmer Rouge genocide in Kampuchea" (p. 13). What Fein meant and was thus referring to is the fact that the Khmer Rouge, which was comprised of a relatively small number of Cambodians, were responsible for the mass murder (or, as it came to be known, "autogenocide") of the much larger group of *non-Khmer Rouge* Cambodians.

In 1980 Irving Louis Horowitz, a sociologist and political scientist, published *Taking Lives: Genocide and State Power*, wherein he argued that genocide is a totalitarian method for gaining national solidarity. Ultimately, he suggested that the definition of genocide be revised as follows: "Genocide is herein defined as a structural and systematic destruction of innocent people by a state bureaucratic apparatus" (Horowitz, 1980, p. 17).

In 1985, Israel W. Charny, a psychologist, developed what he called a humanistic definition of genocide. More specifically, he argued that genocide should be defined as: "the wanton murder of human beings on the basis of any identity whatsoever that they share—national, ethnic, racial, religious, political, geographical, ideological" (p. 4). Many argue that Charny's definition is much too broad to be of use in scholarly research and analysis; certain others, however, agree with Charny that for the purpose of focusing attention on the need to protect *all* victim groups, it is a useful definition. In support of his definition, Charny (1985) argued that "I reject out of hand that there can ever be any identity process that in itself will justify the murder of men, women, and children 'because' they are 'anti' some 'ism' or because their physical characteristics are high- or low-cheekboned, short- or long-eared, or green- or orange-colored" (p. 448).

After developing a number of preliminary working definitions of genocide from the late 1970s onward, Helen Fein, in 1990, settled for the following "sociological" definition: "Genocide is sustained purposeful action by a perpetrator to physically destroy a collectivity directly or indirectly, through interdiction of the biological and social reproduction of group members, sustained regardless of the surrender or lack of threat offered by the victim" (p. 24).

Fein's explanation of her definition is instructive:

> Genocide is sustained purposeful action [thus excluding single massacres, pogroms, accidental deaths] by a perpetrator (assuming an actor organized over a period) to physically destroy a collectivity ['acts committed with intent to destroy, in whole or in part a national/ethnical/racial or religious group:' Art. 2] directly (through mass or selective murders and calculable physical destruction—e.g., imposed starvation and poisoning of food, water, and air—{see Art. 2, a-c}) or through interdiction of the biological and social reproduction of groups members (preventing births {Art. 2, d} and {'forcibly transferring children of the group to another group' Art. 2, e}, systematically breaking the linkage between reproduction and socialization of children in the family or group of origin.
>
> This definition would include the sustained destruction of nonviolent political groups and social classes as parts of a national (or ethnic/religious/ racial) group but does not cover the killing of members of military and paramilitary organizations, the SA, the Aryan Nations, and armed guerrillas (Fein, 1990, pp. 24-25).

In 1990, after examining and critiquing the above definitions as well as others, Frank Chalk, a historian, and Curt Jonassohn, a sociologist, developed their own definition: "Genocide is a form of one-sided mass killing in which a state or other authority intends to destroy a group, as that group and membership in it are defined by the perpetrator" (p. 23). Their rationale for developing yet another definition is as follows: "We have rejected the UN definition as well as others proposed because we want to confine our field of study to extreme cases. Thus, we hope that the term ethnocide will come into wider use for those cases in which a group disappears without mass killing. The suppression of a culture, a language, a religion, and so on is a phenomenon that is analytically different from the physical extermination of a group" (p. 23). (For a more detailed explanation of their work along this line, see the introduction in Chalk & Jonassohn, 1990.)

Fein (1990), for one, has difficulty with some aspects of Chalk's and Jonassohn's definition. For example, she finds the phrase "a state or other authority" too limiting a description of a perpetrator, and the phrase "one-sided mass killing" too unclear when it refers to mass killings which may include various armed elements. Fein also suggests that the definition of "group" is too "open-ended" (Fein, 1990, p. 13).

As the field of genocide studies continues to grow, new and ever clearer distinctions are bound to be made in regard to what does and does not constitute genocide as well as how to define genocide. At the same time, various scholars are bound to disagree over what constitutes a reasonable approach to such issues. In that regard, Charny (1991) cautioned against "obsessive definitionalism" (p. 6), while Fein (1991) voiced a concern about the concept of genocide becoming a "superblanket of generalized compassion" (p. 8).

Quite obviously, if humanity is to develop sound conventions and effective genocide warning systems in order to stave off genocide, then scholars, activists, governmental officials, and others need to come to a consensus in regard to that which does and does not constitute genocide. It should also be noted the misuse of the term (and it is misused and abused on a regular basis by various groups that want to draw attention to the plight of their own people or the focus of their cause) does not assist in either fully understanding or combating actual genocides. (For a discussion of the misuse of the term genocide, see Totten, 1999.)

USING EUPHEMISMS IN THE PLACE OF THE TERM "GENOCIDE" AND/OR AVOIDING THE USE OF THE TERM GENOCIDE

Time and again in the last half of the twentieth century—since, that is, the development and ratification of the UNCG—various governmental and intergovernmental officials—those very individuals who should have been at the forefront of preventive and intervention measures—have purposely used certain euphemisms in place of the term "genocide." Three of the terms used most often in the 1990s as a euphemism for genocide (e.g., during the periods when genocides raged in Rwanda and the former Yugoslavia), were "ethnic cleansing," "ancient ethnic hatred," and "civil war." In the above cases, government officials and bureaucrats in the United States were ordered by their superiors not to use the "g-word."

As for the situation in the former Yugoslavia, Samantha Power (2002) notes that

As pressure picked up, Bush administration [1988-1992] ... officials viewed and spun the violence as an insoluble "tragedy" rather than a mitigable, deliberate atrocity carried out by an identifiable set of perpetrators. The war, they said, was fueled by bottom-up, ancient, ethnic or tribal hatreds ... that had raged for centuries.... Defense Secretary Cheney told CNN, "Its tragic, but the Balkans have been a hotbed of conflict ... for centuries." Bush said the war was "a complex, convoluted conflict that grows out of age-old animosities [and] century-old feuds." [Deputy Secretary of State] Eagleburger noted, "There is no rationality at all about ethnic conflict" (p. 282)....

The Bush administration assiduously avoided using the word [genocide]. "Genocide" was shunned because a genocide finding would create a moral imperative.... Bush told a news conference: "We know there is horror in these detention camps. But in all honesty, I can't confirm to you some of the claims that there is indeed a genocidal process going on there." Policy makers preferred the phrase "ethnic cleansing."

[National Security Adviser Brent] Scowcroft believes genocide would have demanded a U.S. response, but ethnic cleansing, which is the label he uses for what occurred in Bosnia, did not: "In Bosnia, I think we all got ethnic cleansing mixed up with genocide. To me they are different terms. The horror of them is similar, but the purpose is not. Ethnic cleansing is not 'I want to destroy an ethnic group, wipe it out.' It's 'They're not going to live with us'" (pp. 288-289).

Fact upon fact upon fact collected by such human rights groups as Human Rights Watch and Amnesty International, as well as the United States government it should be noted, during the crisis provided ample evidence that contradicted the above assertions and prevarications.

In the case of the 1994 Rwandan genocide, U.N. Secretary General Boutros Boutros-Ghali issued "his most forceful statement on the Rwandan massacres [some two weeks after the genocide began on April 8th, 1994, but] what he had to say simply recycled the mischaracterization of Rwanda as a site of massacres being mutually visited on two ethnic groups by each other, the inevitable result of age old 'ethnic hatreds'" (Barnett, 2002, p. 133). In fact, "it was not until a May 4, 1994, broadcast of *Nightline* that Boutros-Ghali uttered the word genocide in a public forum" (Barnett, 2002, p. 133).

As for the United States, its "official line that the conflict in Rwanda was a civil war formed the basis of its argument that there were not grounds for peacekeepers" to halt a genocide (Barnett, 2002, p. 139). But that was not all. As Samantha Power (2002) notes:

Even after the reality of genocide in Rwanda had become irrefutable, when bodies were shown choking the Kagera River on America's nightly news, the brute fact of the slaughter failed to influence U.S. policy except in a negative way. As they had done in Bosnia, American officials shunned the g-word. They were afraid that using it would have obliged the United States to act under the terms of the 1948 genocide convention.... A discussion paper on Rwanda, prepared by an official in the Office of the Secretary of Defense and dated May 1, testifies to the nature of official thinking. Regarding issues that might be brought up at the next interagency working group, it stated, "1. Genocide Investigation: Language that calls for an international investigation of human rights abuses and possible violations of the genocide convention. *Be Careful. Legal at State was* worried about this yesterday— Genocide finding could commit [the U.S. government] to actually 'do something.'" The Clinton administration opposed the use of the term. On April 28 Christine Shelly, the State Department spokes-person, began what would be a two-month dance to avoid the g-word, a dance that brought to mind Secretary [of State Warren] Christopher's concurrent semantic evasion over Bosnia (pp. 358-359, italics in original).

The above comments are indicative of the power inherent in the term "genocide." They also illuminate the reasons why and the ways in which governmental officials often do all they can to avoid using the term, even when a situation merits it. Finally, the aforementioned passage also speaks to the purposeful obfuscation, disingenuousness, and deception at work when a government does not—for whatever reason—wish to confront the fact that genocide is being perpetrated.

METHODS/STRATEGIES FOR ASSISTING STUDENTS TO GAIN A SOLID SENSE OF THE COMPLEXITIES SURROUNDING THE EFFORT TO DEFINE GENOCIDE

There are, of course, numerous strategies and activities that instructors can use in order to introduce and teach their students about the various issues germane to the definition of genocide. What will be discussed herein are some of the many strategies that the author and/or his colleagues have used over the years with their students.

- Have the students conduct, either individually or in pairs, a study as to how the definition of genocide was decided on by the parties involved in the development of the U.N. Convention on the Prevention and Punishment of Genocide. Have them specifically focus on why and how the U.N. Convention on Genocide has come to be considered a "compromise document." (Among some of the many books that would be useful for this project are: Schabas [2000]; Kuper [1981, 1985]; LeBlanc [1991]; Totten and Jacob [2002]; Whitaker [1985]; and Power [2002].)

- Have the students, individually or in pairs or triads, conduct a study into how legal scholars interpret key terms (e.g., "intent"; "in whole or in part"; "as such"; "sovereignty"; and "internal affairs") in relation to: the definition of genocide; the prevention of genocide; the prosecution of genocide; and how the aforementioned interpretations should impact an individual nation's (other than that of the perpetrator), the United Nations, and the International Criminal Court's obligation to confront and address the perpetration of genocide. (Works that are likely to be helpful in conducting this study are: Schabas [1999, 2000]; van der Vyver [1999]; Fein [1990]; LeBlanc [1991]; Van Schaack [1997]; Triffterer [2001]; Greenawalt [1999]; and Totten and Jacob [2002].)

- Have the students compare and contrast various terms and phrases (e.g., civil war, tribal hatred, ethnic hatreds, "ethnic cleansing," ethnocide, atrocity, massacre, massacres, mass murder, genocidal process, genocidal massacres, selective genocide, genocide) by conducting research into each term and closely examining the various nuances of each. Ultimately, the goal should be to attempt to come to a conclusion as to the exact meaning of each term. (Note: In order for this to be a valuable learning experience, it may be wise for the instructor to model how this should be done by handling at least one of the terms him or herself.) (Some excellent sources for such a study are: Lerner [1994]; Schabas [1999]; Whitaker [1985]; and Power [2002].)

- Have the students conduct research into some of the most thorny issues that generally come into play when a genocide rears its head (e.g., the issue of political will, intent of the perpetrator, sovereignty, internal affairs), why that is often the case, and then have them compare such concerns with the promises made in both the UNCG and the UN Declaration on Human Rights. In order to make this study more manageable for the students, different pairs of students could be assigned different genocides to study. (Among the scores of works that would be useful for this assignment are: Power [2002]; Barnett [2002], Gow [1997]; Melvern [2000]; Forges [1999]; Campbell [2001]; Moore [1998]; Ritter, Roth, and Smith [2002]; Mills and Brunner [2002]; Schabas [1999, 2000]; Van Schaack [1997]; Triffterer [2001]; Greenawalt [1999]; and Totten and Jacobs [2002].)

- Have the students—either individually, in pairs or in triads—compare and contrast various definitions of genocide (e.g., in the U.N. Convention and those developed by various scholars), and come to a conclusion in regard to which one seems to be the most inclusive and practical without being so broad that it is useless. To make this assignment as valuable as possible, the students should be required to provide a detailed discussion as to why and how they came to their ultimate decision. (Note: Instructors may wish to develop a rubric for students which assists and guides them in the type of criteria they should consider in analyzing the various definitions and making their judgments.) (Among the many works that students should find useful for this activity are: Schabas [2000]; Fein [1990]; Kuper [1981, 1985]; Whitaker [1985]; and Totten and Jacob [2002].)

- Students could examine how international law is evolving—even if slowly and unevenly—in regard to that which constitutes genocide by specifically conducting a study into how and why gender-based crimes were largely ignored during the war crimes trials held in Nuremberg and Tokyo after World War II, but how crimes of sexual violence are currently being tried at the International Criminal Tribunal for the former Yugoslavia (ICTY) and the International Criminal Tribunal for Rwanda (ICTR) as violations of the laws or customs of war, genocide, and crimes against humanity. (Two excellent sources for such a study are: Askin [1999] and Chinkin [1999].)
- Examine how the United States (and/or the United Nations or countries other than the United States) described and defined specific genocides while they were being perpetrated, why it did so, and the consequences of doing so. (Three excellent texts to begin such a study are: Power [2002]; Barnett [2002]; and Schabas [1999].)
- Have the students conduct a study into if, when, and to what extent individual states are obligated to attempt to prevent genocide if they have ratified the U.N. Convention on the Prevention and Punishment of Genocide. (Among the many sources that are ideal for use in such a study are: Schabas [1999, 2000]; and van der Vyver [1999].)

CONCLUSION

Defining the term "genocide" is, obviously, not an easy task. It is, though, one that is critically significant to do; and it is only through understanding what genocide is and is not—as well as the strengths and weaknesses of the definition used in the U.N. Convention on the Prevention and Punishment of Genocide—that students can even begin to gain an understanding of the complexities that surround the perpetration, prevention, and intervention of genocide.

REFERENCES

Askin, K. D. (1999). Sexual violence in decisions and indictments of the Yugoslav and Rwandan tribunals: Current status. *The American Journal of International Law, 93*(1), 97-123.

Barnett, M. (2002). *Eyewitness to a genocide: The United Nations and Rwanda*. Ithaca, NY: Cornell University Press.

Barta, T. (1987). Relations of genocide: Land and lives in the colonization of Australia. In I. Wallimann & M. N. Dobkowski (Eds.), *Genocide and the modern age: Etiology and case studies of mass death* (pp. 237-252). New York: Greenwood.

Campbell, K. J. (2001). *Genocide and the global village*. New York: Palgrave.

Chalk, F., & Jonassohn, K. (1990). *The history and sociology of genocide: Analyses and case studies*. New Haven, CT: Yale University Press.

Charny, I. W. (1985). Genocide, the ultimate human rights problem. *Social Education, 49*(6), 448-452.

Chinkin, C. (1999). Women: The forgotten victims of armed conflict? In H. Durham & T. L. H. McCormack (Eds.), *The changing face of conflict and the efficacy of international humanitarian law* (pp. 23-44). The Hague: Martinus Nijhoff.

Dadrian, V. N. (1974, Fall). The structural functional components of genocide: A victimological approach to the Armenian case," In I. Drapkin & E. Viano (Eds.), *Victimology* (pp. 123-135). Lexington, MA: Heath.

Drost, P. (1959). *The crime of state* (Vol. 2). Leyden: A.W. Sythoff.

Fein, H. (1991, February). Genocide: Life integrity violations and other causes of mass death: The case for discrimination—A reply to Israel Charny's critique." *Internet on The Holocaust and Genocide.*

Fein, H. (1990). Genocide: A sociological perspective. *Current Sociology, 38*(1).

Forges, A. D. (1999). *"Leave none to tell the story:" Genocide in Rwanda*. New York: Human Rights Watch.

Gow, J. (1997). *Triumph of the lack of will: International diplomacy and the Yugoslav War*. New York: Columbia University Press.

Greenawalt, A. K. A. (1999). Rethinking genocidal intent: The case for a knowledge-based interpretation." *Columbia Law Review, 99*(8), 2259-2294.

Horowitz, I. L. (1980). *Taking lives: Genocide and state power*. New Brunswick, NJ: Transaction.

Huttenbach, H. (1988). Locating the Holocaust on the genocide spectrum. *Holocaust and Genocide Studies, 3*(3), 289-304.

Kuper, L. (1981). *Genocide: Its political use in the twentieth century*. New Haven, CT: Yale University Press.

Kuper, L. (1985). *The prevention of genocide*. New Haven, CT: Yale University Press.

LeBlanc, L. J. (1988). The United Nations Genocide Convention and political groups: Should the United States propose an amendment? *Yale Journal of International Law, 13*(2), 268-294.

LeBlanc, L. J. (1991). *The United States and the genocide convention*. Durham, NC: Duke University Press.

Lemkin, R. (1973). *Axis rule in occupied Europe: Laws of occupation, analysis of government, and proposals for redress.* New York: Howard Fertig. [Original work published 1944]

Lemkin, R. (2002). Totally unofficial man. In S. Totten & S. Jacobs (Eds.), *Pioneers of genocide studies* (pp. 365-399). New Brunswick, NJ: Transaction.

Lerner, N. (1994). Ethnic cleansing. In Y. Dinstein (Ed.), *Israel yearbook on human rights* (pp. 103-117). The Hague: Martinus Nijhoff.

Melvern, L. (2000). *A people betrayed: The role of the west in Rwanda's genocide.* London: Zed Books.

Mills, N., & Brunner, K. (Eds.) (2002). *The new killing fields: Massacre and the politics of intervention.* New York: Basic Books.

Moore, J. (Ed.) (1998). *Hard choices: Moral dilemmas in humanitarian intervention.* Lanham, MD: Rowman & Littlefield.

Power, S. (2002). *"A problem from hell": America and the age of genocide.* New York: Basic Books.

Rittner, C., Roth, J. K., & Smith, J. M. (Eds.). (2002). *Will genocide ever end?* St. Paul, MN: Paragon House.

Schaack, Beth Van (1997). The crime of political genocide: Repairing the genocide convention's blind spot. *The Yale Law Journal,* 106(7), 2259-2291.

Schabas, W., (1999). *The genocide convention at fifty.* Washington, DC: United States Institute of Peace.

Schabas, W. (2000). *Genocide in international law: The crime of crimes.* Cambridge, UK: Cambridge University Press.

Totten, S. (1999). "Genocide," frivolous use of the term. In I. W. Charny (Ed.), *Encyclopedia of genocide* (Vol. 1, p. 35). Santa Barbara, CA: ABC Clio Press.

Totten, S., & Jacobs, S. (Eds.). (2002). *Pioneers of genocide studies.* New Brunswick, NJ: Transaction.

Triffterer, O. (2001). Genocide, its particular intent to destroy in whole or in part the group as such. *Leiden Journal of International Law, 14*(2), 399-408.

van der Vyver, J. D. (1999). Prosecution and punishment of the crime of genocide. *Fordham International Law Journal, 23*(2), 286-356.

Whitaker, B. (1985, April). *Revised and updated report on the question of the prevention and punishment of the crime of genocide.* (E/CN.4/Sub.2/1985/6, 2 July 1985.)

CHAPTER 5

DEFINING GENOCIDE

Issues and Resolutions

Henry R. Huttenbach

INTRODUCTION: THE PROBLEMS OF DEFINING GENOCIDE

Little progress is made in any study without clarifying the basic terminology, especially the object of the study. Peace studies would not get far off the ground unless the idea of peace—what is meant by "peace"—has been sufficiently agreed on by those conducting an investigation of this otherwise amorphous term. Likewise, biologists would be hard put without a working consensus as to the essence of what is understood by the word "life." The same holds true for those investigating genocide.

Since 1944—when the word was coined by Raphael Lemkin—genocide has become part of the vocabulary both of the general public and of specialists, such as scholars, lawyers, and politicians. One would think that after more than half a century of consideration there would be the semblance of a working definition of genocide. In fact, there is not; repetitive disputations and often bitter disagreements still abound among those engaged in what is generally referred to as Genocide Studies. Indeed, as the 1990s have drastically demonstrated in the well-publicized cases of Bosnia and Rwanda, deep differences abound, whether these incidents

Teaching About Genocide: Issues, Approaches, and Resources, 75–91
Copyright © 2004 by Information Age Publishing
All rights of reproduction in any form reserved.

were genuine cases of genocide or not. To date, they continue to trouble the entire spectrum of interested observers and investigators of genocide. It is particularly confusing for students.

Why is this? What makes identifying an event as genocide so difficult? One would think that recognizing an incident as a genocide or as something else short of genocide would be relatively simple. The reasons for the difference of mind are not as difficult to pinpoint, though they are indicative of the complexities involved, and with which instructors should be familiar. Simply put, they fall into two broad categories: political and conceptual.

From the very beginning, genocide raised political hackles. As the drafters of the 1948 United Nations Genocide Convention (UNGC) soon found out, genocide implied a *state* crime. It was tacitly assumed by the drafters that the crime of genocide basically could only be committed by the state, any lesser organization was considered lacking sufficient violence to marshal the power to exterminate on such a grand scale. Initially, all the drafters of the UNGC had Nazi Germany and the extermination of the Jews of Europe in mind. But, as they thought things through, especially the major powers, they sensed a danger to themselves as they reviewed their own less than innocent pasts. The United States with its morally and legally questionable mistreatment of its indigenous peoples feared possible *post-facto* accusations of genocide in the face of the "disappearance" of many North American Indian tribes. Likewise, Stalin's Soviet Union became nervous over its lethal handling of ethnic minorities in 1944 during World War II, and, above all, of its widespread purges of the anti-Stalin political opposition in the 1930s. The end result was, predictably, a less than satisfactory text with which international lawyers and their tribunals have had to wrestle ever since. On the one hand, they have been relatively content with what constitutes genocidal behavior, but, on the other, are deeply discontent with what genocide *is* as distinct from other mass killings. Hence, they have hesitated to charge anyone with the crime of genocide, often preferring a lesser indictment easier to prove. As we shall see, ad hoc steps to rectify this obstacle have been taken from time to time.

A second problem of identifying a genocide as a genocide arises from the "politics" within the community of scholars of genocide, which has more to do with the chronology of scholarly events. As is well known, genocide studies began under the exclusive (and exclusivist) rubric of Holocaust Studies. As scholars embarked on researching the fate of European Jewry during World War II, they assumed—correctly—that they were engaged in delving into the intricacies of a case of genocide. For decades, there were no scholarly ventures into other genocides, before, during, or after the Holocaust. By default, the "Final Solution" dominated the field, inevitably reducing occasional studies to the catchall "other genocides" and all that this vague phrase implied, namely, the overwhelming *singularity* of the Jewish experience and, by inference, the "lesserness" of other genocidal incidents.

Steadily, as sporadic studies of non-Holocaust genocides appeared, some of those engaged in the on-going work of the Holocaust, such as the early works of Yehuda Bauer, became defensive, seeking to "protect" the Holocaust from "false" comparison. Such comparisons, it was implied, was supposed to "trivialize" the Holocaust, which, many tenacious advocates insisted was "unique." The clamor for the unquestioned *uniqueness* of the Holocaust was based on sheer emotion. Uniqueness, however, is not an academically demonstrable attribute: unprecedence perhaps, but that, too, requires scholarly comparison which all advocates of Holocaust uniqueness could not do, their expertise limited to one case of genocide, the Holocaust.

This "politics" of Holocaust centrism has seriously impaired genocide scholarship and, for purposes here, needlessly delayed the foundation of an inclusive definition which would bring the Holocaust—more appropriately referred to as the Final Solution—under the same conceptual roof alongside all incidents of genocide, regardless of their incidental differences. To do so, however, requires a recognition of a fundamental error in *former* attempts to determine a common denominator for all instances of genocide. That error was the failure to distinguish between two kinds of definitions, the conceptual and the descriptive: one focuses on the central idea—on what genocide *is about*—and the other deals with what constitutes a genocidal act, the specifics of the crime.

CUTTING THE GORDIAN KNOT

Definitions are supposed to clarify and, therefore, help identify. In fact, the search for an adequate definition can often lead to greater confusion and even controversy, bitterly divisive at times. Thus, when a new unnamed phenomenon appears, the first problem begins with the selection of an appropriate and, if possible, value-free appellation. Grappling with genocide is fundamentally no different.

Throughout human history there has been an uninterrupted chain of mass killings. In response to what seemed unprecedented massacres inflicted on European Jews by the Third Reich during World War II, Raphael Lemkin, a Polish lawyer of Jewish background, coined the term "genocide" in order to highlight what the Nazis dubbed the Final Solution of the Jewish Question. In so doing, Lemkin wanted to distinguish between exterminational and nonexterminational massacres committed by Hitlerian Germany. This opened up the problem of determining when and how massacres cease to be large-scale killings and begin to assume genocidal characteristics. But can this distinction be made accurately using academic tools? Or is it largely based on personal impressions (Semelin, 2001)? Prior to having a name of its own, genocide was subsumed in "crimes against humanity" or into a category called "war crimes," as was the case during the post-World War II International Military Tribunal which convened in 1946.

The UNGC of 1948 made genocide—by then an accepted term—a separate criminal category. It tried to solve this problem of specificity and distinction from other crimes by supplying more or less descriptive criteria of what genocidal behavior consists of. In a sense, this has been quite satisfactory, at least to serve as a guideline for those prosecuting genocide. Virtually all courts and tribunals have generally—although not entirely—rested their determination of what makes a genocide and what is or is not genocidal on the 1948 U.N. declaration. Serious deficiencies of the U.N. definition, though, have been found by the 1978 and 1985 reports of the special rapporteurs of the U.N. on its own Genocide Convention. (See Ruhashyankiko [1978] and Whitaker [1985]. See also Lippman [2001].)

Academics—historians, sociologists, psychologists, political scientists, and policymakers—have inundated the field with their conflicting, if not, contradictory and even mutually exclusive definitions of genocide, to the point that the problem posed by this proliferation of definitions is threatening to undermine long-term coherent study of genocide.

Let us begin with the outer extremities of the problem of definition, one I dub the "Katz-Charny Conundrum." In brief, the dilemma of defining genocide is exemplified and distilled by these two polarized views, as follows:

1. In his voluminous 1994 tome, *The Holocaust in History*, Steven Katz developed a methodologically but transparently flawed argument to prove that there is but one bona fide case of genocide, namely, the Holocaust (that is, the "Final Solution of the Jewish Question" as the extermination policy was named by its SS architects). He arrived at this constricting conclusion by examining literally hundreds of instances of mass killing perpetrated over the span of centuries, determining that since none *descriptively* compares with the Final Solution, there remains but one full-fledged instance of genocide.

There are two serious errors here. One is that of false comparison: *first,* by purposely selecting one event to serve as *the* operating paradigm of genocide, Katz could safely conclude that all other events necessarily fail to meet the criteria and standards of his chosen genocide—the Holocaust. Since no two events are ever fully alike, then, logically, all others are automatically disqualified as full-fledged genocides. Second, Katz commits the academic sin of a priori reasoning. He consciously constructed an argument around a conclusion he wanted to prove *prior* to his writing the book, namely, to confirm that the Final Solution is, indeed, unique, standing clearly apart, alone and, therefore, beyond comparison, a sole representative of a class of its own, a super-genocide. All other instances of targeted mass killings of a group Katz classifies, by inference, as near-genocides or lesser genocides. Thus, Katz lays the groundwork for a hierarchy of massacres over which the Holocaust reigns supreme, incomparable, singularly unique, with its own exclusive definition that applies only to itself. Whatever essential knowledge about genocide one needs, Katz concludes, can be gleaned from the Holocaust itself; any insights gained from other (i.e., lesser) genocides are necessarily secondary according to the Holocaust-centric formula posited by Katz.

2.   On the other end of the spectrum of definitions of genocide is that of Israel W. Charny. His definition, if that is what it can be called, is so broad that a wide array of events fit within its amorphous range and, therefore, can be classified as genocides. Charny's much too generous definition is so accommodating that several hundred events are classified as bearing the mark of genocide, including most of those excluded by Katz. This is best illustrated by what Charny refers to as his generic definition of genocide in his two volume *Encyclopedia of Genocide* (Charny, 1999). With but a few exceptions, it is an extremely broad compendium of massacral events, to each of which a genocidal association is attributed. The result is such an extreme universalization of the act of genocide that the word lacks little meaningful core specificity.

Reducing Katz' reasoning diagramatically, two variations of Holocaust "supremacy" can be extracted. Essentially they are the same. The first

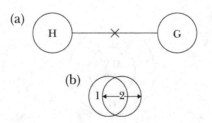

Figure 5.1.   The "Uniqueness" Theory

(Figure 5.1a) highlights the "uniqueness" theory; the second (Figure 5.1b) stresses the Holocaust-centric mode.

In this depiction (Figure 5.1a), the Holocaust (H) stands completely apart from other incidents of genocide (G). It is perceived as "unique," one could say "uniquely unique," a super-genocide that cannot be compared with other, therefore, lesser genocidal incidents.

In this depiction (Figure 5.1b), the Holocaust is recognized as genocide, sharing some features with other genocides (zone 2.) But, above all, it enjoys special distinctive features (zone 1), suggesting it is "more" than a simple genocide. It is "genocide-*plus*."

What this zone of quantitative and qualitative differences consists of has been filled by an endless chain of emotive words, beginning with "unique" itself, a classic case of circular argumentation: something is "unique" because it is! Substitute supportive vocabulary has been used repeatedly: the Holocaust is a "Tremendum"; it is a "watershed" in European history, in world history, in the twentieth century, in modern times. It has been called "unprecedented," and associated with an array of labels: incomparable, inconceivable, unparalleled, indescribable, unspeakable, and so forth. These are all designed to create an aura of mystery to suit various nonacademic, but public (political) purposes. At best they are admissions of emotional shock generated by the event; at worst they obfuscate and serve no intellectual purpose with their purely emotional distractions.

The Charny formula rendered in diagrammatic form appears in Figure 5.2. In this diagram, every violent mass violation of human rights (MVHR—inner circle) is identified as genocidal, as falling within the greater orbit of genocide (outer circle.) Clearly, this seriously blurs the distinctions between two phenomena in the absence of a discriminating definition of genocide.

This approach also has an emotional base to it. Charny is, commendably, at heart, concerned with cruelties to which humanity has been subjected. Genocide is not his top priority. Genocide is but one instance of mass cruelty. In his calculation, all incidents of mass violation of human rights need to be exposed, including genocide. But since genocide is but

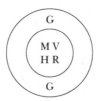

Figure 5.2. The Inclusive Approach

one instance of mass violence it does not, apparently, in his schema, warrant separate treatment. Whereas this approach—strictly humanistic and humanitarian—is morally laudable, it nevertheless is—at least in this author's eyes—of minimal use to those seeking to distinguish *as accurately as* possible genocide from all other criminal acts of mass proportions. It is true that at its heart genocide is also an extreme violation of human rights; but it is also distinctly—but not more!—an event that stands apart if properly conceptualized. Before it can be merged into a greater whole—as Charny does—genocide needs its own qualified definition so an event can be identified as such.

So, how does one extricate oneself from the Katz-Charny Conundrum? The former, Katz, raises the specter of extreme selectivity in his definition, namely: "a genocide is that event that I have chosen." His brand of definition-by-radical-exclusion provides a self-serving methodology that can be applied by any scholar harboring a "favorite" genocide. The latter, Charny, is so indiscriminate (generous?) that virtually all assaults on collective human rights could be perceived as a form of genocide, leaving one with a quandary: unless proven otherwise, any massive, violent event belongs within the parameters of genocide. To repeat, how does one find a middle ground that cuts through this Gordian knot?

In the absence of a satisfactory definition of genocide based neither on nonimpressionistic nor on inflationary criteria, major consequences flow which severely hamper progress in genocide studies. On the monographic level, studies of a single event, claimed to be a genocide by the author, are based, more or less, on a definition satisfactory to that particular scholar. To date, each researcher operates with a definition that suits his or her purposes, namely, to include an event they are studying in the company of other genocides which have been equally arbitrarily or impressionistically labeled genocides by their respective authors.

The result so far has been twofold. First, there are now a series of disputes questioning whether some events are indeed genocides. Thus, the bloody events (dubbed ethnic cleansing) accompanying the dismemberment of Yugoslavia had again and again been treated either as examples of genocide or rejected as falling outside the range of the genocidal. Another example: the decades-long squabble over how to classify the lethal experiences of European Roma and Sinti (Gypsies) at the hands of the Nazis: genocide or not? Prominent scholars, for years using the Holocaust-centric Katz mindset (among them Yehuda Bauer), steadfastly and often vehemently denied that the lethal anti-Gypsy policy of the Nazi regime was genocidal, until recent *volte face* "conversions," not as a result of scholarly reexamination but, one suspects, for expediency: further denial clearly was becoming politically incorrect. In contrast, University of Texas professor Ian Hancock, the foremost U.S. expert on the Gypsy experience, passionately made the (for years, futile) case for the recognition of a Gypsy genocide, the *Porrajmos*. Thus, as events are unsystematically rejected or accepted as genocides, the absence of a more objective definition becomes that much more urgent.

Second, in seeking to examine and then compare genocides, events have first to be classified as such. But in the presence of several competing definitions, it becomes impossible to reach agreement as to which definition should be applied. The absence of a governing definition has led genocide studies to a state of near anarchy, reflecting in part the problem generated both by Katz and Charny. Basically, the lack of consensus as to what makes up a genocide has led individual scholars to tailor their own definition to fit their needs. For example, Vahakn Dadrian, the preeminent expert on the Armenian experience of 1915, has fashioned an à la Katz definition of genocide to fit the specific violent events of 1915 and the subsequent years of mass killing amounting to a genocide, but (again, in this writer's view) is too self-serving (see Dadrian, 1995). Others avoid a formal definition, such as Ben Kiernan (1996) in his studies on Cambodia during the rule of Pol Pot in 1975-1979.

Thus, given 20 hypothetical massacral events, using one definition, the first 10 might qualify; applying another, the second 10 events might be categorized as genocides. Using a third definition of genocide, the middle 10 could be so identified, and so forth (see Figure 5.3)

According to Defintion 2 (D2) and Defintion 3 (D3), events 1-5 are not genocides; yet they are according to Defintion 1 (D1). Conversely, according to D1 and D3, events 16-20 are not genocides, whereas they are according to D2. Thus, without a governing definition there is no reliable way to sort out genocides from nongenocides. In turn, without a precise way of determining what event is or is not a genocide, rational comparison between genocides is impossible. The confusion is obvious. So how to escape this quagmire and arrive at a definition based on a sensible consensus as to *what* genocide is essentially and *of what* it consists?

To begin with: What lies at the epicenter of genocide? What is the common denominator that binds all genocides together? As pointed out ear-

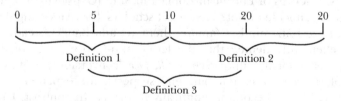

Figure 5.3.   The Holocaust-centric Mode

lier, agreement on a satisfactory definition of genocide still bedevils scholars, though less so trial lawyers.

These are questions that obviously concern us today and will do so long into the future. Today, as the post-Bosnia and Rwanda tribunals—entrusted with the responsibility of trying those apprehended and accused of having committed genocide—proceed, presiding judges, and prosecutors tend to rely relatively uniformly on the spirit, if not the letter, of the 1948 United Nations' Genocide Convention as their basic guide. However, this admittedly pragmatic definition of genocide is more on the order of a *descriptive* formula rather than a *conceptual* definition; nevertheless, it continues to serve as a practical point of departure primarily in the pursuit of justice. Between 1992 and the present, lawyers have patiently honed and refined the U.N. statement, providing future jurists with an accumulation of nuanced interpretations and a body of precedence—case by case—in the hopes of developing legally acceptable formulations, less of the concept of genocide (what one understands it to be *qua* idea) but more of the acts deemed genocidal on which the courts will accept evidence and rest their verdicts.

However, this aggregate approach to a consensus in regard to what genocide as a specific crime is and is not, though practicable in a court setting, is unsatisfactory in the seminar or class room. Whereas the juridical descriptive mode, increasingly empirical, clarifies legal issues of genocide as practiced, it has the opposite effect on the analytic academic mind. In the intellectual quest for the quintessence of genocide, one searches for a *fundamental* concept with which to pinpoint the core meaning of genocide. In their exploration for the quintessence of genocide, academics are still searching for an *Ur-* concept with which to pinpoint the central meaning and significance of genocide. This approach is less concerned with the surface details of what a genocide consists of in action than with the very *idea* of genocide. By means of greater semantic precision it seeks to arrive at a more philosophical determination of genocide, one which has less to do with the lawyers' quotidian concerns for specific evidence of a specific criminal act than one more in tune with an intellectual's and teacher's need for precise abstraction. Scholars of genocide, therefore, before attending to the secondary traits of a specific case of genocide, need to forge an unambiguous conceptual grasp of the *essence* of genocide.

TOWARD A CONCEPTUAL DEFINITION OF GENOCIDE

Originally the term *genocide* enjoyed the advantage of the seeming clarity of a neologism, until its meaning became increasingly obfuscated, buried under countless layers of surface description, ultimately and predictably putting the term's utility into question. So much so, that now another term must be found to determine what precise concept underlies the word genocide. For some years, those using the term genocide have found it to be more and more an empty vessel, a word in search of its meaning. Hence the present state of confusion.

At the heart of genocide lies the existential dimension, the *thought* and the *act* of threatening and endangering a group's very existence. This could be expressed by the term *elimination,* the wish to eliminate a group, except that genocide automatically also raises the thought of mass killing, whereas elimination, etymologically, connotes first and foremost "removal," or, in a genocidal context, "mass expulsion"—but not necessarily mass killing. Mass murder intended and/or committed, however, should be acknowledged in one's basic understanding of genocide, while elimination (at least in English) only secondarily points toward the idea and deed of large-scale killing. A term such as elimination, that only indirectly infers the wide scale destruction of life, misses the sine qua non of genocide, namely, posing a threat to a group's actual physical (biological) survival, in part as a result of a significant loss of life caused by man-made violence. However, a group could be made to disappear culturally by means of forced assimilation; yet most genocides include massacral killings, often in response to opposition to forced cultural conversion. A more satisfactory term to rectify this objection is *extermination.* It strongly suggests broad but focused killing on the order associated with genocide. Interestingly, in 1933, *prior* to the Holocaust, Lemkin had used the term "extermination" in his quest to have it declared a crime while attending the International Conference for the Unification of Criminal Law held in Madrid that year (see Lemkin, 1947), thereby giving a clue as to what idea underlay his newly-minted word—genocide—10 years later. Significantly, the word extermination was also used relatively early to designate genocide, two years before Raphael Lemkin coined the word genocide, by which he meant extermination, the act of killing a group, the act of terminating—ending—existence. The occasion was a virtually forgotten declaration made by the nascent United Nations on December 17, 1942. The statement was issued as a public condemnation of the systematic slaughter of Jews in German occupied Europe (Brecher, 2001). It ran as follows:

German Policy of Extermination of the Jewish Race

The attention of the Belgian, Czechoslovakian, Greek, Luxembourg, Netherlands, Norwegian, Polish, Soviet, U.K., U.S. and Yugoslav Governments and also of the French National Committee has been drawn to numerous reports from Europe that the German authorities, not content with denying to persons of the Jewish race in all the territories over which their barbarous rule has been extended the most elementary human rights, are now carrying into effect Hitler's oft-repeated *intention* to *exterminate* the Jewish people in Europe. From all the occupied countries, Jews are being transported in conditions of appalling horror and brutality to eastern Europe. In Poland, which has been made the principal Nazi slaughter-house, the ghettos established by the German invaders are being *systematically* emptied of all Jews except a few highly skilled workers required for war industries. None of those taken are ever heard of again.... The number of victims ... is reckoned in many hundreds of thousands of entirely inno-cent men, women, and children.

[The signatories] condemn in the strongest possible terms the bestial policy of cold-blooded *extermination*. They declare that such events can only strengthen the resolve of all freedom-loving peoples to overthrow the bar-barous Hitlerian tyranny. They reaffirm their solemn resolution to insure that those responsible for these crimes shall not escape retribution and to press on with the necessary practical measures to this end. [emphasis added]

Notice the express focus on (1) *intention*, (2) on the *systematic*, (3) on *exter-mination*, 4) and on a *group* (the *Jews* of Europe).

Yet, extermination is still too one-dimensional; primarily, if not exclu-sively, it is limited to lethal, physical violence against a defined group, while genocide, as just pointed out, can go considerably further, beyond the mere destruction of the biological life of a targeted group. Genocide can include the wanton pulverization of the corpses (the destruction of the evidence of a crime) and, importantly, the destruction of the entire creative heritage of a people: its literature, its architectural monuments, its arts, its entire legacy, in short, its culture. The possibility of "culture-cide" as an integral part of genocidal intention should not be excluded from the central thought that gives genocide its core meaning.

Indeed, one of the first casualties of genocide or attempted genocide is culture. In extremis, genocide can kill off all the practitioners of a specific culture; for example, Tibetan Lamaist Buddhism has been under existen-tial siege ever since the Chinese invasion in 1955. Yiddish culture is all but extinct as a result of the Final Solution in Eastern Europe. Were it not for the Armenian Diaspora, Armenian cultural life after 1915 could easily have died out. Acts of mass displacement in the name of ethnic cleansing can force refugees in their new environments to lose command of their native language and customs as a result of the erosive consequences of assimilation. In this case, culturecide is an indirect byproduct of genocidal acts.

Like extermination, the term *eradication* also does not go far enough. It does, to be sure, convey the thought of the full physical extermination of a group's biological existence, as well as of its culture as an intended byproduct, leading toward a state of tabula rasa. To *erase*, to wipe out, is indeed a central aspect of genocide, literally and figuratively. The Nazi term *Judenrein* or *Judenfrei* (territory cleansed of Jews) implied both a massive purge of undesired life, of a reviled culture *and* despised memory, thereby implying a clean slate, of a future to be written *without* reference to those erased. But we are still in the realm of pure action, of the descriptive, and insufficiently on the level of the philosophical and the conceptual that is required to supply *the* core meaning or implication of genocide.

Some have suggested the term "extinction"—to render extinct—as the core concept underlying genocide. The problem with this proposal is that it denotes a restricted kind of extermination. When one refers to species and civilizations as becoming or being extinct, it correctly suggests that a future existence has been terminated, but it in no way denies a past existence. They remain, through memory, an integral part of history. However, some genocides are committed to the *explicit* intent of also destroying a collective's past, of denying the group's prior existence, which the concept extinct clearly does not! A more radically precise term is called for.

A more inclusive term that combines the existential destruction of a human collective, including its cultural legacy, is *annihilation*. Its etymology rests on the Latin concept "nihil," namely, nothing. As a verb describing an act, to annihilate unambiguously conveys the concept "to make—to transform—*something*—into nothing." That is, while there once was *a something*, there now is *a nothing*. The idea of *making nothingness* is commensurate in thought with the potentially multidimensional act of destruction that needs to be fully associated with genocide. It allows one to explore the philosophical implications of *nothingness* as a *positive* goal; that is, as a *desideratum* of those committing genocide. Radical genocidal thought and reasoning seek to metamorphose the existential status of a people—its cultural achievements as well as its entire past—from the existent to the nonexistent, to its never having existed. This is one of *the* central aspects of genocide that needs to be fully incorporated into any conceptualization of the whole phenomenon.

Yet, in terms of satisfying a full consideration of all that annihilation implies, the term no longer is powerful enough to encompass the entire spectrum of core implications with which to express genocide in all its fullness, namely, the dimension beyond the philosophical. For that, one must turn to a near synonymous but less frequently used term, *nullification*. It, too, is anchored in the idea of making *nothingness*, but it embodies a far greater emphasis on rendering something into a zero, into an *absolute zero*, the German *das Null*. This makes room for significant expansion of the core idea of genocide; in this case beyond the realm of the purely biological, cultural, and philosophical. While annihilation conveys a strictly rational side of genocidal behavior, as a willed consequence or conscious intention flowing from a well structured syllogism concluding with the fabrication of "absolute nothingness," the word nullification adds a quasi-theological dimension to genocide, an aspect which needs to be included in a full conceptualization of the idea of genocide.

Genocide is made possible by a thought (the desire) and the power to translate it into deed. In the biblical act of creation, the all-powerful creator has arrogated to itself the power both to call something into being, to sustain it *and* to withdraw, to unmake, to decreate its existence. Similarly, by analogy, the *genocidaire* seeks to acquire the power to control the existence of a group, including the ability to obliterate it and every aspect associated with a targeted collective, including its historical existence through memory, as well as any conceivable form of an existential continuity in the future.

Pure genocide, therefore, can be perceived as an act of anticreation, which aims at a totality of extinction so extreme that even the very act of genocide might be denied, its memory fully expunged from future records. In the ultra extreme, genocide—the victims *and* the genocidaires together—will be unacknowledged. The act of complete genocide will become itself a nonact. Genocide of this kind is an act of radical, absolute erasure of every aspect of existence, so radical that, at least in theory or fantasized intention, there will remain not a single shred of evidence of a genocide ever having taken place, as if the terminated group had never existed, consigning it to a timeless, formless, condition of precreation, pushed back to a nonexistence, back to the *Tohu vVohu* of *Genesis*, the unformed void.

This thought and act of extreme nullification as just elucidated lies at the epicenter of genocide, providing the word with a precise but sufficiently elastic conceptual foundation. An event, to be a genocide or, at least, to be genocidal, has to have a direct connection with *an aspect* of the idea of nullification. As illustrated, there is a broad but well defined spectrum of nullification—degrees of intention and consequence—which provide scholars with a sufficient but clearly delineated interpretive leeway to determine whether an event warrants to be labeled a genocide of some kind, ranging from cultural destruction by forced assimilation, through basic biological destruction, to complete nullification which, in turn leads to the ultimate existential *nothingness*, the consignment to the *void* in the name of a Utopian vision which perceives genocide as a means to a *better* world.

RAMIFICATIONS FOR THE CLASSROOM: HOW TO TACKLE
DEFINITIONAL CONCERNS WHEN TEACHING ABOUT GENOCIDE

So, how do we resolve the problem of defining genocide as it relates to the study of genocide within the classroom? To begin with, there is the question of identifying a crime. For example, all homicides are not murder; nor are all murders premeditated. Thus, within the ranges of homicide and murder there are distinct shades determined by intention and circumstances. Similarly, classifying an event as a full genocide or partial genocide must be done with care. Teachers of genocide should warn their students not to be too hasty in rendering judgement. The crime of genocide should not be taken lightly. Concluding that an event falls within the range of genocide is to implicate the event's actor as genocidaires, as criminals. To accuse one of genocide requires proof. Students should be taught to gather sufficient evidence from various sources. They should be made aware of the kinds of sources that are available for use: memoirs, oral testimony, court records, monographs with annotated bibliographies, newspapers, articles, and so forth.

Once they have gathered a reasonable amount of evidence, they are in a position to render a tentative verdict as falling within the range of genocide, depending on the definition they chose. If they rely on an extremely broad definition, it will be easier to conclude that an act of mass killing constitutes genocide. But they need to think and be urged to consider: Is that wise to do? If, however, they select a narrower definition, they may find it more difficult to render a verdict of guilty. Again, they need to seriously consider why they are using a particular definition and the many ramifications of that.

Raphael Lemkin offered a definition drawn from his encounter with the Holocaust. He left unresolved the question of motive. Students should be asked about the relative importance of intention. Genocidal violence can erupt and killings on a genocidal scale can take place without an original intention to wipe out a group. What do we do with unintended genocide? Students should be made aware that in the context of war, unplanned escalation of violence is common.

It might be useful to have the students hold a mock trial, with an accuser, an accused, a jury, witnesses, and a judge. They should focus on one event, which they carefully study together (deciding on which definition of genocide they are going to use to try the case), and then bring it to court. At the end, the whole class, acting as the jury, can deliberate in class and render a verdict. When the exercise is over, students can decide which definition proved the most useful and why others were rejected. That way they can experience the central importance of a definition. Without one, there is no crime and there can be no trial. With an unsatisfactory definition the accused cannot be found guilty. The responsibility for working with a proper definition is a heavy one; indeed, it constitutes a major lesson in life. In the case of genocide, fair justice depends on a carefully wrought definition. After all, one person's terrorist is another person's freedom fighter. The same dilemma can exist, in certain situations, in grappling with genocide. For example, we have been told that in Rwanda in 1994 Hutus committed genocide against Tutsis. That is generally correct. One also learns that Tutsi armies from Tanzania invaded Rwanda and stopped the genocidal carnage. What we do not learn is that the invasion of Rwanda reversed things; now Hutus by the hundreds and thousands were forced to flee into the jungles of east Zaire, most were women and children afraid for their lives. We hear little about the deadly Tutsi actions against these refugees. These Hutu survivors claim a counter-genocide took place. We, in turn, can ask: What definition of genocide did they use to make such a claim?

The crucial importance of definition can be gathered from the Biafran secession war against Nigeria in the mid-1960s. The Biafrans repeatedly accused Nigeria of genocide by starvation—blocking food supplies. The 25 years war of independence of East Timor against Indonesia was dubbed a genocide. The 1972 massacres in Burundi have been classified as genocide. But—in all these cases—according to what standards of a genocide? According to what criteria?

This dilemma provides instructors with the opportunity to demonstrate to their students that the use of accurate definitions is a key to understanding the world of yesterday and especially tomorrow. Tomorrow there will be many more incidents of radical violence. Who is to say it is genocide? Who is to say it is not? The challenge is obvious but not simple.

CONCLUSION:
THE FUTURE OF DEFINING GENOCIDE

This introduction to the problems of defining and teaching about the definition of genocide is by no means the end of the matter. No doubt readers and future scholars will take issue with the definitional approach outlined herein. That in itself is for the good. Dialogue is no enemy of clarity as long as it is not in the service of narrow interests. To perpetuate the Katz exclusionism in the name of promoting a "pet" genocide of one's own should be discouraged as unhelpful. To reward each event of mass killing with the "honorific" status of genocide à la Charny is also a dead-end. One must draw conceptual lines, or else everything is perceived as being the same, which, essentially, is nonsense.

Nevertheless, future events and deeper understanding of those in the past will necessarily force scholars and students to rethink prevailing definitions. The one offered in this chapter will very probably be refined and, finally, replaced by one more satisfactory. That is also as it should be with but one caveat: whatever definition governs should be in the spirit of clarity and accuracy and *not* parochial interests, of which the most serious is the ethnocentric approach to genocide, by which is meant the study of a genocide exclusively by its surviving victims or its ethno-kin. Understandably, initial research on the Holocaust, on the Armenian genocide, and the genocide of the Gypsies was done by the scholars of the respective peoples targeted for extermination. This ethno-scholarship, however, eventually led to a measure of self-promotion. Especially students should be taught not to look at genocide through ethno-prisms.

The danger that this trend might continue is based on the present danger: there are more and more ethnic conflicts being spawned globally, many with genocidal implications. It is this writer's wish that future (and past) genocides be examined by scholars, regardless of their ethnic origins or affiliation. Unless this happens, genocide research and study will be in danger of becoming more apologetic, hagiographic, public relations-minded, and politically correct than sound scholarship. At present, this is dangerous (politically incorrect) territory; but it needs to be said if genocide research is to thrive and definitions of genocide are to retain their credibility and authenticity.

REFERENCES

Brecher, F. W. (2001) David Wyman and the historiography of America's response to the Holocaust: Counter-considerations. In J. R. Mitchell & H. B. Mitchell (Eds.), *The Holocaust: Readings and interpretations* (pp. 353-354). New York: McGraw Hill.

Charny, I. W. (Ed.). (1999). Definitions of genocide. In *Encyclopedia of genocide* (pp. 11-15). Santa Barbara, CA: ABC-CLIO Press.

Dadrian, V. N. (1995).*The history of the Armenian genocide*. Oxford, UK: Berghahn Books.

Katz, S. (1994). *The Holocaust in historical context* (Vol. 1). New York: Oxford University Press.

Kiernan, B. (1996). *The Pol Pot regime*. New Haven, CT: Yale University Press.

Lemkin, R. (1947). Genocide as an international crime. *American Journal of International Law, 41*, 145-147.

Lippman, M. (2001). A road map to the 1948 Convention on the Prevention and Punishment of Genocide. *Journal of Genocide Research*, 4(2), 177-195.

Ruhashyankiko, N. (1978). Special Rapporteur, UN. Escor, 31st Session, UN Doc. E/CN. 4/Sub. 2/416.

Semelin, J. (2001). In consideration of massacres. *Journal of Genocide Research*, *3*(3): 377-391.

Whitaker, B. (1985, April). *Revised and updated report on the question of the prevention and punishment of the crime of genocide*. (E/CN.4/Sub.2/1985/6, 2 July 1985.) Submitted to the United Nations Subcommission on Prevention of Discrimination and Protection of Minorities of the Commission on Human Rights of the United Nations Economic and Social Council in Geneva.

CHAPTER 6

"CASE STUDIES" OF GENOCIDE PERPETRATED IN THE TWENTIETH CENTURY

This section is comprised of succinct but detailed overviews of 10 cases of genocide perpetrated in the twentieth century: the Ottoman Turk genocide of the Armenians, the manmade famine in Ukraine, the Holocaust, the Indonesian massacres of Communists and suspected Communists, the Bangladesh genocide, the Burundi genocide, the Khmer Rouge-perpetrated genocide in Cambodia; the Iraqi genocide of the Kurds; the genocide perpetrated in the former Yugoslavia in the early to mid-1990s, and the 1994 Rwandan genocide. Each essay is written by a scholar who specializes in the study of the particular genocide of which he or she writes.

One might wonder why the aforementioned genocides were selected for inclusion in this book. First, collectively, the genocides included herein were perpetrated over the course of the twentieth century—a century that has been deemed by some to be "the century of genocide." So, readers are able to gain a certain sense as to the chronic problem genocide presented to humanity throughout the entire century. Second, the genocides provide a solid representation of genocides that were perpetrated for different reasons (ideological, retributive, despotic, to name but three). And third, the cases also provide a good sense as to how genocide was perpetrated in different parts of the world (Asia, Europe, the Middle East, and Africa) in very recent times.

Each and every genocide is, in its own way, extremely complex; and that, I believe, demands that educators seek out the most authoritative and accurate discussion of such cases as possible. That is exactly why the scholars whose insights are presented herein were sought out and asked to contribute to this book.

In addition to writing a succinct account of the genocide in which they specialize, each author was also asked to succinctly delineate the key antecedents and events vis-à-vis the genocide. (Note: Some authors chose to include such concerns within the body of their essays, while others chose to highlight such in sidebars.) *The reason for adding this component herein is that far too often students come to understand the whats, hows, whens, wheres but not the whys behind the genocide. In order to truly begin to understand why a genocide was perpetrated, both teachers and students need to fully understand the key antecedents that led up to and, ultimately, culminated in the genocide.*

Finally, each contributor was asked to include, at the conclusion of his/her essay, an annotated list of up to five key works that they deemed important resources for teachers and students.

CASE STUDY 1

THE ARMENIAN GENOCIDE

Richard G. Hovannisian

INTRODUCTION

The Armenian Genocide of 1915 completely altered the course of Armenian history as well as the geopolitical, economic, and ethnographic complexion of the Middle East. The lessons from these crimes remain compelling and need to be passed on to current and future generations. In many ways, the case of the Armenian Genocide has become the prototype of modern premeditated mass killings and their far-reaching consequences.

Civilian populations have often fallen victim to the brutality of invading armies, bombing raids, and other forms of indiscriminate killings. In the Armenian case, however, the government of the Ottoman Empire, dominated by the Committee of Union and Progress or Young Turk Party, turned against a segment of its own population. In international law, there were certain accepted rules and customs of war that were aimed in some measure at protecting civilian populations, but these did not cover domestic situations or a government's treatment of its own people. Only after World War II and the Holocaust was that aspect included in the United Nations Genocide Convention (UNCG). Nonetheless, at the time of the Armenian deportations and massacres beginning in 1915, many governments and statesmen termed the atrocities "a crime against humanity."

OVERVIEW OF THE ARMENIAN GENOCIDE

The Armenians are an ancient people. They inhabited the highland region between the Black, Caspian, and Mediterranean seas for nearly 3,000 years until 1915. Emerging as an organized state in the early centuries of the first millennium B.C.E., Armenia lay on a strategic crossroad between East and West. It was sometimes independent and formidable under national dynasties, sometimes autonomous under native princes who paid tribute to foreign powers, and sometimes subjected to direct foreign rule.

At the turn of the fourth century C.E., after more than a thousand years of polytheism, Armenia adopted Christianity, becoming the first nation in the world to proclaim that faith as the religion of state. Christianity cost the Armenian people dearly, for the tenacity with which they held to the faith exacted from them, down through the centuries and before the genocide itself, virtually millions of lives. Their existence was also made difficult by invasion, draining and devastating the land and compelling many Armenians to seek safety in distant realms. But always,

---

List of Essential Antecedents That Need to Be Addressed When Teaching About This Genocide

- Institutionalized Second Class Citizenship: The Ottoman Empire (14th to 20th century) was a theocratic state in which religious minorities, such as the Armenian Christians, had special taxes and restrictions imposed upon them in exchange for being allowed to practice their "imperfect" religion. (Note: If one refers to the conquest of Constantinople in 1453 as the point at which the empire becomes centralized, then it was the 15th century. However, that came after the Turks had already conquered most of the Balkan lands in the 14th century.)
- Decline of the Ottoman Empire: The Ottoman Empire entered a period of accelerated decline in the 18th and 19th centuries, with extensive territorial losses accompanied by growing domestic corruption and exploitation.
- Unrest among the Subject Nationalities: Parallel with the repressive policies in the empire, the concepts of the Enlightenment and French Revolution stirred unrest among the Balkan Christians. Intellectual revival was followed by political agitation for separation. Most Balkan subject peoples gained independence in the 19th century with external assistance.
- Involvement of the European Powers: The European Powers became increasingly involved in Ottoman affairs. Rivalry among those powers saved the empire from total collapse but still left it susceptible to external political and economic pressures.
- The Plight of the Ottoman Armenians: Most Ottoman Armenians lived in their historic homelands in the eastern Asiatic provinces of the empire. The Armenian peasantry was exposed to intensified exploitation and persecution in the 19th century. The Armenians sought reforms and the safety of life and property by appealing both to the Ottoman sultan and to the European Powers.
- Armenian Resistance Movements: The failure to bring about peaceful reforms prompted some Armenian intellectuals

most Armenians stayed firmly planted on the Armenian Plateau, maintaining their separate ethno-religious identity and culture.

The Turkic incursions into Armenia began in the eleventh century C.E., and the last Armenian kingdom fell three centuries later. Most of the territories that had once formed the ancient and medieval Armenian kingdoms were incorporated into the Ottoman Empire in the sixteenth century. That empire was a theocratic state based on Islamic precepts. The Turkish ruling classes controlled a multinational, multiconfessional realm in which—and this may be a clue to potential genocide—there was a plural, not pluralistic, society in which various groups lived side-by-side yet separate and distinct. They belonged to a common state, but the theocracy was founded on the institutionalized separation of the population into true believers and nonbelievers. The nonbelievers were the *gavurs*, a pejorative term meaning "infidel."

According to the precepts of Islam, tolerance of Christians and Jews, that is, of other monotheists, was to be accorded on condition that they submit to an inferior status of second-class citizenship with certain financial, political, and social disabilities. The testimony of a nonbeliever, for

---

to organize underground political groups for self-defense. The Armenian political parties date to the 1880s, following the failure of Sultan Abdul-Hamid to fulfill his pledge to the European powers in 1878 to implement reforms.

- The Armenians Viewed as a Threat: Armenian strivings for equality were viewed as a dangerous threat by the dominant traditionalist elements in the Ottoman Empire. When forced to promulgate reforms by the European Powers in 1895, Abdul-Hamid reacted by unleashing a series of massacres, causing the death and economic ruin of countless thousands of Armenians.

- The Young Turk Revolution: Some Turkish intellectuals, students, and officers believed that the only way to save the Ottoman Empire was through radical change. In 1908 they succeeded in a near bloodless coup to seize power on the platform of liberty, equality, and justice, and forced the sultan to become a constitutional monarch.

- The 1909 Cilician Massacre: Armenian optimism was soon dashed when traditionalist elements tried to regain power and organized massacres throughout the region of Cilicia, resulting in the deaths of more than 20,000 Armenians and a new chill in Armenian-Turkish relations.

- The Balkan Wars, 1912-1913: The former Christian subject peoples, assisted by Russia, defeated the Ottoman Empire and left it in Europe only with Constantinople/Istanbul and its hinterland. The crisis allowed the radical wing of the Young Turks to seize power and prepare plans to cleanse the Asiatic provinces of the Armenian Christian population as a way of guaranteeing a safe Turkish homeland.

- Turkish nationalism replaced Ottoman multinationalism: The Young Turk dictators adopted the ideology of "Turkism," with the goal of creating a homogeneous Turkic state based on one people and one religion. This became the "warrant for genocide."

example, could not be admitted as evidence against a true believer in an Islamic court. In lieu of military service, because religious minorities were not allowed to bear arms as part of the system of keeping subject groups submissive, a poll tax was imposed on every male child. This was one of the reasons that heads of a household often concealed the true number of family members. There were various other disadvantages, such as special extraordinary taxes, uncompensated labor, and sometimes the need to wear special garb, all in exchange for permission to practice a pre-Islamic "imperfect" religion.

Despite these burdens, most Armenians lived in relative peace so long as the Ottoman Empire was strong and expanding. But as the empire's administrative, fiscal, and military structure crumbled under the weight of

---

Key Events: Armenian Genocide

- World War I erupts in the summer of 1914.
- The Turkish dictators enter into a secret alliance with Germany against the Allied or Entente Powers in August 1914 and then take the Ottoman Empire into the war in October.
- Most able-bodied Armenians are conscripted in a general mobilization and then separated into disarmed labor battalions and massacred during the course of the following months.
- The Turkish Minister of War, Enver Pasha, fails in his attempt to invade and capture the Caucasus region at the end of 1914. This setback, along with the Gallipoli landing of the Allied expedition in 1915, is regarded by some scholars as the trigger for activitating the plan to eliminate the Armenian population under the cover of war.
- The genocide begins with the arrest of Armenian leaders in Constantinople on the night of April 23/24, 1915, and is followed by the deportations and massacres of nearly the entire Armenian population of Asia Minor and the historic Armenian provinces. In general, the adult male population is killed outright, whereas most women and children die in the forced march toward the Syrian desert.
- Many comparisons can be made with other genocides, including the relationship between government and political party; previous demonstrated vulnerability of the targeted group; role of military and special organizations; secrecy and deception; use of technological advances for destruction; denial from the outset.
- The Allied Powers condemn the genocide and pledge punishment for the perpetrators and rehabilitation for the survivors. The defeated Ottoman Empire in 1918 has to face the consequences of Young Turk genocidal policies by initiating courts-martial and condemning chief perpetrators.
- Rise of Turkish resistance movement led by Mustafa Kemal nullifies the process to try the perpetrators of the genocide and opens the way to state denial and suppression of memory, especially after Allied Powers make their peace with the regime and do not want to be reminded of their broken promises.
- The fiftieth anniversary of the genocide in 1965 is a watershed of Armenian activism. Attempts to gain worldwide reaffirmation of the crime are paralleled by intensified denial by the Turkish state and its agencies.
- The Armenian Genocide remains a live issue even after the passage of nearly nine decades and the beginning of a new century and millennium.

internal corruption and external challenges in the eighteenth and nineteenth centuries, oppression and intolerance increased. The breakdown
of order was accelerated by Ottoman inability to modernize and compete
with the West.

By the beginning of the nineteenth century, the Ottoman Empire was
in rapid decline, losing much of its territory in Europe. The concepts
emanating from the Enlightenment and the French Revolution were having an impact on the subject nationalities of the empire, whether Greek,
Serbian, Montenegrin, Romanian, or Bulgarian, and, very belatedly,
Armenian. Perhaps this, too, was one of the contributors to the Armenian
tragedy, in that the Armenians may have stirred too late. Those peoples
who sought emancipation relatively early were able to find European support and ultimately to seize independence, whereas the Armenians
throughout the nineteenth century aspired, not to independence, but
rather to civil rights, equality before the law, security of life and property,
and local self-government, quite some distance from independence.

As Turkish rule weakened and the European powers, for their own selfish reasons, interfered increasingly in the affairs of the Ottoman Empire,
tensions intensified between the various ethno-religious communities,
majorities and minorities, and between the minorities themselves—
Greeks and Armenians, Armenians and Jews, and so forth. More than one
sultan gave in to external pressures and domestic reformers in the nineteenth century to proclaim, against custom and tradition, that all his subjects were equal in his eyes and henceforth would be treated as such. This
was done in an effort to hold the empire together. Unfortunately, most
sultans were not sincere when they issued these decrees under duress.
Moreover, one of the effects of the reform edicts was to anger and arouse
traditional society. For example, if previously a *gavur* came into contact
with a true believer, a first-class citizen, even if poor, humble, and less
educated, a certain demeanor was expected and required. To try to
change that kind of mentality, that type of society, and suddenly to
announce that all were to be equal, when there was no strong, true, sincere governmental support of the declaration, could only lead to trouble.

Armenians came to be portrayed and perceived as an arrogant, schem-
ing element that was conspiring to achieve dominance through the ruse
of equality. And it was not difficult for traditionalist leaders to bring the
masses to regard the specter of equality as being tantamount to exploitation by the *gavur*. It was unfair; it was wrong; it was an attempt ultimately
to usurp the rights and privileges of the true believers. That the European
powers involved themselves in these matters only made things worse.
European pressure for reform was repeatedly applied on the Ottoman
government, but this action was not sustained by effective measures of

enforcement. The result was an even greater suspicion of the subject people.

The Armenian striving to achieve equality through reforms in the Turkish empire was ultimately an utter and dismal failure. Equality through edicts about being the children of a common homeland and of a paternalistic ruler proved to be stillborn. Some Armenian youth gave up hope that reforms could be achieved peaceably. They began to organize underground political parties and encouraged the population to learn to defend itself, but their strength and means were very limited.

When in 1895 the last important sultan of the Ottoman Empire, Abdul-Hamid II (1876-1909), was coerced into signing another reform edict, his real answer to the Westerners who imposed this act on him—and to the Armenians who were seeking assistance and relief from the terrible conditions caused by the breakdown of law and order in the interior provinces—was to unleash a rampage of death and destruction. In October 1895, starting in the port city of Trebizond on the Black Sea and spreading in the winter months to every province of historic Armenia and into Cilicia along the Mediterranean Sea, there erupted mayhem lasting for up to a week during which hapless Armenians were cut down wherever they were found. Armenian shops were looted; Armenian homes were burned; Armenian villages were pillaged. Thousands of terrified people fled to the mountains or abroad, and still other thousands were forcibly converted to Islam. The number who died was placed minimally at about 100,000, although most sources report the number at 200,000, and some as many as 300,000.

Here is a key question to be considered: How are the massacres of 1895-1896 that claimed so many Armenian lives to be interpreted? Was it, in fact, the beginning of the end for the Armenian people? Should the Armenian Genocide be regarded as starting in 1915 or rather as being a continuous process from 1895 to the end of World War I in 1918 and even beyond? This issue requires further thought and analysis.

Whatever the answer, in the Armenian case there was a very important qualitative and quantitative difference between 1895 and 1915. The sultan, however oppressive, however sinister, however paranoid, probably did not conceive realistically of an empire without Armenians. The Armenians had a place and a function in his realm. They simply had to be taught a lesson; they needed to be intimidated back into complete submission. The Armenians should be impoverished somewhat, and their concentrations in their historic provinces should be diluted. Certain demographic changes were in order. What better response to Western meddling?

While Abdul-Hamid's actions in 1895 may be classified as genocide according to the United Nations Convention on the Prevention and Punishment of the Crime of Genocide, in the narrower sense in which many Holocaust and Armenian Genocide scholars interpret the term as implying the attempted total annihilation of a people, it may be more proper to define the massacres of 1895-1896 as pogroms, albeit the term was not then used for the Armenians. Even though there was much bloodshed and certainly the intent was to kill an ethnic or religious group, at least in part (the U.N. definition reads "in whole or in part"), there was a beginning and an end to the violence. After several days, when the mobs had done their work, regular army units appeared to establish a degree of order.

The sultan could not allow the entire country to get out of hand. The intended message had been given, and it was time for the government to bring a halt to the pillage and plunder. For the Armenians, it seemed that they had sustained but survived yet another in a long series of calamities. Thus, once more the challenge was to reconstruct and go forward.

Abdul-Hamid was not trying to bring about drastic changes in society. Rather, he was desperately attempting to preserve a system that was unsalvageable, a foundering ship of state that was being sunk by external volleys and internal disintegration. Pogroms—massacres—were his misguided and vain response to the critical problems besetting the empire.

If this interpretation is accepted, then it is obvious that there was a fundamental difference between 1895 and 1915. In 1908, Abdul-Hamid, the old sultan, was overthrown and sent into exile the next year by the Committee of Union and Progress (CUP) or, as they were commonly referred to, the Young Turks, a political movement that held forth the vision of a new Ottoman Empire based on constitutional government and the principles of equality, fraternity, and justice. It is beyond the scope of this overview to explain in detail where that experiment went wrong. That said, it is worth noting that in his comparative study of the Armenian Genocide and the Holocaust, genocide scholar Robert Melson maintains that the Turkish genocide of the Armenians stemmed from a revolution that went sour, that failed to achieve its anticipated objectives (see Melson, 1992).

In the Ottoman Empire, the hopes placed on constitutional government in 1908 soon dimmed, partly because of European exploitation and self-interest and partly because of internal discord. By 1913, that which had started as a democratic revolution culminated in a dictatorship of the ultra-rightwing faction of the Young Turk Party. It was that extremist element that took Turkey—the Ottoman Empire—into World War I as an ally of the German Empire. A fundamental calculation was that the anticipated triumph of the Central Powers against Great Britain, France, and the Turk's old nemesis, Russia, would allow for Turkish annexation of territories that had been lost to the tsars in one war after another. Moreover, there was the scheme of creating a new Turkish realm, no longer based in Europe, but rather extending eastward toward the original Turkic homelands in central Asia.

Various pan-Islamic and pan-Turkic concepts were at work, but an overriding theme of Young Turk ideologues was the unification of the Turkic-speaking peoples within a common framework. Ideology in the case of the Armenian Genocide was very important, perhaps not a singular explanation, yet nonetheless a critical justification of radical measures against the targeted group.

If Adolf Hitler, whose ideology or objective was to establish a new world order based on a racial formula in which there was no room for Jews, the Young Turk ideology to create a new regional order without Armenians was similarly at work. Armenians were regarded as being an alien element unwilling to assimilate. The tenacious Armenians had existed as a subject people for centuries and had clung to their ethno-religious identity.

The Armenian Question relating to the need for measures to safeguard the lives and properties of the Armenians in the provinces had become an international issue since 1878 and had allowed for intermittent European intervention. It was feared that sooner or later the Armenians would try to follow the example of the former subject European Christian nationalities to establish a separate state, thereby becoming a major barrier to any and all pan-Turkic objectives. Thus, the time had come to supplant the old, tired concept of Ottomanism—that is, a society with Turks, Kurds, Greeks, Armenians, Arabs, Jews, and others all living side-by-side—with that of a modern state or empire anchored in a single ethnicity and a single religion.

Although many Young Turk leaders were agnostics or atheists, they exploited religion and the traditional precepts of Islam to spread fear and suspicion of the Armenian people. Previously, under the sultans, the loyalty of the masses was directed toward the person of the sultan—to God and Suzerain—but now with the sultan discredited, the Young Turks made the state the new focus of allegiance. This is clearly reminiscent of Nazi ideology. Among the new Turkish intellectuals and ideologues, such as Zia Gokalp, are heard poetic lines of exaltation of the state:

> I am a soldier, it is my commander.
> I obey without question all its orders.
> With closed eyes I carry out my duty.
> (Quoted in Heyd, 1950, p. 124)

So the state is above all else, and for its sake anything is possible.

Genocide scholar Helen Fein (1979) has rightly observed: "The victims of twentieth century premeditated genocide—the Jews, the Gypsies, the Armenians—were murdered in order to fulfill the state's design for a new order.... War was used in both cases ... to transform the nation to correspond to the ruling elite's formula by eliminating the groups conceived as alien, enemies by definition" (pp. 29-30).

The Genocide

When was it determined that the solution to the Armenian Question was to be found in the elimination of the Armenian people? The mass arrests, the segregation of Armenians in the Turkish army into labor battalions before they were killed, and then the decrees of deportation came after the Turkish armies had suffered major setbacks on the battlefield.

The Young Turk leader and Minister of War, Enver Pasha, seized the first opportunity to strike against Russia on the Caucasus front to break through to Baku and the Caspian Sea. He ordered the campaign against the advice of his general staff and military commanders, who warned that the Armenian Plateau was impassable during the winter blizzard conditions and that the Turkish army would sustain terrible casualties as much from exposure as from combat. But driven by his ideology and so fixated on achieving his objective, Enver dismissed this counsel and took personal command. The misadventure led to the loss of an entire army corps.

Some would say there is a definite connection between Enver's frustration and embarrassment and the decision to implement a genocidal campaign against the Armenians. Without the active and sincere cooperation of today's Turkish scholars and the Turkish government, however, the answer to such questions will remain circumstantial and cannot be known with indisputable certainty. Even then, the precise decision-making processes may never be established unless the secret records of the Young Turk inner circle are revealed and made accessible.

Although the genocide of the Armenian people and the destruction of millions of persons in Central and Eastern Europe during the Nazi regime a quarter of a century later each had particular and unique features, there were some striking parallels. The similarities include the perpetration of genocide under the cover of a major international conflict (World War I and World War II, respectively), thus minimizing the possibility of external intervention; conception of the plan by a monolithic and xenophobic clique; espousal of an ideology giving purpose and justification to racism, exclusivism, and intolerance toward elements resisting or deemed unworthy of assimilation; imposition of strict party discipline and secrecy during the period of preparation; formation of extralegal special armed forces to ensure the rigorous execution of the operation; provocation of public hostility toward the victim group and ascribing to it the very excesses to which it would be subjected; certainty of the vulnerability of the targeted groups; exploitation of advances in mechanization and communication to achieve unprecedented means for control, coordination, and thoroughness; and the use of sanctions such as promotions and incentive to loot or, conversely, the dismissal and punishment of reluctant officials and the intimidation of persons who might consider harboring members of the victim group.

It is especially important to note that the perpetrators were confident of the vulnerability of their intended victims. After Kristallnacht in 1938, had anyone intervened on behalf of the Jews? After the massacres of 1895-1896 and once again in 1909, when more than 20,000 Armenians were massacred in the region of Cilicia, had anyone intervened? The inaction of the world community provided the obvious answer.

The involvement of the armed forces and the creation of special organizations to oversee the genocidal operations are another important parallel. During the Holocaust there were the SS and the *Einsatzgruppen*, whereas during the Armenian Genocide there was the *Teshkilati Mahsusa*—the Special Organization. It was an organization whose ostensive purpose was to further the war effort but whose secret mandate was to supervise the destruction of the Armenian people and to make certain that recalcitrant officials would be forced to cooperate or removed from office and punished. The Special Organization recruited hardened criminals and tribesmen into killer battalions. These fell on the deportee caravans, usually in places of no escape, such as mountain passes and river crossings.

There were, of course, many Turks and other Muslims who felt that what was happening was an affront to God and to humanity. A significant number tried to protect their friends and neighbors at some risk to themselves. Unfortunately, denial of the genocide for more than eight decades has not allowed the Armenians to honor those who attempted to help, yet nearly every survivor story entails some kind of intervention that made possible escape from certain death. Intervention was not necessarily altruistic. It often entailed the desire to acquire a maid, a servant, field hands, or even girls to provide personal pleasures. Nonetheless, someone intervened to pluck these people from the death caravans. It is also true that many Armenians survived only by forfeiting their identity. They were registered as Muslims and given Turkish names. They forgot or dared not use their native language, and little children even lost the memory of parents as they were absorbed into the larger new society that was being created.

One must ask: If the intent of genocide is to destroy the targeted group, why then does the perpetrator differentiate in the process of annihilation? Why was it that the Armenian men, in city after city, town after town, and village after village, were roped up, usually by fours, and taken to the nearest killing field, to the nearest river crossing, to the nearest mountain gorge, and killed outright—shot, axed, stabbed, hacked to death? In that crude form of killing, there would be two, three, or four men who emerged from the bloody heaps as living witnesses of what had occurred. If the intent was to destroy the Armenians, then why the belabored process of taking hundreds and thousands of women and children and forcing them to march to death, rather than killing them in the same way as the men? And yet in most areas this is just what happened. Women, children, and the elderly were driven toward the Syrian desert with little food or water and tormented all along the way. Perhaps, it was thought that any potential resistance could be obviated by wiping out the male population, who were also viewed as the primary perpetuators of the race. Moreover, since antiquity there was a twisted code of conduct that often spared women from direct killing but in fact subjected them to even greater misery and agony. This was fully manifest in the case of the Armenian Genocide.

There were many choiceless choices that had to be made during the Armenian Genocide, as in all genocides. Women who were interviewed when they were in their eighties, who presumably should have sublimated or at least reconciled themselves to a distant past, still sobbed in anguish as they spoke of having two children on the deportation route when they were prodded on by bayonets to ford a fast-flowing river. They could carry only one child without being swept away. Which one to take and which to leave? And how to leave? Hence, the choiceless choice, as one child was placed under a tree or near a boulder. The last sound to haunt the mother for the rest of her life was the child's cry not to be abandoned. And what of the abducted teenage girls, who gave birth to one or two babies during their years of captivity? Following the defeat of both Germany and Turkey in World War I, relief agencies and relatives came to rescue the girls, who now were faced with the choiceless choice of either abandoning their bastard children who were their flesh and blood or else renouncing their faith, family, and nationality to live out their years with an imposed and completely different identity.

As for the aftermath of the Armenian Genocide, the trauma is especially enduring because of the refusal of the perpetrator regime or its successors to acknowledge the crime and engage in acts of contrition and redemption. On the contrary, the Turkish government has engaged for decades in an unrelenting campaign of denial and suppression of memory. Denial has taken on the new and sinister forms of rationalization, relativization, and trivialization. Soon, the last Armenian survivor will pass from the scene. Then the perpetrator side can say: "Were you there? Did you see it? Is your testimony allowable in court as first-hand evidence?

You weren't there. You're not an eyewitness. You are imagining and fabricating."

This challenge makes it all the more important that the Armenian Genocide, its effects and implications, be integrated into collective historical memory and made a part of the permanent record of humankind.

ISSUES AND QUESTIONS TO CONSIDER WHEN TEACHING ABOUT THE ARMENIAN GENOCIDE

The Armenian Genocide eliminated a people from its homeland and wiped away most of the tangible evidence of its 3,000 years of material and spiritual culture. The calamity, which was unprecedented in scope and effect, may be seen as the culmination of the ongoing persecutions and massacres of Armenians in the Ottoman Empire, especially since the 1890s. Or it may be placed in the specific context of modern nationalism and the great upheavals that brought about the dissolution of a multiethnic and multireligious empire and the emergence in its place of a Turkish nation-state based on a monoethnic and monoreligious society. The approaches are not mutually exclusive and should be examined in the context of the plight of the Armenians in the nineteenth century and their ultimate elimination from the Ottoman Empire in the first part of the twentieth century.

A critical issue in the Armenian case that has general application is the way traditional-bound societies react to change or attempted change. If the Armenian quest for equality and security in the Ottoman Empire was viewed by the dominant element as a serious threat to its accustomed way of life, one need look no further than the reactions in the United States to the civil rights movement in the second half of the twentieth century to see certain comparisons in the strong, sometimes violent, response to impending change. There are, of course, fundamental differences that must be noted as well. If in the U.S. case the government intervened to enforce legislation and change, in the Armenian experience, the sultan's government was directly complicit in obstructing the very reforms to which it had acceded, at least on paper.

One might consider whether there was anything that the Armenian people or their leaders could have done to escape their fate in face of an emerging militant nationalism espoused by the Turkish rulers. Was there any real way for the Armenians to have kept their identity, their religion and culture, and still survived in the changing geopolitical, ethnic, and economic environment? Could they or should they have avoided intellectual and political currents that emanated in Europe and gradually made their way eastward? In what ways did their own cultural, educational, and economic progress affect their relations with the dominant group and impact on the course of their history?

It is important also to consider the role of foreign governments that intervened from time to time in the Armenian Question. What circumstances could have made the results of external intercession more favorable? And what was the role of bystander governments during the period from the 1890s to the 1920s? How did the demonstrated vulnerability of the Armenian people make the perpetrators all the more audacious?

A common feature of most genocides is denial by the perpetrator side. In the Armenian case, the question should be raised as to why, long after the Ottoman Empire has been succeeded by the Republic of Turkey, does there continue to be such adamant rejection and denial of the truth. Were there conditions that made the aftermath of the Armenian Genocide radically different from the post-Holocaust period? And why do powerful countries such as the United States participate in trying to cover up or obscure the magnitude and significance of the Armenian Genocide while fully recognizing the crimes of Nazi Germany and the genocidal policies of that regime?

Students should also consider the effects of the trauma and of post-traumatic stress, the ways in which survivors live with painful memory and react to denial, and how the trauma manifests itself in subsequent generations. As for legal recourse, one may ask how victim groups, especially those that are also dispossessed of their goods, properties, and even homeland, can place their case before national and international bodies that tend to be made up of mutually-protective nation-states? Might the outcome for the Armenian victims and survivors have been different if the international tribunals that now operate in the Hague and elsewhere been empowered at the time? Finally, how is it possible to seek legal recourse and to have truth prevail over perceived national interests? Is it possible to liberate history and human rights from politics?

CONCLUSION

The late Terrence Des Pres (1986), author of *The Survivor: An Anatomy of Life in the Death Camps*, has captured the importance of remembering:

> Milan Kundera ... has written that "the struggle of man against power is the struggle of memory against forgetting."... National catastrophes can be survived if (and perhaps only if) those to whom disaster happens can recover themselves through knowing the truth of their suffering. Great powers, on the other hand, would vanquish not only the peoples they subjugate but also the cultural mechanism that would sustain vital memory of historical crimes.... When modern states make way for geopolitical power plays, they

are not above removing everything—nations, cultures, homelands—in their paths. Great powers regularly demolish other peoples' claims to dignity and place, and sometimes, as we know, the outcome is genocide.

In a very real sense, therefore, Kundera is right: Against historical crimes we fight as best we can, and a cardinal part of this engagement is "the struggle of memory against forgetting" (p. 10-11).

Memory will prevail when crimes against humanity such as the Armenian Genocide become an undisputed integral part of the collective historical record. In that endeavor, the roles of education and the educator are critical.

Recommended Reading on the Armenian Genocide

Balakian, P. (1997). *Black dog of fate.* New York: Basic Books.
   This semiautobiographical memoir of an Armenian American on a journey of self-discovery, including the revelation of the "forgotten" or "suppressed" genocide, is effectively written and can have broad classroom application.

Dadrian, V. N. (1999). *Warrant for genocide: Key elements of Turko-Armenian conflict.* New Brunswick, NJ: Transaction Books).
   The author draws on much of his previous work in the field to synthesize critical factors contributing to the Armenian Genocide.

Hovannisian, R. G. (Ed.). (1986). *The Armenian genocide in perspective.* New Brunswick, NJ: Transaction Books.
   Scholars in the fields of history, political science, sociology, theology and ethics, literature, and psychiatry offer a multidisciplinary perspective on the Armenian Genocide.

Hovannisian, R. G. (Ed.). (1998). *Remembrance and denial: The case of the Armenian genocide.* Detroit, MI: Wayne State University Press.
   Several contributors to this volume examine the ways in which memory of the Armenian Genocide is maintained and expressed, and others analyze the changing strategies and moral implications of the phenomenon of denial.

Miller, D. E., & Miller, L. T. (1993). *Survivors: An Oral History of the Armenian Genocide.* Berkeley: University of California Press.
   This insightful study is based on oral history interviews with Armenian survivors, relaying first-hand experiences and categorizing and analyzing survivor reactions ranging from resignation to rage.

REFERENCES

Des Pres, T. (1986). Remembering Armenia. In R. G. Hovannisian (Ed.) *The Armenian Genocide in Perspective* (pp. 9-17). New Brunswick, NJ: Transaction.

Fein, H. (1979). *Accounting for genocide: National responses and Jewish victimization during the Holocaust*. New York: The Free Press.

Heyd, U. (1950). *Foundations of Turkish nationalism: The life and teachings of Ziya Gokalp*. London: Luzac.

Melson, R. 1992. *Revolution and genocide: On the origins of the Armenian genocide and the Holocaust*. Chicago, IL: University of Chicago Press.

CASE STUDY 2

THE SOVIET MANMADE FAMINE IN UKRAINE

James Mace

INTRODUCTION

The manmade famine of 1932-1933 in the Ukrainian Soviet Socialist Republic (S.S.R.) claimed the lives of at least 4.8 million people (estimates vary upward to over twice that number, for the Soviet statistics of the period are notoriously unreliable) because of an official Soviet policy of seizing grain from the Ukrainian countryside in amounts clearly in excess of what was actually available, oppressing all manifestations of Ukrainian national and territorial political self-assertion, and bringing in tens of thousands of new officials to rule the republic in a manner seen fit by the central Soviet authorities in Moscow. A major point of interest is that the Ukrainian Famine shows that genocide need not necessarily flow directly from ethnic, national, or religious antagonisms (little of these were involved) but purely from concerns of the centralization of political power in a Communist (Stalinist) totalitarian system by manipulating categories of class conflict such that they assume new, essentially national, content.

Teaching About Genocide: Issues, Approaches, and Resources, 111–118

OVERVIEW OF THE
SOVIET MANMADE FAMINE IN UKRAINE

The issue of genocide in the Soviet Union is intricately and integrally tied to the policies of social engineering carried out under the leadership of Joseph Stalin from the late 1920s until his death in 1953. The main transformations of this period included the forced collectivization of agriculture on the basis of the liquidation of the kulaks as a class, rapid industrialization made possible by the lowering of real labor costs through the drastic reduction in the living standards of free workers and the extensive use of forced labor, the absolute standardization of all spheres of intellectual activity and their strict subordination to state priorities, and the integration of a large and varied collection of national and religious groups into a Russocentric political structure.

The manmade famine of 1932-1933 poses particular problems from the standpoint of internationally accepted definitions of genocide, since its focus was geographic rather than discriminatory against specific groups within a given area, and it was clearly not an attempt to destroy all members of a given group. Rather, its national and ethnic target (that is, genocidal nature) must be inferred from the clarity in which it was geographically focused against areas containing target populations and from the particularly harsh policies of the Soviet authorities in the national sphere as applied to the main victimized group, the Ukrainians.

After a brief period of liberalization in Soviet policies toward rural areas in spring 1932, the central authorities—led by Communist Party General Secretary Joseph Stalin, Union of Soviet Socialist Republics (U.S.R.R.) Prime Minister Vyacheslav Molotov, Minister of Agriculture Lazar Kaganovich, and others—despite clear warnings from the Communists of Ukraine at the July 1932 Third All-Ukrainian Party Conference that the demands made by Moscow could not be met—decreed a series of policies of unprecedented severity. Among them were the U.S.S.R. law of August 7, 1932, providing prison sentences of up to 10 years or execution for gleaning harvested agricultural fields. When the harvest came in, the Communist authorities mobilized all available personnel to seize as much grain as possible in a vain attempt to meet the obviously unrealistic demands for grain. The official press each day carried exhortations for stricter enforcement of the grain procurements against "kulaks" (wealthier peasants), who were supposed to be hoarding the nonexistent grain. On October 12, Moscow made a major assault on the republic's relative self-assertion by replacing three of its seven oblast (regional) Communist Party secretaries and at the end of October placing Molotov in direct charge of the grain seizures in Ukraine. Kaganovich was given the same task in the nearby largely Ukrainian Kuban district of the North Caucasus. These commissions acted with unprecedented severity.

On November 18, U.S.S.R. Premier Molotov forced a resolution through the Communist Party of Ukraine Central Committee authorizing the seizure of all "bread resources" from collective and individual farms (at that time still about 20% of the total agricultural land area), except for a seed reserve later ordered seized on December 24th, if they had not fulfilled their quotas. It also authorized the imposition of "fines in kind" (that is, in such products as potatoes, beans, and meat). On December 6, "maliciously" delinquent villages were subject to the sanctions of being placed on a "black board," which meant the closing of all local stores and physical removal of all goods contained therein, a "thorough purge" (that is, arrest and repression) of the officials in charge, and physically cordoning off the area to prevent people from seeking food elsewhere.

On December 14, the top Communist leaders of the Ukrainian S.S.R, North Caucasus Territory, and Western (Smolensk) oblast were called to Moscow. While the resultant decision signed by Stalin and Molotov shows no particular criticism of the Smolensk leader, those of Ukraine and the North Caucasus were scathingly criticized for their failure to seize the required amounts. Without mentioning the fact that Molotov and Kaganovich had been directly in charge of the campaign of seizures, the failures were attributed to the alleged fact that the authorities in both areas had been deeply infiltrated by "hostile elements," largely but not exclusively Ukrainian nationalists, who had been deliberately assisting the hoarders and sabotaging the grain seizures. This was in turn attributed to the allegedly artificial character of the Ukrainization policy of fostering the development of Ukrainian culture in the North Caucasus and the "mechanistic" way it had been carried out in Ukraine. This resulted in the immediate closure of all Ukrainian institutions in the North Caucasus and the Ukrainization policy becoming a dead letter in Ukraine. The decision authorized the mass exile of specific Ukrainian Cossack settlements from the North Caucasus to the Far North of the Soviet Union along with the arrest of designated officials from a number of Ukrainian S.S.R. *raions* (districts equivalent to small U.S. counties) to be imprisoned for five to ten years in concentration camps and authorizing the death penalty for especially evil "enemies with Party cards in their pockets." Thus a campaign of terror was begun against local officials in the Ukrainian S.S.R. and North Caucasus in order to stimulate them to even greater exertions in seizing foodstuffs. One is struck by the degree in which the whole campaign was micromanaged such that Stalin and Molotov, the two most powerful men in the Soviet Union of the period, were directly designating by name which district officials were to get what punishment. This was obviously designed as an example of what they expected from subordinates in finding similar malefactors on their own.

This campaign culminated when, on January 24, 1933, the All-Union Communist Party (of Bolsheviks) issued a decree removing a number of top oblast officials of the Communist Party (Bolshevik) of Ukraine (CP(b)U), replacing them with nominees from Moscow led by Pavel Postyshev (officially the Second Secretary of the CP(b)U but clearly giving the orders), who immediately announced that the grain seizures would continue unchanged and initiating a wave of removals and repression against tens of thousands of communist party political and cultural figures associated with the republic's previous self-assertion. These latter measures, not in themselves genocidal, are nevertheless important in placing in context and making understandable actions that were clearly genocidal, most notably in this case, deliberately creating conditions of life calculated to make it impossible for members of the group to survive.

In terms of human suffering—largely captured through eyewitness accounts recorded in emigration and later in Ukraine but confirmed by now published secret police reports to the relevant authorities—the apogee was reached in the spring of 1933 with a daily ritual of horse-drawn carts rounding up bodies in the villages for burial in mass graves, some villages dying out entirely, and hundreds of official cases (and probably many more) of people being driven to cannibalism, of homeless people—especially children—being rounded up and dumped outside of town or of children being placed in homes where they were left to starve to death.

Having accomplished its most plausible political purpose, the destruction of the Ukrainian S.S.R. as a barrier to the direct authority of Stalin and those around him, the famine in Ukraine basically ended with the harvest of 1933, followed by a policy of strict centralization in ideology, politics, and culture. Indeed, since permissible new ideas, policies, and cultural initiatives could only be enunciated in Moscow and adapted in republics like the Ukrainian S.S.R., the Ukrainian language and culture became derivative and second-rate, because if one sought new ideas and methods, they could come only from Russia in the Russian language.

ANTECEDENTS

The background to the famine extends to the nineteenth century, when the Ukrainians, living mainly but not only in the Russian Empire, evolved into a national movement quite characteristic of peoples that had for one reason or another lost their traditional elites and been pushed back to traditional villages—beginning with cultural efforts and then extending to political demands as at least some elements of the group became literate and urbanized. This movement faced harsh official repression, with education and publishing in Ukrainian being banned in decrees of 1863 and 1876. Still, the movement had clearly developed a mass character by the time of the Russian Revolutions of 1917, when socialist Ukrainian national leaders soon showed their ability to quickly mobilize tens of thousands of supporters for urban demonstrations through an already developed system of rural cooperative societies. The Ukrainian socialists, however, were plagued by a lack of practical political experience, no clear idea of what a legitimate state structure was or how to build it, a preference for autonomy over independence, and, above all, their weakness in urban areas. They did establish, however, an autonomous (and in January 1918) independent Ukrainian People's (or National) Republic, which rapidly came into armed conflict with the Soviet Russian government established by Bolshevik (soon Communist) leader Vladimir Lenin in November 1918. While elections to the All-Russian and All-Ukrainian Constituent Assemblies in late 1917 and early 1918 showed overwhelming popular support in the areas inhabited predominantly by Ukrainians for the Ukrainian socialists, their failure to organize a standing military left them vulnerable to attack by Soviet Russia. Turning first to Germany and later to Poland for assistance in their Liberation Struggles, the wars of the Russian Revolution ended with the bulk of the Ukrainian lands going to the Communist-dominated Ukrainian S.S.R., which became a founding member of the Union of Soviet Socialist Republics in 1922-1923.

In the early Soviet period, known as War Communism, Lenin's Communists, who wanted to bring about "communism" immediately by nationalizing all industry and forcing the peasantry into Party-dominated communes, also had little patience with national aspirations of any type. They considered all such hopes "bourgeois nationalism" and considered any expression of them a sign that someone was not really "socialist" but in fact "counterrevolutionary." Thus they had no qualms about attacking the nascent Ukrainian People's Republic in the name of a "fraternal" Ukrainian S.S.R., the latter of which had neither an army nor policies of its own.

Ironically, the situation changed after Ukraine had lost its formal independence. In 1921 the Communists proclaimed a New Economic Policy (NEP) of allowing private peasant agriculture and small-scale private industry. Continued armed resistance in the non-Russian countryside of Ukraine and Central Asia led the Communists in 1923 to proclaim a policy of "taking root" or indigenization (and Ukrainization in Ukraine), which attempted to give the non-Russian Communist regimes a veneer of national legitimacy by means of directing local Russian-speaking officials to learn the local language and culture, actively recruiting members of the local national group into Communist Party and Soviet state work, tolerating and financially supporting the national intelligentsia and encouraging the growth of a Communist trend within it, along with strengthening the stability of the non-Russian republics in general. As the struggle for Lenin's (who died in 1924) succession absorbed the energies of the top Communist leadership in Moscow, the Communists of Ukraine (most of whom were not ethnic Ukrainians) were able to assume more and more authority, since the CP(b)U was by far the largest single subdivision of the All-Union Communist Party and the Ukrainian S.S.R. the largest non-Russian republic within the U.S.S.R. By the mid-1920s, Ukrainian Communists were making significant demands for recognition of their national character in politics, culture, and the economy.

Once Stalin had completed his struggle for supremacy in Moscow in the late 1920s, he turned to a policy of withdrawing the concessions implemented in the early 1920s and restoring the original Bolshevik program of War Communism. In December 1929, a policy of the "complete collectivization of agriculture on the basis of the liquidation of the kulaks as a class" (collectivization meant a modified form of the communes that peasants saw as little different from serfdom) and rapid industrialization. This break with the peasantry as such meant that trying to placate the non-Russian, still overwhelmingly rural nations like the Ukrainians, lost its political necessity. But in the intervening years a strong local Communist elite with significant bases of support had evolved, and this meant Stalin had to discredit that elite in order to destroy it. A series of show trials of "bourgeois nationalists" began with the trials of the Union for the Liberation (1930) and Ukrainian National Center (1931) to encompass 15 publicly announced conspiracies of "wreckers" (saboteurs) by the end of the decade. The destruction of Ukraine's Communist administrative elite, begun in 1932-1933, was accompanied by the wholesale repression of Party members who had earlier taken part in what had been called the Ukrainian cultural process. The powers of the republic were gutted.

CONCLUSION

The destruction of the old leadership of the Communist Party of Ukraine was completed in 1938 when, out of 102 members of its Central Committee, the head of state was demoted to an assistant at the Museum of the Revolution in Leningrad (now St. Petersburg), the premier committed suicide, and 96 were arrested and ultimately executed. Only two individuals, each of whom had no political significance, were allowed to remain. In place of the traditional Ukrainian identity, a new, broader Soviet identity was imposed, creating a fundamental discontinuity in the development of the Ukrainians, their partial deliberate destruction as such, which allows us to speak of Ukraine today as a postgenocidal society.

Postscript

Students might be asked to consider the problem of absolute power without moral responsibility motivated by some overweening ideal (whether social progress, racial, or ideological purity) as something that fosters genocide and massive violations of human rights. They could also posit such questions as: How has the basic concept inherent in UN Convention on the Prevention and Punishment of Genocide—that national, ethnic, racial, and religious groups have a right to survive without somebody attempting to violently get rid of them—sought to prevent such human suffering? and Could the formulation and international acceptance of the idea of genocide that took place in 1944-1948 have helped save the Ukrainians?

Recommended Reading on the
Soviet Manmade Famine in Ukraine

Conquest, R. (1986). *The harvest of sorrow: Soviet collectivization and the terror-famine*. New York: Oxford University Press.

This book constitutes the first full history of the Soviet manmade famine in Ukraine. In this book, Conquest, a senior research fellow and scholar-curator of the East European Collection at the Hoover Institution, Stanford University, focuses on dekulakization, collectivization, and the imposition of a "terror-famine" between 1932 and 1933 on the collectivized peasants of the Ukraine and certain other areas.

Conquest, R., Dalrymple, D., Mace, J., & Novak, M. (1984). *The man-made famine in Ukraine*. Washington, DC: American Enterprise Institute.

This booklet contains a discussion between four scholars (two of them, Conquest and Mace, experts on the famine) that was held on the fiftieth anniversary of the Soviet manmade famine in Ukraine.

Mace, J. E. (1988). Genocide in the U.S.S.R. In I. W. Charny (Ed.), *Genocide: A critical bibliographical review* (pp. 116-136). New York: Facts on File.

A succinct but informative overview of the manmade famine in Ukraine. It also includes an annotated list of key works on the famine.

U.S. Commission on the Ukraine Famine. (1988). *Investigation of the Ukrainian Famine, 1932-1933*. Washington, DC: U.S. Government Printing Office.

    Written by James Mace, when he was director of the U.S. Commission on the Ukraine Famine, this is an invaluable text that is comprised of eight highly informative chapters: (1) Non-Soviet Scholarship in the Ukrainian Famine; (2) Post-Stalinist Soviet Historiography on the Ukraine; (3) Soviet Press Sources on the Famine; (4) Soviet Historical Fictions on the Famine; (5) The Famine Outside Ukraine; (6) The American Response to the Famine; (7) Summary of Public Hearings; and (8) Oral History Project. It also includes a "Glossary of Terms," and a lengthy section of translations of oral histories (pp. 235-393).

CASE STUDY 3

THE HOLOCAUST

Michael Berenbaum

THE HOLOCAUST

The Holocaust is the systematic, state-sponsored murder of six million Jews and millions of other people by the Germans and their collaborators during World War II.

The word holocaust is derived from the Greek word *holokauston*, a translation of the Hebrew word *Olah*, meaning a burnt sacrifice offered whole onto the Lord. It was the name given to this genocide because in the ultimate manifestation of the Nazi killing program—the death camps—Jews were murdered in gas chambers and their bodies were consumed whole in crematoria and open fires. The Germans called the murder of the Jews euphemistically but all too accurately "The Final Solution to the Jewish Question." Defining the Jews as a question invited a solution, and the annihilation of men, women and children is all too final. Immediately after the war, the survivors called the murders, the *churban*, a word that evokes the destruction of the First and Second Temples in Jerusalem in the years 586 B.C.E and 70 C.E., respectively. In contemporary Hebrew, the term used is *Shoah*, also meaning destruction but restricted solely to the events of World War II.

Holocaust scholar Raul Hilberg (1985) has suggested that there were six stages to the destruction of the European Jews: (1) definition,

Teaching About Genocide: Issues, Approaches, and Resources, 119–132
Copyright © 2004 by Information Age Publishing

(2) expropriation, (3) concentration, (4) deportation, (5) mobile killing units and (6) death camps.

German law *defined* Jews in 1935 in the so-called Nuremberg laws, biologically based on the religion of their grandparents. Two or more Jewish grandparents put a person at risk. Three or more and they were defined as full Jews, much to their peril. From 1933-1945, the Germans promulgated some 400 pieces of law relating to *expropriation*. They isolated the Jews, confiscated their property and their possessions, and ultimately deported them from their homes. In Jewish history, this could be viewed as a process of disemancipation, depriving Jews of their rights as citizens, the loss of their civil rights and civil liberties. Expropriation was designed to make it ever more difficult for Jews to live in Germany or later in German-occupied territories. The impoverished community deprived of their rights was expected to leave. The early goal of German policy was forced emigration. Yet, there were two flaws with this plan. The Third Reich continued to expand, and as it expanded—to Austria and part of Czechoslovakia in 1938, to Poland in 1939, to Western Europe including France, Belgium and the Netherlands in 1940, to Greece, Yugoslavia and the Soviet Union in 1941—more and more Jews came under its control. Secondly, few countries were willing to accept Jews. The British did not permit massive immigration to Palestine and the United States would not

---

List of Essential Antecedents That Need to Be Addressed When Teaching About This Genocide

- The origin of the Jews with the Biblical Hebrews; that the story of the Exodus and the revelation at Sinai are essential to Jewish consciousness and that Jews were exiled from their land in 586 B.C.E. and 70 C.E.
- The religious roots of antisemitism within Christianity, most especially that Jesus was a Jew, that crucifixion was a Roman punishment, that Jews were held responsible for the crucifixion of Jesus of Nazareth and for denying his messianic mission, and that two traditions in Christianity—supercessionism and the teaching of contempt—were both deeply harmful to Jewish-Christian relations.
- The three forms of antisemitism—religious, political, and racial. Religious antisemitism was rooted in a religious antagonism toward Judaism and could be resolved by conversion; political antisemitism was the use of hatred of the Jews to advance one's political agenda and could be resolved by emigration or expulsion; racial antisemitism is the hatred of Jews for their bloodlines, their very being. The Nazi form of resolution was annihilation.
- The Armenian genocide is an antecedent to the Holocaust. It shared much in common with the Holocaust, both were undertaken by the state against ethnoreligious communities partially integrated into society; and differs in important ways in that Jews were stigmatized as deicides and the attack against them was global in scope.
- Extreme Nationalism—Nazism was a most extreme manifestation of German nationalism. The Nazis believed that Aryans were the master race and

admit Jews except in conformity with preestablished national quotas, and even then, the full number of Jews allowed by law were permitted to enter in United States in only 2 of the 12 years of Hitler's rule.

Unable to leave, Jews were *concentrated* in ghettos pending a decision as to their fate. To the Germans, the ghetto was an interim arrangement. To the Jewish inhabitants, it was the form that life would take until the Germans were defeated or the war came to an end. Jews were *deported* from small towns to larger ghettos, some were shipped from Germany to ghettos in the east and from western European cities to transit camps. As a result, deportation occurred at several times during the destruction process. In 1941, a decision was made at the highest level of the German government to annihilate the Jews. The actual systematic killing began as *mobile killing units*—supported by local gendarmerie, police units, native antisemitic elements of the populations and the German army—murdered Jews in towns, villages, and cities throughout the German-occupied Soviet Union. The shooting and killing shook many of the murderers. Killing fields marred the landscapes of urban areas in German-occupied territories of the Soviet Union. Ponar was the site of destruction for 60,000 Jews of Vilna, Babi Yar for 33,771 Jews of Kiev in September 1941; Rumballa, for 25,000-28,000 Jews from Riga in November 1941. The *Einsatzgruppen*, or mobile killing units, murdered more than 1.3 million Jews

that Germany was destined to rule the world and dominate culture. They not only excluded the outsider—the Jews—but viewed the Jews' annihilation as essential to the national salvation of the German people.

• Industrialization—The destruction of the Jews followed many of the principles of industrialization in the sense that an assembly line was introduced wherein the victim upon arrival faced selection. Those chosen to live were sent to work, and those who were selected to die faced the confiscation of all possessions, then a process of undressing, gassing, recycling—gold teeth were taken, hair was shaved—and cremation. Body fat was even used as fertilizer.

• Social Darwinism—Social Darwinism was based on a reading of Charles Darwin's idea of survival of the fittest. Social competition and struggle were to eliminate those unfit and permit the most fit to dominate. In the case of Nazism, the struggle between the German and Jew for dominance was a fight unto the death for survival.

• Totalitarianism—Hitler and his followers sought total control not only of the apparatus of government but of society as a whole. From March 1933 onward, Hitler's decree had the force of law, and from 1934 onward he was not only the head of government but the head of state.

Note: The last three antecedents to the Holocaust—extreme nationalism, industrialization, social Darwinism and totalitarianism—are often taught in explaining the Holocaust—as they should be—and consequently, they are present but not particularly emphasized in this essay.

between 1941-43. Special units returned in 1943 to dig up the bodies and to burn them and thus to eradicate evidence of their crime. The goal, then, was no longer forced emigration but systematic annihilation, the murder of all Jews, a new configuration for the human species, the Final Solution.

Death camps were the essential instruments of the "Final Solution." Whereas *Einsatzgruppen* killers were sent to their victims, in the death camps the process was reversed; the victims were mobile and the killing centers stationary. The death camps became factories producing dead corpses, effectively and efficiently at minimal physical and psychological costs to German personnel. At Chelmno, the first of the killing centers, mobile gas vans were used. Elsewhere, the gas chambers were stationary and linked to the crematoria where bodies were burned. Sobibor, Treblinka, and Belzec were dedicated to murder alone. Auschwitz was three camps in one, a prison camp, a slave labor camp, and a death camp, called Auschwitz II or Birkenau.

Death camps basically constituted the "industrialization" of killing. The death camp was in essence a killing factory using the elements of assembly-line mass production perfected in the nineteenth and early twentieth century industrialization process. Those parts of the body that were of value were recycled; gold was taken from the victims' teeth and hair was shorn and used to stuff mattresses and to line submarines. Human fat was *not* used for soap, as myth has it, but ashes were used for fertilizer. The three camps of Auschwitz were linked, as slaves were worked until they were no longer useful and those unable to work were sent to Birkenau for gassing. Leading German corporations in Auschwitz III made massive corporate capital investments. They expected the camps to last for years and to recoup their investments.

While Jews were the primary victims of Nazism and were central to Nazi racial ideology, other groups were victimized as well, some for what they did do, some for what they refused to do, and some for what they were.

Political dissidents, trade unionists, and Social Democrats were among the first to be arrested. Additionally, German and Austrian male homosexuals were arrested and, like the others, were later incarcerated in concentration camps, where some 5,000 died. There was no systematic persecution of lesbians. Jehovah's Witnesses were also a problem for the Nazis because they refused to swear allegiance to the state or to register for the draft and would not utter the words, "Heil Hitler." Twenty thousand in number among Germany's more than 65 million people, many were incarcerated. They could be freed from the concentration camps if they renounced their faith. Few did. Their pacifism was so complete that they could serve as barbers and valets for the SS, even in the camps.

Poles, too, were a particular target of Nazi persecution in the aftermath of the German invasion: Polish priests and politicians were murdered, Polish leadership was decimated and the children of the Polish elite were kidnapped and raised as "voluntary Aryans" by their new German "parents." The rest of the Polish population was to be trained to be subservient to the Germans.

Along with the Jews, two other groups were targeted for systematic murder. Beginning in 1939, the Germans initiated a plan of "euthanasia," a euphemistic name for the murder of mentally retarded, physically disabled, and emotionally disturbed Germans who were seen as a contradiction to the ideal of Aryan supremacy held by the Nazis. The gas chambers and crematoria were first developed for their murder; a signed order by Adolf Hitler initiated this process. These "medicalized" killings came to a formal halt in 1941, after the Catholic Church and the parents of the victims protested. In fact, they were driven underground where they continued informally in concentration camps. Systematic murder intensified from "euthanasia" which killed tens of thousands, to genocide where slaughter was in the millions.

The other group singled out by the Nazis were the Roma and the Sinti, known collectively as the Gypsies, who were systematically killed in the gas chambers alongside the Jews in death camps such as Auschwitz.

By war's end six million Jews were dead, two out of three European Jews were dead, one in three of all Jews in the world.

Covering 21 countries and a period of 12 years, the Holocaust varied country by country, region by region, year by year. We know where it ends; with liberation in May 1945, with the rebuilding of the lives of the Holocaust survivors in its aftermath, a process than continues to our own days; and with the rebuilding of Jewish life in Israel, in the United States and in other countries to which survivors emigrated, a process that also continues to our days; and with an understanding of the Holocaust as a central event in twentieth century Western civilization, a process that is also ongoing.

WHERE DOES THE HISTORY OF THE HOLOCAUST BEGIN?

Does the history of the Holocaust begin with the rise of Adolf Hitler and the Nazi party or with the German defeat and the victorious Allies' punitive responses to the German defeat in World War I—the penalties of the Versailles treaty that demanded harsh reparations from the German nation? Does it begin with the emancipation of the Jews and their integration into German society and with the dream that Jews could be part of rather than apart from the German nation? Does it begin with the statement of Konstantin Pobedonostev, the Supreme Prosecutor of the Holy Synod and one of the most influential advisors the Russian czar who in May 1882 promulgated the infamous one third Laws—one third of the Jews will be killed, one third of the Jews will leave the country, and one third of the Jews will be assimilated in the population—which the czar approved? By doing so, he thus formulated—and anticipated—the history of the Jews over the next 70 years. Does it begin with the Armenian genocide, the first major genocide of the twentieth century and with the failure of the Western world to remember the Armenians, the very words that Hitler is reported to have invoked on the eve of the German invasion of Poland? Or does it begin even earlier with the history of antisemitism and its evolution from religious antisemitism to political antisemitism and finally to the singularly distinct Nazi contribution of racial antisemitism that was expressed in what Daniel Jonah Goldhagen (1996) described as "exterminationist" antisemitism or as Raul Hilberg (1985) said so very simply: "The missionaries of Christianity had said in effect: You have no right to live among us as Jews. The secular rulers who followed had proclaimed: You have no right to live among us. The German Nazis at last decreed: You have no right to live" (p. 9).

There are cogent, intellectually compelling historical reasons to include each of these issues as antecedents to the study of the Holocaust and thus to ensure that our students understand that the Nazi policies did not emerge creation *ex nihilo*, but within a specific context that require some background and some understanding

THE JEWS

Teaching the Holocaust is not a course about Jews; it is about their victimization. But there is an essential moral task and intellectual task that can be accomplished within the framework of teaching the Holocaust. The moral task is: to remember, emphasize and teach that before the Jews were victims, they were people. If we know them as people, we will care more about their victimization. And if we know their history, we will also understand facets of their brutalization and annihilation that were directly linked to their historical experience, such as, the burning of synagogues and Torah scrolls, shaving beards of Orthodox Jews, and so forth.

WHAT DO STUDENTS NEED TO KNOW ABOUT THE JEWS?

It is essential to situate the Jews in Europe, defining the origins of Judaism with the Biblical Hebrews and the Patriarchs Abraham, Isaac, and Jacob, establishing the formative experience of slavery and the Exodus from Egypt, defining the monotheism of the Jews and the covenant at Sinai, and providing a sense of the commandments with an emphasis on circumcision and linking the Biblical Hebrews to the land of Israel. The theme of exile and return in Judaism should be emphasized with mention of the first exile and more detailed concentration should be placed on the second exile, which left Jews living in exile, without a country of their own, without sovereignty and without an army.

Because neither the situation of Jews in twentieth-century Europe nor the death camps and the Holocaust itself can be understood apart from the vexed and painful relationships between Jews and Christians since the birth of Christianity—and the hostility of Christianity toward Judaism—particular emphasis should be given to Jesus as a practicing Jew and his rootedness in Judaism. The story of the crucifixion should be presented as understood by contemporary historians and endorsed by Vatican II, which depicts crucifixion and *the* crucifixion as a Roman crime, an act committed by Pontius Pilate and not by the Jewish community of his era and de-emphasizes the later account of the crucifixion that are found in the Gospels. Students should have a passing familiarity with the theological debate between Christianity and Judaism over the Messianic mission of Jesus of Nazareth and with the break between Christianity and Judaism after the death of Jesus. They should be aware of the problem of supercessionism within Christianity, the view that Christianity had come to fulfill Judaism, the New Testament to complete the Old without leaving room to the religiously legitimate ongoing life of the Jewish people and of Judaism as a religion—as well as with the teaching of contempt for Judaism that was prevalent in the thought of some major Christian thinkers.

My sense is that this may be the only time in a student's education in which this information is conveyed and it can be done so in a respectful and dispassionate manner, not to challenge the religious beliefs of either faith but to present their disagreements in context. We should also indicate that there was a social dimension to the religious dissent and move quickly through some mention of the expulsions in the Middle Ages and also to the antisemitism of Martin Luther and finally to the struggles of the Enlightenment and to forms that Enlightenment and post-Enlightenment Judaism took and the forms of post-Enlightenment antisemitism.

It is also important to provide detailed information on the inner forms of Jewish life in the areas occupied by the Germans depicting the diverse Jews who came under German control and who were sent to their death—

from the fiercely pious to the passionately secular, Zionist and Bundist, Socialist and Communist, and all those in-between. This should include specific attention to the inner life of these Jews in all of its complexity and in all of its contradictions and conflicts, including Sephardic Jews from North Africa to Greece and Bulgaria who were also murdered in the Holocaust.

It is also proper to debunk the all too prevalent myth of Jewish powerlessness. In his work *Power and Powerlessness in Jewish History*, David Biale (1987) has summarized the Jewish predicament: "From biblical times to the present day, Jews have wandered the uncertain terrain between power and powerlessness, never quite achieving the power necessary to guarantee long-term security, but equally avoiding, with a number of disastrous exceptions, the abyss of absolute impotence. They developed the consummate skill of living with uncertainty and insecurity" (p. 210).

ANTISEMITISM

A consideration of antisemitism is essential to understanding Nazism, the targeting of the Jews and thus, indispensable in a study of the Holocaust.

One must begin with religious antisemitism. As uncomfortable as this may make certain teachers, especially those teaching in schools with fundamentalist students, the roots of Nazi antisemitism are to be found in Christian antisemitism. Thinkers as diverse as Holocaust scholars Raul Hilberg, Yehuda Bauer, David Bankier, Omar Bar-Tov, and Daniel Goldhagen all emphasize the Christian roots of antisemitism, and major Christian scholars have for three generations stood four square behind a reconsideration of Christian antisemitism. Recent events such as the visit of Pope John Paul II to Jerusalem in 2000 and the renunciation of Martin Luther's teaching on the Jews by the American Lutheran Church in 1994 have made teaching this history of antisemitism far easier because teachers can invoke not only the authority of dissenting and formerly avant garde Christian thinkers but even slow to change Christian Churches which have taken monumental steps to ensure that this form of antisemitism cannot recur and will not receive the veneer of respectable Christianity. Pope John Paul II (2000) put it succinctly: "Antisemitism is anti-Christian."

Again, it is imperative that the religious roots of antisemitism be taught, and in doing so, it must include the role of Jews in Christianity, the tradition of contempt, the problem of supercessionism and the transformation of the portrayal of the crucifixion is the Gospels. One must emphasize that religious teaching had social consequences and remember the admonition of Abraham Joshua Heschel (1965): "Speech has power and few men realize that words do not fade. What starts out as a sound ends in a deed" (p. 81).

Students also need to learn that the Enlightenment did not end the problem of antisemitism, but that Enlightenment thinkers often attacked Judaism and Jews to delegitimate Christianity, masking their critique of the latter to vehemently attack the mother religion. At the least, mention must be made regarding the emerging role of Jews as citizens of a "secular" state and the problem of heterogeneous society and the sense of displacement that some sensed by the greater integration of the Jews.

The political use of antisemitism must also be explored, most especially in the form it took in Vienna, the city so formative to Hitler's own development. Vienna was a cosmopolitan city in which Jews flourished when Hitler lived there. It was the home of Sigmund Freud and Theodore Herzl, the founder of Zionism; and yet, it was also the home of several movements that impacted Hitler, including political antisemitism, which, under the leadership of Karl Leuger, exploited antisemitism for political gain and the concept of Social Darwinism, which viewed society as the battleground in the struggle for survival of the fittest. Tolerant and open, there was also an undercurrent of the hatred and antagonism that were to fuel Hitler's view of the world. In the Vienna of his youth, Hitler may well have learned how to utilize hatred to propel his movement.

Racial antisemitism must be seen in its context as part of the eugenics movement as an outgrowth of some particular readings of Darwin's survival of the fittest, which gave a scientific cover to deep-seated hatreds. The Nazis exploited that science, trained, and educated scientists who endorsed the Nazi killing programs—beginning with the T-4 murders of those deemed to be incurably ill (the mentally and physically handicapped) by Nazi medicine from 1939 onwards. It was in these programs that the role of the gas chambers came to the fore, that key elements of the killing process were refined, essential personnel trained, and the role of the physician firmly established.

Surely, the full range of Nazi racism must be explored, but the particular focus on and the passionate animus toward the Jews must be understood. It was the foundation for much that was soon to happen.

ANTECEDENTS

We must also explore other antecedents. Hitler and the National Socialists fostered an atmosphere of extreme nationalism, *Deutschland uber Alles*, "Germany above all." They redefined the German nation to exclude the Jews, even those who had dwelled in Germany for generations, and to include the *volkdeutsch*, people of German ancestry whose family had lived elsewhere for generations. They introduced the concept of social Darwinism, viewing the world as a social struggle among inferior and superior races, viewing Germany as being engaged in a national struggle for its own survival, which required the destruction of the Jews. They ruled by the Fuhrer's decree, which had the force of law. Germany became a totalitarian society; all power was concentrated in one political party—opposition parties were outlawed in 1933, freedom of the press and freedom of assembly were curtailed, there was no separation of powers between the executive, the legislative and the judiciary. As Nazi rule evolved, all power, indeed all legitimacy, was concentrated in Adolf Hitler. During the years 1933-1939, even before the start of World War II, Hitler and his followers had created a Nazi society and a police state, in which the party and its leader dominated all aspects of society. Even the army swore allegiance directly to the Fuhrer and not to the constitution or the state.

GENOCIDE

It is important to note that the study of other genocides and/or a comparative study of genocide need not diminish the uniqueness of the Holocaust; rather, such efforts can provide greater understanding of its singularity. As to the argument that no comparisons can be made to the Holocaust, medieval theology taught us how to approach God without diminishing the uniqueness of the Divine by *via positiva* and *via negativa*.

In the *Encyclopedia of Genocide*, Robert Melson (1999) offers several parallels between the Holocaust and Armenian Genocide and several important distinctions. Melson (1999) regards the Holocaust and the Armenian genocide as the "quintessential instances of genocide in the modern era" (p. 69). Mass murder were the "products of state initiated policy whose intentions were the elimination of the Armenian community from the Ottoman Empire and the Jews from Germany and Europe and beyond" (Melson, 1999, p. 69). Furthermore, Melson argues, both victimized groups were "ethnoreligious communities that had been partially integrated and assimilated into native societies" (Melson, 1999, p. 69). In that regard, both were "total domestic genocides" (p. 69). Thirdly, Armenians and Jews were unmistakably communal or ethnic groups, not political groups or classes. Finally, both mass murders were the products of modern ideologies in circumstances of revolution and war.

Melson also delineates three significant distinctions. While both groups were regarded as inferior, only the Jews were stigmatized as deicides—murderers of the Christ. He concurs with Holocaust scholars Richard Rubenstein and Yehuda Bauer that an essential component of the Holocaust was the religious dimension of the hatred of the Jews. Secondly,

Armenians were a peasant, agrarian society while living on their own land. Jews were urbanized and, with the exception of the small Zionist groups who sought the Jewish future in the Jewish homeland, the over-whelming majority of the murdered Jews were seeking to continue to dwell and to be included into the countries in which they lived. *Finally—and most importantly—the Nazi racial ideology was global in scope, seeking to murder all Jews everywhere.* There was no plan for the global elimination of all Armenians everywhere. In fact, even during the height of the slaugh-ter, Armenians continued to reside in Istanbul and other Turkish cities (Melson 1999, p. 70).

There are several direct links between the Holocaust and the Armenian genocide: Hitler's statement to the German high command in 1939 (see below); the invention of the term "genocide" by Raphael Lemkin—with the Armenian experience clearly in mind—when in 1933 he submitted a draft proposal to the League of Nations for an international convention on barbaric crimes and vandalism; the use of photographs as a form of protest by Armin Wegner who photographed and distributed for publica-tion pictures of the Armenian genocide and three decades of pictures of the Holocaust. (As my late former colleague and noted historian Sybil Milton has shown, Wegner documented and circulated his pictures of the Armenian and the Jewish experience to call attention to both events, to protest both killings.); the modeling of behavior in such instances as when the resistance fighters in Warsaw and Bialystock saw their deeds as a con-temporary Musa Dagh, the site of Armenian resistance during the Arme-nian genocide, south of the coastal town of Alexandretta and west of ancient Antioch (Franz Werfel's (1934) important novel *Forty Days at Musa Dagh*, which was widely read in Europe, served to transmit the model of Musa Dagh to Jews in Warsaw and Bialystock who invoked that experience as a guide to their behavior under very different circumstances more than a quarter of a century later.); and the actions of Henry Morgenthau Jr., the distinguished son of a distinguished father who saw his father's open opposition to the Armenian genocide when he served as Ambassador to Turkey during World War I as the model of what he should do during the Holocaust. Morgenthau Sr. had risked prestige and position when, as Ambassador to Turkey during World War I, he vehemently protested the fate of the Armenians. At the fateful moment on January 13, 1944, when his staff confronted him with a report entitled "On the Acquiescence of the American Government to the Murder of the Jews," Morgenthau Jr., too risked career and position to confront President Franklin Delano Roosevelt. Their meeting in turn led to the creation of the War Refugee Board, one of the few American efforts on behalf of rescue.

The oft-reported story of Hitler's address to the German High Command in Obersalzberg on the eve of the Polish invasion bears repetition. Hitler told his officers:

> I have issued the command—and I'll have any body who utters but one word of criticism executed by a firing squad—that our war aim does not consist in reaching certain lines, but in the physical destruction of the enemy. Accordingly, I have placed my deathhead formations in readiness—for present only in the East—with orders to send to death mercilessly and without compassion, men, women, and children of Polish derivation and language. Only thus shall we gain the living space (*Lebensraum*), which we need. Who, after all, speaks today of the annihilation of the Armenians? (quoted in Office of the United States, Chief of Counsel for Prosecution of Axis Criminality, Nazi Conspiracy and Aggression, 1946, Vol. 7., p. 753)

Contrary to some claims, I regard the statement in its entirety as reliable. This version of Hitler's speech is traceable to Louis Lochner, the famous American correspondent in Berlin, who received a copy of the notes of the meeting from Hermann Maass, who received it from Hans Oster, a key assistant of Admiral Wilhelm Canaris, director of the Abwehr, the counterintelligence department of the German high command. The information was sent to London on August 29th. (In 1944 Maass was a leader in the conspiracy against Hitler.)

Gerhard Weinberg (personal communication, 1990) has said that this speech was "clearly designed to reassure his [Hitler's] listeners and assuage any doubts they might have" (n.p.). The audience, Weinberg said, "consisted of men who had themselves as adults lived through the events to which Hitler referred" (Weinberg, 1990). This was not the first time that Hitler had invoked the Armenian experience. He had previously used it as an example of how common massive resettlement had been in history.

The implications of Hitler's reference to the Armenians are clear: had the world remembered the Armenian experience, German offices might have been more reticent to carry out such a campaign of terror and destruction. Remembrance might have protected future generations.

A Convention for the Prevention of Crimes of Genocide was adopted by the United Nations on December 9, 1948. Lemkin had a major hand in drafting the Genocide Convention, which was designed to overcome the claims of the Nuremberg defendants that they had violated no law. The Convention specifically defines the various aspects of Nazi genocide as criminal. It prohibits the killing of persons belonging to a group (the final solution); causing grievous bodily or spiritual harm to members of a group; deliberately enforcing upon the group living conditions which could lead to complete or partial extermination (ghettoization and starvation); enforcing measures to prevent births among the group (sterilization); forcibly removing children from the group and transferring them to another group (the "Aryanization" of Polish children).

The adoption of the Convention was followed by the adoption of a Universal Declaration of Human Rights. In 1949, the United Nations adopted the Geneva Convention on the Laws and Customs of War, enunciated the rights of prisoners of war and the conduct of armies toward the populations they control. Yet articulation of the crime is no guarantor of its prevention.

I suspect that many teachers will read this discussion of antecedents, roll their eyes and say "what does he know?" I get three hours to teach the Holocaust while reading *Night* or *The Diary of Anne Frank* in a literature course. Trust me, I do know as I always listen to teachers when we meet. My response: Do not make the perfect the enemy of the good. Do the best that you can with the time allotted you. I was asked in this short essay to be prescriptive of the desirable not descriptive of the possible.

Recommended Reading on the Holocaust

Berenbaum, M. (Ed.). (1997). *Witness to the Holocaust: An illustrated documentary history of the Holocaust in the words of its victims, perpetrators and bystanders*. New York: HarperCollins.

This volume includes documents related to a wide array of issues, including the following: the first regulatory assault against the Jews, early efforts at spiritual resistance, the Nuremberg Laws, *Kristallnacht*, the beginning of ghettoization, the Judenrat, the *Einsatzgruppen*, Hitler's plan to exterminate the Jews, the Warsaw Ghetto Uprising, and the Nuremberg Trials.

Hilberg, R. (2003). *The destruction of the European Jews—college student edition*. New Haven, CT: Yale University Press.

This is the college student edition of Hilberg's highly acclaimed three-volumes set, *The Destruction of the European Jews*. Commenting on the three-volume set, Holocaust scholar David Wyman asserted that it constitutes "the standard text in the field."

Langer, L. L. (1991). *Holocaust testimonies: The ruins of memory*. New Haven, CT: Yale University Press.

A sustained analysis of the unique ways in which oral testimony of survivors contributes to humanity's understanding of the Holocaust.

Rubinstein, R. L., & Roth, J. K. (1987). *Approaches to Auschwitz: The Holocaust and its legacy.* Atlanta, GA: John Knox Press.

Using a multidisciplinary approach in this philosophical history of the Holocaust, the authors analyze the roots of the Holocaust, the power of the Nazis in Germany and abroad, the reactions and responses of the Christian community, the business world, and the literary community; and the "silence of God"; among other issues.

Wiesel, E. (1969). *Night.* New York: Avon.

A searing account of Wiesel's experiences during the Holocaust, including the murder of his mother and youngest sister in Auschwitz, and his father's death at the end of a death march.

REFERENCES

Biale, D. (1986). *Power and powerlessness in Jewish history.* New York: Schocken Books.

Goldhagen, D. (1996). *Hitler's willing executioners.* New York: Knopf.

Heschel, A. J. (1966). *The insecurity of freedom.* New York: Noonday.

Hilberg, R. (1985). *The destruction of the European Jews: Revised and definitive edition.* New York: Holmes and Meier.

Melson, R. (1999). The Armenian genocide and the Holocaust compared. In I. Charny (Ed.), *Encyclopedia of genocide* (Vol. 1, pp. 69-70). Santa Barbara, CA: ABC-Clio.

Milton, S. (1989). Armin T. Wegner: Polemicist for Armenian and Jewish rights. *Armenian Review, 42*(4), 17-40.

Office of the United States, Chief of Counsel for Prosecution of Axis Criminality. (1946). *Nazi conspiracy and aggression* (Vols. 1-9). Washington, DC: U.S. Government Printing Office.

Pope John Paul, II. (2000, March 23). The pope's speech at Yad Vashem, Jerusalum.

Werfel, F. (1934). *The forty days of Musa Dagh.* New York: Viking.

CASE STUDY 4

THE INDONESIAN GENOCIDE OF 1965-1966

Robert Cribb

INTRODUCTION

Between October 1965 and March 1966, perhaps half a million members of the Indonesian Communist Party (PKI) and its affiliated organizations were killed in Indonesia. The killings have been remarkably little researched, but we know that the perpetrators included units of the Indonesian army, religious militias, village mobs, and men of violence. The genocide dramatically transformed the Indonesian political landscape, destroying what had been the world's largest communist party in a noncommunist country and paving the way for General Suharto's seizure of power. Under Suharto's long rule, Indonesia gave up the activist, nonaligned foreign policies and the leftist domestic policies of former President Sukarno and adopted a combination of political repression and economic development which prevailed until the Asian economic crisis finally toppled Suharto in 1998.

Teaching About Genocide: Issues, Approaches, and Resources, 133–141
Copyright © 2004 by Information Age Publishing
All rights of reproduction in any form reserved.

133

OVERVIEW

The killing of communists began in about the third week of October 1965 and continued until approximately March 11, 1966, when the party was formally banned. By that time a high proportion of the party's active leaders and cadres had been killed, a few had been captured for show trial purposes and an unknown number escaped abroad or went underground. A very large number of people with party links—1.8 million according to one official estimate—were detained for longer or shorter periods during and after the killings. After the main killings ceased, moreover, there was a related episode of massacre in West Kalimantan (Borneo) in 1967 and a brief period of communist-led guerrilla warfare in southern East Java.

The immediate trigger for the killings was an ambiguous coup in Jakarta in the early hours of October 1, 1966. Military units led by junior army officers kidnapped and killed six leading generals, including the army commander General Ahmad Yani. The junior officers described their action as a move to forestall a military coup that they believed was planned for October 5th. They themselves, however, proceeded to seize power in Jakarta in the name of an apparently fictitious Revolutionary Council. The actions of the plotters, however, seem to have been very poorly planned, and their movement was defeated within 24 hours by the swift action of the senior surviving general, Suharto (many Indonesians have only one name), who headed the army's crack Strategic Reserve. Suharto then built on this success to remove President Sukarno from power and to establish a long-lasting authoritarian government generally called the New Order.

Most observers are now confident that the junior officers who carried out the coup were working in collusion with other forces, but opinions differ greatly on what these forces were and on what the coup was intended to achieve. The official view of Suharto's regime was that the coup was devised and coordinated by the PKI as a grab for power. Other observers have suggested that President Sukarno might have inspired the action (key military figures in the coup were from his presidential guard). Some time later, speculation began to grow that either Suharto himself or the United States' Central Intelligence Agency (the CIA), or perhaps the two of them in collusion, had engineered the coup, using the junior officers as dupes to implicate the PKI and to provide a pretext for military rule. The evidence, however, is still inconclusive and to some extent contradictory and unreliable; it is possible, even likely, that more than one plot was being carried out on the morning of October 1, 1965.

Whatever the truth of the matter, the coup of October 1st was widely and immediately interpreted as the work of the PKI. Under Sukarno's "Guided Democracy" (1957-1966), the army and the communist party had emerged as the two principal political organizations in the country and the only serious contenders for power following the eventual departure of the ailing Sukarno. The PKI had wide, though unmeasured popular support, especially in densely populated Java. It had abjured the idea of armed struggle in the early 1950s and had opted instead for a march through the institutions of the state. In 1965, therefore, it had no active guerrilla units, but rather a considerable degree of influence in the civilian bureaucracy, a lightly trained youth militia, a powerful ideological voice in national affairs and an uncertain sympathy in parts of the armed forces, especially the airforce and the Central Java division of the army. As Sukarno's health showed signs of failing before the PKI could consolidate these limited advantages, the idea that the PKI might stage a preemptive action seemed plausible.

On the other hand, at the time of the coup the army itself was far from united. No officers or units openly supported the PKI, but some were clearly sympathetic to the Left. Many more believed passionately in Sukarno's spirited radical nationalism and were suspicious of the pro-Western orientation of General Yani. Indonesian military doctrine, hammered out in the 1950s, prescribed a so-called "middle way" in which the army would fully participate in national politics but would not seek to rule in its own right. In these circumstances, it is not impossible that a limited action to remove Yani and his associates from the scene might have achieved guarded but widespread acceptance. Although the military as a whole was much more powerful than the PKI, Suharto's success in seizing power was not at all a foregone conclusion to many observers. Suharto's position remained uncertain until at least mid-1966 and he did not take over as acting president until March 1967.

Destroying the PKI, therefore, was a necessary part of Suharto's drive for power, but it seems improbable that this need extended to massacring half a million people. It is likely that the killing of a few thousand leaders would have been sufficient and, indeed, there are strong indications that United States embassy officials at the time supplied the army with a list of some 5,000 PKI leaders who were to be "removed from political circulation." Explaining the scale of the killings in the context of the Indonesian politics in 1965-1966 and of the broader history of the archipelago is the principle challenge for analysts.

The pattern of the killings was very uneven across the Indonesian archipelago, but it closely followed the distribution of PKI influence. More specifically, the killings were worst in Central and East Java, Bali, North Sumatra, serious in southern Sumatra, parts of West Java, Aceh, North Sulawesi, West Timor and Flores, and apparently little more than incidental in other regions. The massacres took place overwhelmingly outside the cities; those who remained in the cities were generally detained rather than killed. The timing was different too—killing began in Aceh in the third week of October and in East Java soon after, but did not become serious in Central Java until mid-November and did not begin in Bali until early 1966. Massacres in West Kalimantan in 1967 are sometimes considered part of a series with the earlier killings, sometimes as a separate reflection of the military regime's abundantly demonstrated willingness to use force against dissent.

The army played a key role in the killings in most areas but especially so in Central Java. There, the PKI had won an extensive following in rural areas by supporting poor peasants against village and regional elites. The PKI Party had become the only real advocate defending poor peasants against exploitation and abuse of power by those elites and had supported the redistribution of land. As a result, whole villages and even whole districts were predominantly communist, depending on the issues available to the party and the vigor with which local elites resisted it. After the coup, the region seems to have polarized quickly into communist and noncommunist zones, and both sides began sporadic killing of their opponents within their own zones while launching raids into neighboring enemy zones. This stalemate lasted more than a month. Then the army's elite paracommando unit, the RPKAD, entered the province and systematic killings began. The paracommando units moved against the fortified communist villages one by one, usually prevailing easily over the poorly armed PKI supporters. In these clashes normally all the defenders were killed—those not committed to the communist cause had fled or been expelled.

In most parts of Indonesia, the army and police also rounded up party members and leaders of affiliated organizations (the Indonesian Peasants' Front, the Indonesian Women's Movement and so on) by name, using lists compiled by local anticommunist officials. Teachers were often especially targeted because they were the most important conduits for modern, often leftist, ideas in remote villages. Artists and craftsmen were also targeted. Those detained were generally kept in local jails or in makeshift camps before being handed over to civilian killing squads.

These killing squads were a feature of the massacres in many regions. They were often recruited from local anticommunist youth groups and were generally trained, equipped, and controlled by the military. Their gruesome task was to take PKI members to remote places and kill them. Most anecdotes describe the killings as being done with machetes and rope, though firearms were sometimes used. The victims were sometimes made to dig their own graves in forests or rice fields, but sometimes their bodies were simply tipped into rivers. Folklore from Java includes many stories of people who stopped eating fish during these months because they were assumed to have consumed human flesh. The early killings seem to have taken place secretly, but as the killers gained confidence they exulted in brutality, sometimes nailing body parts to the front doors of houses or piling bodies on bamboo rafts which were then floated downstream. The civilian killing squads also took part in raids on communist villages. Engaging the broader public in the killings seems to have been a deliberate military strategy, perhaps partly for logistical reasons, partly to strengthen the impression that the killings were indeed a reflection of popular anger against the PKI, and partly to leave blood on as many hands as possible. There are reports from several regions of family members of communists forced to kill their own relatives to prove their anti-communist credentials.

In East Java and Bali, civilian killing squads were particularly important. In both regions, the party already had powerful local enemies in the form of Muslim traditionalists (in East Java) and Hindu religious leaders (in Bali). East Java was divided culturally between traditionalist Muslims generally called *santri* and followers of a distinctively Javanese form of Islam called *Kejawen*, which drew heavily on Hindu, Buddhist, and animist beliefs and practices from before the coming of Islam. Some communities still regarded themselves as Hindu. *Kejawen* was perhaps even stronger in Central Java, but there it was a belief system encompassing both elites and masses. In East Java, by contrast, the Islam-Kejawen division tended to follow class lines, with village elites generally *santri* in orientation. The *Kejawen* community, therefore, was the PKI's natural class constituency in the province. Even before the emergence of PKI influence, there was a deep ideological dimension to the conflict as well. The followers of *Kejawen* feared the Muslims as fanatics, while the Muslims saw the eclectic magical beliefs of the *Kejawen* followers as sorcery. To their minds, the link between atheistic communism and *Kejawen* was to be expected. The class divisions in rural East Java were not wide—there was no class of large land-owners—but the differences within village communities between those who had enough land to rent out and those who were tenants was often felt acutely and the PKI had vigorously exploited these divisions in a land reform campaign in 1963-1964.

When the failure of the October coup signaled that the PKI might be eliminated from national politics, local Muslim militias in East Java, drawn especially from the youth organization Ansor, began to attack communists and communist-influenced villages. The subsequent arrival of army units in the province led to a further wave of killings, based on lists as in Central Java, but much of the work of destroying the PKI had already been done in many parts of the province. Although the killings in Bali began later, they, too, seem to have been prompted above all by local antagonisms, rather than army prompting.

The pattern of the killings varied in other regions. In West Java and Aceh, where the army had recently suppressed a deeply rooted Islamic rebellion, the military did not dare to arm civilian militias and thus carried out most of the killings itself. In North Sumatra, the main victims were Javanese unionists who had been active on military-controlled plantations and the killings had an ethnic tinge because many of the killers were drawn from the indigenous Malay and Batak communities. In parts of eastern Indonesia, the victims included isolated mountain tribes which had been unwittingly "recruited" by PKI activists and who suddenly found themselves attacked by their traditional enemies from the lowlands.

ANTECEDENSTS/KEY ISSUES

Like all genocides, the mass killing in Indonesia was a complex, multifaceted event, and the causative forces combined in different ways in different parts of Indonesia to produce very different outcomes. In general, however, it is necessary to focus on four major factors:

1. A deep-seated ideological distaste for communism in many sections of Indonesian society, especially within sections of the army and especially amongst Muslims. This distaste arose partly from the PKI's association with atheism, which was anathema to Muslims, partly from a nationalist perception that the PKI served foreign powers (especially China), partly from the real threat a PKI government might have posed to power and privilege, and partly from the party's reputation for ruthlessly pursuing its goals. Under all these circumstances, demonization of the PKI easily developed.
2. The PKI's own involvement in social conflict throughout Indonesia. In order to win popular support and to raise levels of political consciousness, the party diligently took sides on a wide range of social conflicts, including such issues as land, money-lending, culture, and religion. For each issue that it won supporters, it also earned enemies. The party effectively aligned itself with some of the oldest and deepest enmities in the archipelago, and when the protection of law and order was suddenly suspended in 1965, the party was targeted as the modern embodiment of those ancient antagonisms.
3. The peculiarly tense political environment of late Guided Democracy. With President Sukarno's personalist Guided Democracy clearly unsustainable in the long term, Indonesia seemed to be approaching a crucial moment in its history, when its future—communist or noncommunist—would be decided. The tension created

by this sense of destiny was exacerbated both by terrible economic decline and by an opaque political discourse in which no one publicly discussed either the future or the present crisis. This atmosphere fed the already well-developed pattern of rumor-mongering in Indonesia, so that people were willing to believe the worst of their opponents and to take revenge not just for what they were said to have done but also for what they were rumored to be planning.

4. The role of men of violence. For centuries, unofficial men of violence have existed on the margins of politics in the Indonesian archipelago. They have acted both as independent gangsters and as temporary agents of those in power. Their style is above all one of fearsomeness, of creating a paralyzing sense of fear in their victims. They are capable of resorting to extreme violence, but the combined force of state power and public hostility generally prevents them from doing so. News of the October 1965 coup, however, created an atmosphere of impunity, in which the violent capacities of these gangsters was no longer repressed. Recruited into the civilian militias which carried out the bulk of the massacres, they wrought a terrible toll on life.

In analyzing the killings, four further issues need to be addressed. First, although the killings have sometimes been described as a kind of collective "running amok" by observers, amok is now regarded as a clinical state induced by an extreme sense of shame or loss of honor. The perpetrator of amok is generally a lone, male individual who has lost all possibility of an honorable future; running amok is characteristically a terrible form of suicide in which the perpetrator also kills those who are affected by his dishonor (mainly his family) and those who have caused or witnessed it (members of the community) in order to be killed himself. *There was no element of amok in the 1965-1966 killings.*

Second, there is a widespread belief that Chinese Indonesians constituted many, or even most, of the victims of the Indonesian massacres. This belief is also entirely mistaken. Although there have been repeated outbreaks of violence against Chinese Indonesians, both because of their dominance in the economy and because their loyalty to Indonesia is doubted, they were not targeted in 1965-1966 because they were generally not politically engaged. In Java, moreover, Chinese traders had been expelled by law from the countryside in the 1950s, and they were thus concentrated in the towns and cities, where relatively few killings took place. Chinese traders were widely found in rural areas outside Java, and a great many of these were driven from their homes with threats of violence, but the number killed seems to have been only a minute proportion of the overall death toll.

Third, although the number of victims of the killings in Indonesia is conventionally stated as half a million, the evidence for any figure is extremely flimsy. No official records of the killings were kept, there has been no large-scale collection of testimonies from witnesses, survivors, or perpetrators, almost no bodies have been exhumed, and census records in a mobile society are virtually useless for this purpose. All figures rest primarily on the judgment of informed observers and may thus be influenced by political interests.

Finally, to consider the Indonesian killings as genocide requires a definition of genocide which does not limit the victim group to those defined in racial, ethnic, or religious terms. The Indonesian killings were overwhelmingly political and would not normally be considered as genocide under the 1948 United Nations Convention on the Prevention and Punishment of Genocide. Nonetheless, Indonesian national identity has always been defined in terms of beliefs and values—as, indeed, it is in the United States—rather than in terms of ethnicity. The communists and their enemies, in effect, sought to create very different kinds of Indonesian identity for the peoples of the archipelago. In this respect, the destruction of the communist version of the Indonesian identity has much more in common with the destruction of an ethnic group than first appears.

Recommended Reading on the Indonesian Genocide of Communist and Suspected Communists (1965-1966)

Aveling, H. (Ed.). (1975). *Gestapu: Indonesian short stories on the abortive communist coup of 30th September 1965* (Southeast Asian Studies Working Paper No. 6). Honolulu: University of Hawaii .

A gripping collection of fictional short stories written soon after the killings, describing the period from a variety of points of view.

Cribb, R. (Ed.). (1990). *The Indonesian killings of 1965-1966: Studies from Java and Bali* (Papers on Southeast Asia No. 21). Clayton, Victoria, Australia: Monash University Centre of Southeast Asian Studies.

A collection of analytical chapters and translated documents. Sums up the state of knowledge of the killings in the early 1990s.

Cribb, R. (1997). The Indonesian massacres. In S. Totten, W. Parsons, & I. W. Charny (Eds.), *Century of genocide: Eyewitness accounts and critical essays* (pp. 236-263). New York: Garland.

A succinct analysis of the killings, accompanied by extracts from three eyewitness accounts.

Hefner, R. (1990). *The political economy of mountain Java*. Berkeley: University of California Press.

Chapter 7 provides a detailed description and analysis of the violence in one region of East Java.

Hughes, J. (1967). *Indonesian upheaval*. New York: David McKay.

A generally reliable and detailed journalist's account of the killings and the political events which preceded them.

CASE STUDY 5

THE BANGLADESH GENOCIDE

Rounaq Jahan

INTRODUCTION

The birth of Bangladesh in 1971 was a unique phenomenon. It was the first nation state to emerge after waging a successful liberation war against a postcolonial state, Pakistan. The nine month long liberation war in Bangladesh drew worldwide attention because of the genocide committed by Pakistan which resulted in the killings of approximately three million people and raping of nearly a quarter million girls and women. Ten million Bengalis reportedly took refuge in India to avoid the massacre of the Pakistan army and 30 million people were displaced within the country (Mascarenhas, 1971; Payne, 1973).

In addition to providing an overview of the genocide, a list of antecedents and key issues that need to be addressed when teaching about this genocide is also provided. The conclusion attempts to draw some lessons from the 1971 Bangladesh genocide.

OVERVIEW OF THE GENOCIDE

The conflict between the ruling elites in Pakistan and the Bengali nationalists that resulted in the genocide of 1971 started soon after the creation

Teaching About Genocide: Issues, Approaches, and Resources, 143–153
Copyright © 2004 by Information Age Publishing

of Pakistan in 1947. When India was partitioned on the basis of religion, creating two new independent states (a Hindu majority in India and a Muslim majority in Pakistan), Bengali Muslims who lived in the eastern part of India voluntarily joined fellow Muslims who lived 1,000 miles apart in western India and spoke different languages, to form one state. However, very soon it became apparent to the Bengali Muslims that despite their numerical majority (54% of the population) in the new state, they were being economically exploited and excluded from exercising political power (Jahan, 1972). Their language and cultural expressions were threatened as Pakistani ruling elites, who were predominately Urdu speaking, looked on Bengali language and culture as too "Hindu leaning" and made repeated attempts to cleanse it from Hindu influence.

From the beginning, the Bengalis demanded democracy (free and regular elections, a parliamentary form of government, freedom of political parties, free and uncensored use of the media, and so on), but the ruling elites in Pakistan thwarted every attempt at instituting democracy in the country. Elections were repeatedly postponed and the country was ruled by civil bureaucratic or military dictators. In 1969, a popular movement in both East and West Pakistan toppled the military rule of Ayub Khan, which forced the ruling elites to agree to hold a countrywide democratic

List of Essential Antecedents That Need to Be Addressed When Teaching About This Genocide

- In 1947 when India was partitioned on the basis of religion creating two independent states—Hindu majority India and Muslim majority Pakistan—the majority ethnic-linguistic group of Pakistan, the Bengalis, who lived in East Pakistan, were excluded from state power by a small civil military bureaucratic elite who lived in West Pakistan and represented minority ethnic-linguistic groups.
- The non-Bengali ruling elite inherited the British colonial image of the Bengalis as a nonmartial race. They considered the Bengalis as racially inferior and under the influence of Hindu culture.
- The Bengalis resisted the attempts of non-Bengali ruling elites to suppress Bengali language and culture and establish a nondemocratic theocratic state.

- To exclude the majority Bengalis from exercising state power, national elections were repeatedly postponed and the military took control in 1958.
- The Bengalis launched a radical autonomy movement in the 1960s during the rule of the military dictator General Ayub Khan.
- Ayub Khan fell from power in 1969 as a result of popular movements both in East and West Pakistan. His successor General Yahya Khan agreed to hold democratic national elections.
- The first free and fair national election, held in 1970, resulted in an overwhelming electoral victory for the Bengali nationalist party, the Awami League, which was then poised to form a government both at the center and in the province of East Pakistan.
- To prevent the Bengalis from assuming state power, on March 1, 1971, General Yahya Khan canceled the scheduled March 3, 1971 session of parliament. To

election in 1970. The first free national election, held two decades after the birth of Pakistan, resulted in a sweeping victory for the Bengali nationalist party, the Awami League. The election results gave the Awami League not only control over their own province, East Pakistan, but also a majority nationally and a right to form the federal government.

Again, though, the ruling elites in Pakistan took recourse to unconstitutional measures to prevent the Bengalis from assuming state power. On March 1, 1971, Pakistan's president, General Yahya Khan, postponed indefinitely the scheduled March 3rd session of parliament which threw the country into a constitutional crisis. The Awami League responded by launching an unprecedented nonviolent, noncooperation movement, which put the whole administration in East Pakistan into a virtual stand still. Even the government bureaucracy complied with the noncooperation movement.

The Yahya regime initiated political negotiations with the Bengali nationalists but at the same time flew in thousands of armed forces from West to East Pakistan, consolidating preparations for a military action. On March 25, 1971, General Yahya Khan abruptly broke off the negotiations and unleashed a massive armed strike against the population of the capital city, Dhaka. In two days of uninterrupted military operation, hundreds

protest this undemocratic act, the Bengalis started a nonviolent, noncooperation movement under the leadership of the Awami League and its leader Sheikh Mujibur Rahman.

• Yahya Khan began political negotiations with Sheikh Mujibur Rahman but at the same time flew in thousands of armed forces from West to East Pakistan. On March 25, 1971, Yahya suddenly broke off negotiations and unleashed a military operation against unarmed Bengali citizens in the capital Dhaka and in the various garrison cities, killing, within twenty four hours, thousands of ordinary citizens and destroying houses and property. In the ensuring nine months the Pakistani military engaged in a genocide against the Bengalis killing nearly three million people and raping a quarter million girls and women. Ten million people became refugees in neighboring India.

• Following the armed crackdown and acts of genocide, the Bengalis declared independence on March 26,1971 and launched a war of national liberation that lasted nine months. The Pakistan army surrendered on December 16,1971.

• During the nine month long national liberation war, the Pakistani military targeted certain groups as their special enemies—students and intellectuals, Awami Leaguers and their supporters, the Bengali members of the armed forces and the police. Though the Hindus were specially targeted, the majority of the victims were Muslims—ordinary villagers and slum dwellers—who were caught in the military's sweeping sprees of killing and destruction. Women constituted a particular victim group as thousands were raped either in front of their families or in military camps where they were kept after abduction.

of ordinary citizens were killed, houses and property were destroyed, and the leader of the Awami League, Sheikh Mujibur Rahman, was arrested.

The genocide in Bangladesh, which started with the Pakistani military operation against unarmed Bengali citizens on the night of March 25, 1971, continued unabated for nearly nine months until the Bengali nationalists, with the help of Indian army, succeeded in liberating the country from Pakistani occupation forces on December 16, 1971. The atrocities committed by the Pakistan army were widely reported by the international press during 1971 (see, for example, Mascarenhas, 1971; Schanberg, 1971; *Newsweek*, August 2, 1971; and *Time*, August 2, 1971). From the eye witness accounts documented during and immediately after the genocide in 1971 and 1972 and those published in the last 30 years, it is possible to analyze the major features of the Bangladesh genocide (why and how it was committed, who were involved in the crimes, who were the victims, and the world response).

WHY WAS THE GENOCIDE COMMITTED?

The genocide in Bangladesh caught the outside observers as well as the Bengali nationalists by surprise. After all, the Bengali nationalists were essentially waging constitutional peaceful movements for democracy and autonomy. Their only crime, as U.S. Senator Edward Kennedy observed, appeared to have been to win an election (Malik, 1972). Perhaps, the main reason behind the atrocities was to terrorize the population into submission. The military commander in charge of the Dhaka operations reportedly claimed that he would kill four million people in 48 hours and thus have a "final" solution of the Bengal problem (Jahan, 1972).

But the atrocities created a completely opposite effect on the Bengalis who immediately rose in revolt, and chose the path of armed struggle to resist armed aggression. Though their initial armed resistance failed, the Bengali nationalists were not prepared to give up the liberation struggle. Instead of direct confrontation, the liberation fighters chose the course of guerrilla warfare. To retaliate against the guerrillas, the Pakistan army embarked on a strategy of destruction of entire areas and populations where guerrilla actions were reported. Massive killing, looting, burning and raping took place during these "search and destroy" operations.

But the reasons behind the genocide were not simply to terrorize the people and punish them for resistance; there were also elements of racism in this act of genocide. The Pakistan army, consisting of mainly Punjabis and Pathans, had always looked on the Bengalis as racially inferior (a non-martial, physically weak race, not interested or able to serve in the army).

This image of the Bengalis, created by the British colonialists, was accepted as a given by the Pakistani ruling elites. A policy of genocide against fellow Muslims was deliberately undertaken by the Pakistanis on the assumption of racial superiority and cleansing the Bengali Muslims of Hindu cultural-linguistic influence.

HOW THE GENOCIDE WAS COMMITTED

On March 25, 1971, when the Pakistani government initiated army action in Bangladesh, a number of sites and groups of people were selected as targets of attack. In Dhaka, the university campus, the headquarters of police and Bengali paramilitia, slums and squatter settlements, and Hindu majority localities were selected as special targets. The Pakistani ruling elites believed that the leadership of the Bengali nationalist movement came from the intellectuals and students; that the Hindus and the urban lumpen proletariat were the main supporters; and that the Bengali police and army officials could be potential leaders in any armed struggle.

When the news of the Dhaka massacre spread and the independence of Bangladesh was declared on March 26, 1971, in the name of Sheikh Mujibur Rahman, leader of the Awami League, spontaneous resistance was organized in all the cities and towns of the country. This first phase of the liberation war was, however, amateurish and uncoordinated and lasted only approximately six weeks. By the middle of May, the Pakistani army was successful in bringing the cities and towns under their control though the villages remained largely as "liberated" areas. As mentioned earlier, the army's campaign against the cities and towns not only led to massive civilian casualties but it also resulted in large-scale dislocation of people.

During the second phase of the liberation war (from mid May to September) the Pakistan army essentially dug in strong holds in the cities and carried out periodic operations in rural areas in order to punish the villagers for harboring freedom fighters. The army also engaged in large scale looting and the raping of girls and women. In fact, systematic and organized rape was the special weapon of war used by the Pakistan army during the second phase of the liberation struggle. Thus, while young able bodied males were the victims of murder during the first phase, girls and women became the special targets of Pakistani aggression during the second phase. During army operations, girls and women were raped in front of close family members both to "punish" them and to terrorize them. Girls and women were also abducted and repeatedly raped and gang-raped in special camps run by the army near army barracks. Many of the rape victims were either killed or committed suicide. Altogether, it is estimated that approximately a quarter million girls and women were raped during the 1971 genocide.

Throughout the liberation war, thousands of able-bodied young men were arrested, tortured, and killed—either because they were suspected of being freedom fighters, had the potential of being freedom fighters, or for actually being freedom fighters. Ultimately, cities and town were virtually bereft of young males who were either arrested, took refuge in India, or joined the liberation war. In the rural areas, another group of Bengali men (those who were coerced or bribed to collaborate with the Pakistanis) fell victim to the attacks of Bengali freedom fighters during the second phase.

The third phase of the liberation struggle (from October through mid-December) saw intensified guerrilla action and finally a brief conventional war between Pakistan and the combined Indian and Bangladeshi forces, which ended with the surrender of the Pakistan army on December 16, 1971. The Pakistan government engaged in its most brutal and premeditated murder campaign in the last week of the war when their defeat was virtually certain. At that point in time, the Pakistanis decided to kill well-known intellectuals and professionals in each city and town to deprive the new nation of its most talented leadership. Subsequently, between December 12th and December 14th, a selected number of intellectuals and professionals were picked up from their houses and murdered.

The victims of the 1971 genocide were, thus, first and foremost Bengalis. Though Hindus were especially targeted, the majority of the victims were Bengali Muslims (ordinary villagers and slum dwellers) who were caught unprepared during the Pakistan army's sweeping spree of killing and destruction. The Pakistani ruling elites also identified certain groups as their special enemies (students and intellectuals, Awami Leaguers and their supporters, and Bengali members of the armed forces and police). As mentioned previously, women constituted a particular victim group, and as a result thousands of women were suddenly left defenseless and to fend for themselves as widows and rape victims.

WHO COMMITTED THE GENOCIDE?

The Pakistan government was primarily responsible for the genocide. It decided not to let the Awami league and Sheikh Mujibur Rahman form the federal government. It opted for a military solution and made the decision to unleash a brutal military operation to terrorize the Bengalis.

When Bangladesh was liberated and the Pakistan army surrendered, the Bangladesh government declared its intention to hold war crimes trials for members of the Pakistan army. Ultimately, specific charges were brought against 193 officers. Bangladesh, however, later gave up the idea of the war crime trials in exchange for negotiated settlement of outstanding issues with Pakistan, including the return of the Bengalis held hostage in Pakistan, repatriation of the non-Bengali Biharis from Bangladesh to Pakistan, a division of assets and liabilities, and formal recognition of Bangladesh.

The Pakistani military leaders, however, were not the only culprits. The political parties such as the Pakistan People's Party (PPP) also played an important role in instigating the army to take military action in Bangladesh. There were also Bengali collaborators within the Pakistani regime. Many of the Islamist political groups opposed to the Awami League such as the Muslim League and the Jamaat-e-Islami collaborated with the army. Some Bengali intellectuals were also recruited to do propaganda in favor of the Pakistanis. The non-Bengali residents of Bangladesh (the Biharis) were yet another group that collaborated. The latter acted mainly as informants and also participated in riots in Dhaka and Chittagong. It needs to be noted, though, that Biharis were also victims of Bengali mob violence.

THE WORLD RESPONSE TO THE GENOCIDE

World response to the genocide can be classified in two categories—official and nonofficial. At the official level, world response was—understandably—determined by geopolitical interests and major power alignments. Officially, India was sympathetic and supportive of the Bangladesh cause from the beginning. India's major superpower ally, the Soviet Union, supported the Indian backed cause. All the Eastern bloc countries naturally were also supportive of Bangladesh.

Pakistan and her allies were predictably opposed to Bangladesh. It also launched a propaganda campaign to deny the fact of genocide. Islamic countries were generally supportive of Pakistan. So was China. The official policy of the United States was to "tilt in favor of Pakistan" because Pakistan was used as an intermediary to open the door to China.

At the nonofficial level, however, there were great outpouring of sympathy for the Bangladesh cause worldwide because of the genocide. The Western media, particularly in the United States, Great Britain, France, and Australia, kept Bangladesh on the global agenda all through 1971. Well-known Western artists and intellectuals also came out in support of Bangladesh. In the United States, citizen's groups and individuals successfully lobbied the U.S. Congress to halt military aid to Pakistan.

Despite U.S. President Richard Nixon's administration's official support of Pakistan government, influential senators and congressmen spoke out strongly against the genocide. Members of parliament in Great Britain and many Western countries throughout Europe were also highly critical of the Bangladesh genocide.

India, officially and nonofficially, played a critical role in mobilizing support for Bangladesh. The genocide and the resultant influx of 10 million refugees in West Bengal and neighboring states created spontaneous nonofficial sympathy. Officially, India sponsored a systematic international campaign in favor of Bangladesh. And finally in December 1971, when the ground was well prepared, Bangladesh was liberated as a result of direct Indian military intervention.

The world's sympathy for the 1971 genocide was also demonstrated by the tremendous relief and rehabilitation efforts mounted by the United Nations and private voluntary organizations in Bangladesh. Even before the liberation of Bangladesh, large-scale relief efforts were undertaken by the world community to feed the refugees in camps in India. However, despite the international community's generous response in providing humanitarian aid, there was very little support for the war crime trials that Bangladesh proposed to hold. The Indian army very quickly moved Pakistani soldiers from Bangladesh to India in order to prevent reprisals or mob violence against the Pakistan army. India and other friendly countries were also in favor of a negotiated package to settle all outstanding issues between Pakistan and Bangladesh—including the issue of war crimes. Though public opinion favoring war crime trials against the Pakistan army was high in Bangladesh, Sheikh Mujib's regime had to finally give up on the trials. This decision, though, created a sense of betrayal and mistrust and left a deep scar in the national psyche. Plain and simple, the victims of the 1971 genocide wanted the criminals tried and brought to justice.

LONG RANGE IMPACT OF THE GENOCIDE

A major impact of the genocide was the introduction of violence in Bangladeshi society and politics. Prior to 1971, Bengalis were a relatively peaceful and homogeneous community with a low level of violent crime. They were highly faction ridden and politicized but differences and disputes were settled through negotiations, litigations and peaceful mass movements. After the Pakistani armed attack, Bengalis took up arms and for the first time engaged in armed struggle. This brought a qualitative change in people's attitude to conflict resolution. The nonviolent means of protest and conflict resolution were discarded in favor of armed violence.

The genocide, looting, burning, and raping brutalized the Bangladeshi society. After witnessing so much blood shed and violence people developed a higher degree of tolerance toward violence.

Students and youth who became familiar with the use of arms did not give them up after 1971. Continuous armed conflict between rival student groups have made the college and university campuses one of the most dangerous places in the country.

The role of Bengali collaborators who helped to facilitate the genocide created deep division and mistrust in the otherwise homogeneous Bengali social fabric. After the birth of Bangladesh, the whole country appeared to be divided between the freedom fighters and collaborators.

The issue of collaborators also created deep division within the armed forces. Between 1975 and 1981 the various factions of the armed forces staged numerous bloody coups and counter coups, which resulted in the murder of virtually all the military leaders who participated in the liberation war.

It is also important to note that the Hindu community never felt safe again in Bangladesh. In fact, many of them decided not to return to Bangladesh after 1971. And even after Bangladesh was liberated, there was a steady migration of young Hindus to India.

Women's status was also altered as a result of the genocide. The sudden loss of male protection forced thousands of women to seek wage employment outside home. Violence against women became more widespread and common.

CONCLUSION

The genocide and the liberation war has been kept alive primarily through creative arts (theatre, music, literature, and paintings). Many eyewitness accounts and personal diaries of the genocide, which are extremely evocative, have been published in Bangladesh.

While the genocide and the liberation war have not been forgotten by the people of Bangladesh, the collaborators have been gradually rehabilitated through state patronage. Since the 1975 army coup and the overthrow of the Awami League regime, many of the collaborators, who had been opposed to the Awami League, joined the political parties floated by two military leaders, Ziaur Rahman (1975-1981) and Ershad (1982-1990). These two military leaders tilted the country toward Islamic ideology, allowed religion-based parties to function, and appointed a few well-known collaborators to their cabinets. The gradual ascendance of the Islamist forces in the country became even more evident after the return of democratic elections in the 1990s. Following the 1991 election, the

Bangladesh Nationalist Party (BNP) succeeded in forming a new government (1991-1996) with the support of the fundamentalist party, *Jamaat-e-Islami*. In the 1996 election, however, *Jamaat* lost to the Awami League when the latter won and formed a new government (1996-2001). But then, the Islamists became part of the ruling coalition government after the 2001 election when the BNP formed an electoral coalition with the Islamist parties.

The control of state power by the collaborators of the 1971 genocide led the victims of genocide to take direct political action in the 1990s. In 1991, they launched a mass movement and established a citizens' committee for the express purpose of ridding the government of the "Killers and Collaborators of 1971." This civic organization galvanized the support of the intellectuals and youth, and also received the support of the Awami League. The citizen's committee organized a public trial where children, wives, and other relatives of the victims of genocide presented testimony against the *Jamaat-e-Islami Party* and its leader Golam Azam. Ultimately, then, the genocide and the collaborators' issue, which were gradually sidestepped after 1975 when the military took control of state power, was brought back to the center stage of the political arena.

Finally, it must be duly noted that the genocide has created a deep trauma in the national psyche. It creates fear, suspicion, and mistrust. The Bengalis are still suspicious of all foreign powers, including India, which helped to liberate the country. There is not only constant fear of foreign aggression, there is also mistrust concerning the presence of foreign agents and collaborators. The deep animosity between the freedom fighters and collaborators makes national consensus building efforts almost impossible. Indeed, creating a civil society in Bangladesh is terribly difficult when the issue of genocide divides the nation so deeply.

Recommended Reading on the Genocide in Bangladesh

Jahan, R. (1997). Genocide in Bangladesh and eyewitness accounts. In
    S. Totten, W. S. Parsons, & I. Charny (Eds.) *Century of genocide: Eyewitness accounts and critical views* (pp. 291-316). New York: Garland.
    This essay on the Bangladesh genocide is accompanied by six, short
    first-person accounts of the genocide.

Kuper, L. (1981). *Genocide: Its political use in the twentieth century.* New
    Haven, CT: Yale University Press.
    Kuper succinctly discusses various aspects of the Bangladesh genocide.

Kuper, L. (1985). *The prevention of genocide.* New Haven, CT: Yale University Press.

Just as he did in his earlier book, *Genocide: Its Political Use in the Twentieth Century,* Kuper succinctly discusses various aspects of the Bangladesh genocide throughout this book.

MacDermot, N. (1971, April) Crimes against humanity in Bangladesh. *The International Lawyer,* 7(2), 476-484.

The author, then Secretary General of the International Commission of Jurists, discusses the 1971 mass killings in Bangladesh, and in doing so discusses whether what took place could be deemed genocide or not and why.

Payne, R. (1973). *Massacre.* New York: Macmillan.

The author describes the sequence of events of the 1971 Bangladesh genocide, and proposes a schema of such "massacres."

REFERENCES

Coggin, D., Shepard, J., & Greenway, D. (1971, August 2). Pakistan: The ravaging of golden Bengal. *Time,* pp. 24-29.

Jahan, R. (1972). *Pakistan: Failure in national integration.* New York: Columbia University Press.

Jenkins, L., Clifton, T., & Steele, R. (1971, August 2). Bengal: The murder of a people. *Newsweek,* pp. 26-30.

Malik, A. (1972). *The year of the vulture.* New Delhi, India: Orient Longman.

Mascarenhas, A. (1971). *The rape of Bangladesh.* New Delhi, India: Vikas.

Payne, R. (1973). *Massacre.* New York: Macmillan.

Schanberg, S. H. (1971, October). Pakistan divided. *Foreign Affairs, 41,* 125-135.

CASE STUDY 6

THE BURUNDI GENOCIDE

René Lemarchand

INTRODUCTION

Contrary to a widespread opinion, the first recorded case of genocide in the Great Lakes region of Central Africa occurred not in Rwanda, but in neighboring Burundi, Rwanda's "false twin," 22 years prior to the 1994 bloodbath. The scale and targeting of the massacres in Burundi, not to mention their purposefulness, leaves no doubt as to their genocidal character. From May to July 1972, 200,000 to 300,000 Hutu lost their lives in an orgy of killings triggered by an abortive Hutu insurrection. Though largely forgotten in the West, the events of 1972 remain deeply etched in the collective memory of the Hutu people, not only in Burundi but also among the older generations of Hutu in Rwanda.

The Burundi carnage—the worst in a series of massacres of Hutu populations since independence in 1962—is a crucial element in the regional historical context of the 1994 Rwanda genocide, as well as a major watershed in the convoluted sequence of events that led to the collapse of the Burundi state after the assassination its first elected Hutu president in 1993. It has drastically reconfigured the country's ethnic map, driving a deep wedge between Hutu and Tutsi; the immediate consequence was the physical elimination of the entire pool of educated Hutu elites, and of all Hutu officers and troops. It has paved the way for the rise of an all-Tutsi

Teaching About Genocide: Issues, Approaches, and Resources, 155–167
Copyright © 2004 by Information Age Publishing

state protected by an all-Tutsi army. Tens of thousands of Hutu refugees also fled their homeland to seek asylum in neighboring states. Significantly, it was among the refugee community of Tanzania that the most radical, bitterly anti-Tutsi sentiments took hold of the hearts and minds of a growing number of Hutu politicians. Their continuing impact on the destinies of Burundi—and of Rwanda—can hardly be overestimated.

Foreshadowing the hand-washing response of the international community to the 1994 Rwanda bloodbath, nothing was done by the United Nations or anybody else to bring the bloodshed to an end. Despite the alarming cables sent to Washington by the U.S. Embassy, the State Department was emphatic in its "desire to avoid any indication of the U.S. taking sides in this current tragic problem."[1] In 1972, as in 1994, the killings went unabated while the by-standers looked the other way.

List of Essential Antecedents and Events That Need to Be Addressed When Teaching About This Genocide

Key Antecedents

- The Hutu revolution in Rwanda (1959-1962) provided the nascent Hutu elites in Burundi with a republican model to emulate, and the Tutsi with the nightmare scenario of a Hutu-dominated state to be avoided at all cost.
- The influx of tens of thousands of Tutsi refugees fleeing the Rwanda revolution acted as a powerful vector of ethnic polarization. Their mere presence was enough to alert their Burundi kinsmen of the danger involved in the rise of a Hutu political consciousness. During the genocide many Tutsi refugees from Rwanda became actively involved in the killings.
- The void left by the assassination of Prince Rwagasore in 1961 paved the way for a bitter struggle for the leadership of the ruling *Union pour le Progres National* (Uprona) between the Hutu and Tutsi candidates in 1962-1963, a key factor in the rise of Hutu-Tutsi antagonisms.
- The wholesale elimination of Hutu politicians in the wake of the aborted Hutu-led coup in 1965, followed by another

massive purge of remaining Hutu politicians and army officers in 1962, was a critical element behind the deepening Hutu-Tutsi crisis preceding the genocide.
- In late 1971 and early 1972 a bitter intra-Tutsi feud pitted the ruling Tutsi-Hima faction against the Tutsi-Banyaruguru. Sensing the weakening of the Tutsi-dominated state apparatus, and realizing that force was now the only option, it was at this moment that a group of Hutu politicians in exile decided to organize a rural insurgency.
- Perceived as a mortal threat to their hegemony, the April 1972 Hutu insurrection gave the Tutsi-Hima hard-liners in the army and the government the pretext for the indiscriminate mass slaughter of Hutu populations.

Key Events

- July 1, 1962, Burundi becomes independent as a constitutional monarchy.
- On January 18, 1965, Pierre Ngendadumwe, Burundi's first Hutu prime minister is assassinated by a Tutsi refugee.
- Hutu candidates emerge triumphant from the legislative elections of May 1965, with 23 seats out of a total of 33 in the legislative assembly, but they are robbed of their victory when king (*mwami*) Mwambutsa appoints one of his

OVERVIEW

The Burundi genocide did not come about like a bolt out of the blue; nor was the Hutu insurgency that triggered it totally unforeseen. Hutu-Tutsi tensions have been a central feature of Burundi politics ever since the country's independence in 1962, yet they never reached the degree of intensity experienced by Rwanda during and after the 1959-1962 Hutu revolution. Unlike its neighbor to the north, Burundi acceded to independence as a constitutional monarchy, with both Hutu and Tutsi elements holding important positions in the government and the army. The turning point in the escalation of the Hutu-Tutsi conflict came in May 1965, when, shortly after the victory of Hutu candidates in the legislative elections, King Mwambutsa, yielding to the pressures of the Tutsi minor-

courtiers (Leopold Bihumugani) as prime minister.

- At dawn on October 19, 1965, a group of Hutu gendarmerie and army officers attack the royal palace but fail in their attempt to seize power. On October 21, thirty-eight Hutu officers are arrested and shot; on October 28, ten leading Hutu politicians are tried and immediately executed. In subsequent weeks 86 death sentences were handed down by improvised military tribunals, virtually decapitating the emergent Hutu leadership. The *mwami*, panic-stricken, leaves his kingdom to seek refuge in Switzerland.

- Thousands of Hutu, accused of being "enemies of the *mwami*" are rounded up and executed in and around the royal capital Muramvya in October and November 1965.

- On March 24, 1966, King Mwambutsa abdicates in favor of his son Charles Ndizeye, who reigned from July 8 to November 28 under the dynastic name of king Ntare.

- On November 28, 1966 Captain Michel Micombero, a Tutsi-Hima from Bururi, abolishes the monarchy, proclaims the First Republic and appoints a predominately Tutsi government.

- In September 1969 allegations of an impending Hutu-engineered coup lead to the arrest and execution of scores of influential Hutu personalities in the army and the government. It was at this point that some of the Hutu personalities later to be involved in the 1972 insurrection decided to leave the country.

- In July 1971 charges of conspiracy are brought against a group of leading Tutsi politicians from the north, all of Banyaruguru origins, and on January 14, 1972 a military tribunal issues nine death sentences (four officers and nine civilians) and seven life sentences against Banyaruguru elements.

- On April 29, 1972, a Hutu insurrection breaks out in the south, resulting in thousands of deaths among Tutsi civilians. Later the same day, King Ntare was executed in Gitega at the request of President Micombero, under the pretext that he was plotting against the government.

- The Hutu insurrection served as the triggering factor for one of the most appalling bloodbaths recorded in the annals of an independent African state, only exceeded in scale and savagery by the 1994 Rwanda genocide. Anywhere from 150,000 to 300,000 may have died in the course of the repression.

ity, decided to appoint one of his closest courtiers to the post of prime minister, thus effectively robbing the Hutu of their landslide victory at the polls. On October 18, 1965, Hutu anger broke out in an abortive coup directed against the king's palace, followed by sporadic attacks against Tutsi elements in the interior. Panic-stricken, the king fled the country, never to return.[2] In reprisal Tutsi units of the army and gendarmerie arrested and shot 86 leading Hutu politicians and army officers. Furthermore, after the discovery of an alleged Hutu plot in 1969, some 30 Hutu personalities, civilian and military, were arrested and immediately executed. Exclusion of the Hutu from political participation thus made recourse to force the only viable option.

But if the Hutu insurgency of April 1972 must be seen in the context of the growing polarization of ethnic ties, its timing draws attention to the violent intra-Tutsi squabbles and maneuverings that preceded the uprising and may have prompted the few remaining Hutu leaders to exploit the situation to their advantage. During the months preceding the slaughter, the country seemed to be tottering on the brink of anarchy. The long simmering struggle between Tutsi-Hima and Tutsi-Banyaruguru was threatening to get out of hand.[3] The country was awash with rumors of plots and counterplots, leading to the arrest and bogus trials of scores of Banyaruguru politicians, while the ruling clique, headed by President Michel Micombero, consisting principally of Tutsi-Hima from the Bururi province, saw its legitimacy plummet. Nothing could have done more to solidify Tutsi solidarities than the looming threat of a violent Hutu uprising.

The insurrection broke out on April 29, spreading indiscriminate anti-Tutsi violence in several localities in the south, notably Rumonge, Nyanza-Lac, and Bururi, bastion of the ruling Hima faction. Armed with machetes, bands of insurgents proceeded to kill every Tutsi in sight, including women and children. Countless atrocities were reported by eye-witnesses, including the evisceration of pregnant women and the hacking off of limbs. In Bururi, all military and civilian authorities were killed. A short-lived republic, bearing the mysterious name of "République de Martyazo" was proclaimed in Vyanda. Although no one knows for sure how many were involved, the insurgents could not have numbered more than a few thousand (and not 25,000 as the government subsequently claimed). A French pilot who flew helicopter missions on behalf of the Burundi army, put their number at one thousand, "including the majority of committed or conscripted Hutu, Zairian Mulelistes in the middle, and the Hutu organizers at the top" (U.S. Embassy cable, 1972). Despite persistent reports of a Muleliste presence among the insurgents, questions remain as to their numbers and motives.[4] As for the number of Tutsi civilians killed, their number is equally difficult to assess, the figures ranging from 2,000 to 5,000.

The ensuing repression, soon to reach genocidal proportions, got under way almost immediately, and lasted well into the months of August and September. No region was spared. Bujumbura and other cities were thoroughly "cleansed" of every literate Hutu, and not a few illiterate. In a matter of weeks virtually all the Hutu elites and potential elites, including university students and secondary school children, were either dead or in flight. Institutions of higher learning and secondary schools (*athénées*) were particularly hard hit. According to a document of missionary origins,[5] by July, 250 Hutu students out of 350 enrolled at the University of Bujumbura were reported to have "disappeared," while about 60 were said to have been killed by other students; out of 700 Hutu students enrolled at the Bujumbura *athènèe*, 40% were reported missing. At the Kiremba school, "out of 350 Hutu students at least 100 were executed, as well as three or four instructors—among the victims were boys aged 13 or 14." Altogether, the number of Hutu students "executed or in flight" was estimated at 1,450.

The army, meanwhile, was thoroughly purged of Hutu elements. During the night of May 22, 150 Hutu soldiers were ordered to be killed by their commanding officers; another 40 were executed on May 27. A total of about 700 Hutu soldiers were exterminated during the early stages of the repression; to make up for these losses, some 800 Tutsi recruits, most of them from Bururi, were conscripted into the army in early June. By then the Burundi army was—and remains to this day—an all-Tutsi army.

Cables from the U.S. Embassy to the State Department leave no doubt as to the scale of the bloodbath. Here are some examples:

- "No respite, no letup. What apparently is genocide continues. Arrests going on around the clock." (May 26)
- "The liquidation of Hutu goes on apace. Catholic missionaries are increasingly disgusted. Stories which can only be called sickening reach us every day; many Hutu are being buried while still alive. [One informant] calculates that between 1,400 and 1,500 Hutu males were killed in reprisals in Rutovu alone. The most normal means of execution there has been sledge hammers. In the region between Mwaro and Bukirasazi the army was called in and killed all Hutu males it could find; Tutsi civilians killed the women and children. The death toll is in the thousands." (June 20, 1972)
- "In two days following July 14 three new ditches filled with Hutu bodies [were spotted] near Bujumbura airport." (July 21)

- "Repression against Hutu is not simply one of killing. It is also an attempt to remove them from access to employment, property, education, and the general chance to improve themselves." (July 25)
- "We have clear report mass graves near the airport were again utilized at the beginning of the week." (August 11)

The manhunt continued through much of September, adding scores of refugees to the tens of thousands already seeking asylum in neighboring states.

Though much of the "cleansing" was done by Tutsi units of the army and gendarmerie, the youth militias, better known as the *Jeunesses Révolutionnaires Rwagasore* (JRR), served as a major auxiliary force. Sometimes joined by "volunteers" from the Tutsi refugee population from Rwanda, thousands of them roamed the countryside in search of Hutu suspects. In Muramvya, panic struck the local population when a thousand JRR militants suddenly appeared and proceeded to round up and arrest scores of Hutu. Many of the summary executions conducted in the Bujumbura prison were the work of JRR units, and so also the dumping of corpses in mass graves. On May 17, the U.S. Embassy routinely reported, "many arrests are now being carried out by youth militias. Each night trucks carry bodies to a mass grave near the airport, particularly from the prison, where Hutu are believed to be clubbed to death." In early June, the Swiss honorary consul and Dean of Social Sciences at the University of Bujumbura, Francois Bonvin, described the situation as "dramatic": "The events have truly made more than 100,000 victims; the number of refugees is estimated at between 50,000 and 100,000. The University is decimated" (U.S. Embassy cable, June 12). As subsequent events showed, Bonvin's estimates erred on the conservative side.

Difficult though it is to discern a chain of command in the planning and organization of the massacres, no one bears a heavier responsibility than Artémon Simbananiye, who served as minister of foreign affairs in the Micombero government before being appointed minister plenipotentiary and roving ambassador (and who, today, peacefully lives in Bujumbura as a born again Christian). It was Simbananiye who set in motion and supervised the genocidal machine, while at the same time informing foreign governments that the "real" genocide was the one being committed by Hutu against Tutsi, with the active encouragement of "imperialist and reactionary forces."

There can be little question that in the minds of most Tutsi, the insurrection was seen as posing a mortal threat to their survival as a minority. Even moderates, as well as Tutsi-Banyaruguru, closed ranks behind the Micombero government in its ruthless attempt to restore "peace and order." But there was a great deal more at stake in the massive repression unfolding in every part of the country. The underlying objectives of the government can be broadly described as follows: (a) to insure the long-term stability of the state by the wholesale elimination of all Hutu elites and potential elites; (b) to transform the instruments of force (i.e., the army, the police, and the gendarmerie) into a Tutsi monopoly; (c) to rule out the possibility of a restoration of the monarchy (hence the killing of King Ntare, in Gitega, in early May); and (d) to create a new basis of legitimacy for the Hima-dominated state by projecting an image of the state as the benevolent protector of all Barundi against their domestic and external enemies.

Critical to the restoration of state legitimacy was the diffusion of an inversionary discourse aimed at shifting the onus of genocide to the insurgents. Through a variety of official channels, including the White Paper issued by the government, the point that comes across again and again is that the rebels had committed genocide against the people of Burundi; in putting down the rebellion, the state prevented the insurgency from taking an even bigger toll.

The inability or unwillingness of the international community to see through the humbug of official media and take heed of the many "warning signs," all wrenchingly clear, is little short of astonishing. To take the full measure of Western indifference one can do no better than quote from the extraordinarily guarded tone, verging on a tacit approval of the killings, of the letter of the diplomatic corps delivered to President Micombero on May 30, at the initiative of the papal nuncio: "As true friends of Burundi we have followed with anguish and concern the events of the last few weeks. We are thus comforted by your formation of groups of wise men to pacify the country, and by the orders that you have given to repress the arbitrary actions of individuals and groups, of private vengeance and excesses of authority. With all our heart we hope that your laudable initiatives will have the cooperation of all. We assure your Excellency that the governments and organizations that we have the honor to represent to you will do everything to assist those who have suffered and those who suffer still, at the same time support your efforts to promote the peace, unity and progress in Burundi and all its inhabitants" (quoted in U.S. Embassy cable, May 30). By then the "excesses of authority" *of the Micombero government* had sent well over 100,000 Hutu to their graves.

Hardly more edifying was the response of U.N. Secretary General Kurt Waldheim to the carnage. Following the visit of a U.N. Special Mission to Burundi from June 22-28, headed by I. S. Djermakoye, Under Secretary General and Special Advisor on African Affairs, Waldheim expressed his "fervent hope that peace, harmony. and stability can be brought about successfully and speedily, and that Burundi will thereby achieve the goals of social progress, better standards of life and other ideals and principles set forth in the Charter of the United Nations" (Teltsch, 1972, p. 1).

The cynicism behind such pious hopes is a devastating commentary on the role of the United Nations during the genocide. It has an all too familiar ring. In 1972, as in 1994, the United Nations sat on its hands as hundreds of thousands of innocent Africans were being slaughtered.

THE ISSUES/ANTECEDENTS
(THE BASIS FOR A COMPARATIVE STUDY OF THE 1972
BURUNDI GENOCIDE AND THE 1994 RWANDA GENOCIDE)

In what sense does the Burundi bloodbath bear comparison with the Rwanda genocide? In what ways is it different? What are the connecting links between the two? To what extent did it shape the subsequent course of events in Burundi?

That both states shared the characteristics of hierarchically structured plural societies constitutes a major point of convergence between them. It brings to mind Leo Kuper's (1981) argument that such societies are particularly vulnerable to genocidal violence, especially where exclusion creates the seeds of violent retribution (p. 57). The threats posed to their ruling ethnocracies by members of the politically subordinate group meant that entire communities suffered the stigma of guilt by ethnic association and therefore could be targeted for physical elimination. To be sure, whereas in Burundi the victims were Hutu, in Rwanda they were overwhelmingly Tutsi, notwithstanding the tens of thousands of Hutu killed in Rwanda by the genocidaires. Furthermore, even though the number of lives lost may never be known, the scale of the killings in Rwanda—between 600,000 and 800,000—is far above the number of Hutu victims in Burundi, estimates ranging from 100,000 to 300,000, with 200,000 probably closer to the truth. One was a "total" genocide (1994 Rwanda), the other a "selective" genocide (1972 Burundi). The fact remains that in each case the roots of the killings stemmed from much the same social context of plural societies marked by ethnic exclusion.

Both genocides are best described, in Helen Fein's terms (1990) as "retributive genocides" (pp. 28-29), in that they occurred in reaction to perceived threats to the ruling elites. The initial shock and anguish caused by the indiscriminate killing of thousands of Tutsi civilians quickly gave way to the conviction that what was at stake was their survival as an ethnic minority. As one cable from the U.S. Embassy reported, "all moderate Tutsi we have had contact with are saying quite openly—it is us or them" (June 11). There is a striking parallel here with Rwanda's "security dilemma" in the hours following the crash of President Habyalimana's plane in 1994, when the choices facing the Hutu elites were phrased in much the same terms, and given wide diffusion by the Hutu-dominated media.

In both instances, the seriousness of the threat was greatly magnified by the weakness of the state. By the same token, the magnitude of the threat offered a unique opportunity to legitimize mass murder in the name of security. In Burundi, as noted earlier, the conflict between Tutsi-Hima and Tutsi-Banyaruguru had reached unprecedented intensity, undermining both the legitimacy of the Hima-dominated state and its capacity to rule. In Rwanda, intra-Hutu divisions were even more profound, resulting in political assassinations, chronic rural violence and widespread looting of property. In such exceptional circumstances, nothing short of exceptional means were "required" to deal with the enemies of the state. In Burundi as in Rwanda, mass murder was more than a means of coming to terms with a perceived threat; for the ruling elites in each state it was also a strategy for strengthening their vacillating hold on the state.

Where the case of Burundi departs most conspicuously from that of Rwanda is in their radically different outcomes. In Burundi, the Tutsi genocidaires were able literally to get away with murder, and reap the full benefits of their successful trial of strength against the "enemies of the nation." For the next 16 years—until the massacres of Ntega and Marangara in 1988—the state remained a monopoly of the Tutsi minority. South Africa under apartheid offers the closest parallel to the political and economic domination exercised by Tutsi oligarchy during these years. Only if one remembers the extent and duration of Tutsi hegemony can one begin to understand the murderous impulse behind the assassination in 1993 of Melchior Ndadaye, Burundi's first elected Hutu president. Then, as now, for many Tutsi hard-liners, the prospect of sharing power with the Hutu was simply unacceptable. One wonders, however, whether the endless cycles of ethnic violence unleashed by Ndadaye's death is a more acceptable alternative for both Hutu and Tutsi.

If the legacy of Tutsi rule weighs heavily on the future of the country, so does the memory of the 1972 blood bath. Time and again, observers have been struck by the sheer "docility" with which Hutu victims allowed themselves to be taken to their graves. This meekness belongs to the past. In the collective memory of the Hutu, what happened in 1972 is the ever present scenario that keeps reminding them of the cruelties they have suffered at the hands of the Tutsi. It shapes their tendency to react violently in moments of crisis, as happened in 1993 after the assassination of Ndadaye, and provides Hutu extremists with the ideological map they need to recast their precolonial history in the mold of an unending Hutu-Tutsi struggle. While much of the recent history of Burundi is indeed written in blood, what most Hutu remember is that it is Hutu blood.

Compared to the orgy of media attention attracted by Rwanda, the Burundi genocide received minimal coverage in the Western press. Today it has fallen into virtual oblivion. That a tragedy of such magnitude should have gone almost unnoticed is all the more difficult to comprehend in the light of the enormous impact of the Rwanda genocide on public opinion. For this anomaly a number of reasons come to mind. The arcane quality of Burundi politics is hardly sufficient as an explanation. A more important consideration is that many observers, including some well-known journalists and academics, have tended to endorse uncritically the official version of events conveyed by the Burundi media, thereby hoping to obtain the friendship and favor of Burundi officials as well as continued access to the field. The extreme ambivalence of the international community concerning the extent and responsibility of the killings is yet another element to bear in mind. Underlying all of this, however, is the fundamental fact that in the 1970s, human rights issues did not attract as much as a fraction of the frenzied attention they now command from policy makers, academics, and journalists. There was no such thing as a "CNN effect"—only a handful of courageous newspapermen covered the events in Burundi at the time, most of them ignorant of the most elementary facts of Burundi politics, and whose travels and contacts were closely monitored by the Burundi authorities. Nor was there the equivalent of a Romeo Dallaire (a brigadier general in the Canadian army and the U.N. force commander in Rwanda prior to and during the 1994 genocide) on the ground to alert international public opinion about the horror unfolding before his eyes.

If there is such a thing as a conspiracy of silence surrounding the Burundi tragedy, this is nowhere more evident than in Bujumbura: to this day, Burundi officials have consistently denied that anything like a genocide happened in 1972; the only officially recognized genocide is that of the thousands of Tutsi killed by Hutu in 1993, in the wake of Ndadaye's assassination. To suggest that the events of 1972 might conceivably provide part of the explanation for the atrocities committed against Tutsi in 1993 violates all the rules of political correctness.

Many observers have noted the impact of Ndadaye's assassination on the sharp radicalization of anti-Tutsi sentiment in Rwanda. Few have paid attention to the complex causal connections between the Burundi and Rwanda genocides. And yet, for those familiar with the regional roots of the Rwanda bloodbath, there is more than a coincidence between the events of 1972 and the violent anti-Tutsi backlash in Rwanda in 1973, leading to the seizure of power by a new and more radical generation of Hutu elements from the north, under the leadership of Juvenal Habyalimana. Again, it is worth noting that it was among the Hutu refugees of the 1972 bloodbath, in Tanzania, and not in Rwanda, that emerged for the first time a stridently anti-Tutsi ideology, rooted in a blatantly biased "mythico-history" of Hutu-Tutsi relations.[6] Because ethnic memories transcend geographical boundaries, it is impossible to grasp the regional underpinnings of the 1994 Rwanda genocide unless we appreciate the fact that the horrors of 1972 are permanently lodged in the collective consciousness of all Hutu, in Rwanda and in Burundi.

CONCLUSION

Official denials that a genocide of Hutu ever happened in Burundi is not the least of the obstacles that stand in the way of a normalization of Hutu-Tutsi relations. A persuasive case can be made for giving proper recognition to the claims of ethnic memory. Repossessing the past, for many Hutu, is a precondition of peace. Effacing the horrors of 1972 from the historical record, while giving ample space to those committed by Hutu against Tutsi in 1994, can only strengthen the conviction of Hutu extremists that recourse to violence is the price to be paid for recovering their past, and through it their collective identity.

Recommended Reading on the Genocide in Burundi

Lemarchand, R. (1995). *Burundi: Ethnic conflict and genocide*. Washington, DC: Cambridge University Press and Wilson Center Press.
   An in-depth discussion of the historical roots of the Hutu-Tutsi conflict, with a detailed analysis of the 1972 bloodbath.

Lemarchand, R. (1997). The Burundi Genocide. In S. Totten, W. Parsons, & I. W. Charny (Eds.), *Century of genocide: Eyewitness accounts and critical views* (pp. 317-333). New York: Garland.
   Examines the causes and impact of the Burundi genocide. Includes a set of first-person accounts by survivors of the genocide.

Lemarchand, R., & Martin, D. (1974). *Selective genocide in Burundi* (Report No. 20). London: Minority Rights Group.
   The first serious effort to come to grips with the roots and scale of the Burundi genocide.

Ntibantuganya, S. (1999). *Une démocratie pour tous les Burundais (A democracy for all Burundais)*. Paris: LíHarmattan.
   An outstanding semiautobiographical account by Burundi's former president, a Hutu; especially arresting are the author's personal memories of the 1972 killings and his subsequent exile to Rwanda.

Sommers, M. (2001). *Fear in Bongoland: Burundi refugees in urban Tanzania.* New York: Berghahn Books.

A valuable ethnographic inquest into the attitudes, social organization and survival strategies of Hutu refugees in Dar-es-Salaam.

Weissman, S. (1997, July-August). Living with genocide. *Tikkun,* 53-57.

A powerful critique of Western donors' responses to the 1972 genocide and subsequent crises.

NOTES

1. State Department cable to U.S. Embassy in Bujumbura, June 20, 1972. I am grateful to Michael Hoyt, who served as Deputy Chief of Mission in Bujumbura at the time of the genocide, for giving me access to Embassy and State Department cables from April to July 1972. The entire set of communications to and from the U.S. Embassy will be published in a forthcoming book by Michael Hoyt. A complete set is now available from my own collection at the University of Florida.

2. The throne was left vacant until July 1966, when the king's younger son, Charles Ndizeye (later known by his dynastic name of Ntare) was called on to replace his father. His reign, the shortest in the history of Burundi, lasted from July 8 to November 28, 1966, when the monarchy was abolished by President Michel Micombero. Fearing that he might exploit the situation to restore the monarchy, Micombero had him killed in May 1972. Though seen as neither Hutu nor Tutsi but *mwami* (king) of all Barundi, Ntare was among the first victims of the genocide. See Lemarchand (1995, pp. 58-75). See also the excellent background discussion in Kiraranganiya (1985).

3. The distinction between Tutsi-Banyaruguru and Tutsi-Hima refers to a major difference of status among Tutsi: the Banyaruguru, meaning "those from above," were considered closer to the monarchy and hence enjoyed considerably more prestige in the traditional pecking order than the Hima, generally viewed by others with much disdain. Although the Hima are found in large numbers in the south, most notably in the Bururi province, and Banyaruguru in the north, the Hima-Banyaruguru cleavage has nothing to do with geographical distribution.

4. The term "Muleliste" refers to the leader of the 1964 Kwilu insurrection in the Congo, Pierre Mulele, and became widely used to designate the Congolese insurgents operating in eastern Congo during the 1964-1965 rebellion.

5. Quelques données sur les arrestations et dèparts dans les ècoles secondaires et supèrieures du Burundi. Evènements du 29 avril 1972 et les mois de mai et juin. Liste arretèe le 2 juillet 1972. Typescript, available from the author's collection, University of Florida.

6. On the concept of "mythico-history," see Malkki (1990).

REFERENCES

Fein, H. (1990, Spring). Genocide: A sociological perspective. *Current Sociology*, *38*(1, Special Issue).

Kiraranganiya, B. (1985). *La vèritè sur le Burundi* (The truth about Burundi). Serbrooke, Canada: Editions Naaman.

Kuper, L. (1981). *Genocide: Its political use in the twentieth century.* New Haven, CT: Yale University Press.

Lemarchand, R. (1995). *Burundi: Ethnic conflict and genocide.* Washington, DC: Cambridge University Press and Wilson Center Press.

Malkki, L. (1990). *Purity and exile: Transformations in historical-national consciousness among the Hutu refugees in Tanzania.* Chicago: University of Chicago Press.

Teltsch, K. (1972, July 29). Killings go on in Burundi, UN statement suggests. *New York Times*, p. 1.

CASE STUDY 7

THE CAMBODIAN GENOCIDE

Craig Etcheson

OVERVIEW OF THE CAMBODIAN GENOCIDE

Between 1975 and 1978, the Cambodian communist movement known as the "Khmer Rouge," or Red Khmer, carried out what was arguably the most brutal revolution of the twentieth century. Scholars now believe that between 2.2 and 2.5 million people lost their lives in less than four years, up to half of them by execution. That amounted to somewhere between one quarter and one third of the entire population of the country. The fact that the vast majority of the victims were from the same ethnic and linguistic group as the perpetrators makes this terrible crime all the more unprecedented. The country was virtually destroyed by its own people, and for a time, there was some question as to whether Cambodia would continue to exist at all.

The Cambodia we see on contemporary maps is all that remains of a once-great empire called Angkor. Eight hundred years ago, the Angkor Empire encompassed what today is southern Vietnam, southern Laos, much of Thailand and parts of Myanmar (formerly known as Burma). The god-kings who ruled these lands built religious monuments which still remain some of the most impressive architectural wonders of the world. After the peak of the Angkor Empire in the thirteenth century, Cambodia's faster-growing neighbors to the east and the west—Vietnam

Teaching About Genocide: Issues, Approaches, and Resources, 169–179
Copyright © 2004 by Information Age Publishing
All rights of reproduction in any form reserved.

and Thailand—began to nibble away at the empire, continually shrinking it. But even as Cambodia's empire shrunk, imperial attitudes remained deeply ingrained in Cambodian culture, and the country continued to be ruled by monarchs who wielded life-and-death powers over their subjects, right up into the twentieth century. As a result, Cambodia's kings and their royal courts lived in luxury and opulence, while the Cambodian people toiled away in lives of desperate poverty.

During the 1940s and 1950s, a group of Cambodian leftists began to revolt against the social and economic inequality in Cambodia. Many of them were from elite backgrounds, educated abroad in Thailand, Vietnam or France. They believed that communism offered a way to change Cambodia, to end the privileges of the elite and improve life for ordinary

List of Essential Antecedents That Need to Be Addressed When Teaching About This Genocide

- Economic dislocations exacerbate tensions in society. Economic problems, including low productivity, declining average land holdings, ill-advised central policies, and war-induced strains in commodity markets combine to create a falling standard of living and rising popular discontent.
- Triggering effect of war in Vietnam. The civil war in neighboring Vietnam, greatly intensified by U.S. intervention, begins to take a toll on Cambodia as the Vietnamese parties to the war intrude onto Cambodian soil, and the United States begins to pursue its enemies into Cambodia, destabilizing Cambodia's society and politics.
- Ethnic cleavages in society. Long-standing grievances over territory lost to Cambodia's expanding neighbors Vietnam and Thailand fuel an undercurrent of racism that is particularly virulent against Cambodian citizens of Vietnamese ethnic origin. An ethnic chauvinism emerges in the communist party, according to which everyone must become like the ethnic Khmer majority.
- Inflammatory rhetoric used by political leaders. Cambodians on both sides of the civil war denigrate the "Other" as

"unbelievers" and "parasites," thus dehumanizing individuals and contributing to the perception that the "enemy" is not worthy of existing.

- Religious differences. Viewing religion as an archaic impediment to their modernist project, the Khmer Rouge move to eradicate all vestiges of the religions practiced by Cambodian people, especially Buddhism and Islam.
- Utopian ideology. Devising a twisted version of Marxist-Leninist-Maoist theory, the Khmer Rouge dream they can instantly create a classless society by eliminating every social class except poor peasants. All who resist are executed.
- Urban-rural conflicts in society. Social and economic divisions grow between Cambodia's urban population and the vast majority of Cambodians who are rural dwellers, and become critical when the Khmer Rouge blame the city people for all of Cambodia's problems.
- Division of Society into "New" and "Old" People. The "Old People," or those who lived in communist-based areas prior to the end of the civil war, are distinguished from the "New People," who lived in the last enclaves of the previous regime. The New People are singled out for especially harsh treatment, and are readily suspected of treasonous thoughts or activities.

people. But through the 1960s, Cambodia's communists had little success in persuading Cambodians that they should rise up against the country's ruler, Prince Norodom Sihanouk. Nine out of ten Cambodians were farmers, and many were conservative, simply wanting to be left alone to grow their crops, raise their families, and get on with life. But events in the outside world were conspiring to make that impossible.

In neighboring Vietnam, the United States was engaged in a brutal war, attempting to prevent communist forces from dominating all of Vietnam, north and south. Communist North Vietnam aimed to drive the United States out of the south and to reunify north and south Vietnam under a single leadership. The Vietnamese communists were using Cambodia's border areas to supply their forces and take sanctuary from combat in South Vietnam. Unable to tolerate this, the United States began attacking communist forces across the border in Cambodia, which until then had been neutral in the Vietnam War. At about this same time, key members of Cambodia's ruling class began to have serious doubts about the leadership of Prince Sihanouk. In 1970, he was ousted in a coup d'état. These events combined to plunge the previously peaceful Cambodia into all-out war.

Cambodia's new leaders quickly allied themselves with the United States. The ousted Sihanouk joined forces with his previous enemies, the Khmer Rouge, who until then had been weak and ineffective. And now that Cambodia had abandoned its official policy of neutrality in the Vietnam War, North Vietnam and its communist allies in China and the Soviet Union began to assist Sihanouk and the Khmer Rouge. In a matter of weeks, the entire country was engulfed in a war which was to last five years, and cost hundreds of thousands of lives.

Sihanouk wanted revenge against those who had overthrown him, and he was willing to work with the Khmer Rouge to get that revenge. Since he was popular with the Cambodian people, the Khmer Rouge did all they could to convince the latter that they were engaged in battle for the purpose of restoring Sihanouk to power. In reality, the Khmer Rouge were out to destroy everything that Prince Sihanouk represented and to establish a communist Cambodia. They kept tight control over the revolution's military forces, and by the time they won the war in April 1975, Prince Sihanouk was reduced to a powerless figurehead. Henceforth, the Khmer Rouge said, Cambodia would be ruled by and for the poorest people.

The revolution would change almost everything in Cambodia, everything in its 2,000 year-old way of life. Immediately on victory, all the cities were emptied of people, and everyone was put to work in the fields as agricultural slaves. Money and markets were abolished. All aspects of life were communalized, right down to clothes and choice of marriage partners. No one could leave their assigned work site without permission.

Children were taken from their parents and taught that their only real family now was the communist party. Religion was forbidden. Anyone who resisted in any way was subject to immediate execution.

Key Events: Cambodian Genocide

- 1945—Restoration of French Colonial Power in Indochina. France attempted to reassert its colonial control over Vietnam, Laos, and Cambodia in the wake of World War II. This act gave rise to a variety of indigenous independence movements, one of which would eventually evolve into the Communist Party of Cambodia.
- 1955—Assumption of Political Power by Prince Norodom Sihanouk. King Norodom Sihanouk abdicates the throne, and, as a "Prince," forms a political party with the aim of totally dominating political power. He represses all competing political parties, frustrating the democratic aspirations of many Cambodians and driving political resistance underground.
- 1963—Rise of Saloth Sar as the General Secretary of the Communist Party of Cambodia. Saloth Sar, later known to the world under the name Pol Pot, becomes the top leader of Cambodia's communist revolutionaries, and begins to place cadres loyal to him in key positions.
- 1968—Samlot Tax Rebellion Spurs Communists to Adopt Armed Struggle. A local tax rebellion in western Cambodia brings brutal repression by the Cambodian government. The violence inspires the communists to declare armed rebellion and begin preparing for war.
- 1968—Expansion of Vietnam War into Cambodia. The United States begins sporadic bombing of Cambodian territory, targeting Vietnamese revolutionaries hiding there. This policy exacerbates strains on Cambodia's economy, society, and polity that result in a coup d'état. As U.S. bombing intensifies after 1970, it helps to radicalize Cambodia's conservative peasantry.
- 1970—Coup d'état Against Prince Norodom Sihanouk. Royal rivals and disgruntled military officers oust Sihanouk and shatter the delicate neutrality he had been pursuing. Russia, China, and Vietnam throw their support to Cambodia's revolutionaries, while the United States adopts the new Cambodian regime as a client and expands its involvement in the war.
- 1975—Cambodia's Communists Win Civil War. The Khmer Rouge seize state power and implement extremely radical policies designed to instantly create a one-class society. Authority to execute actual and suspected opponents of these policies is delegated to local cadres, and the killing spreads rapidly.
- 1979—Vietnamese Invasion Brings Down the Khmer Rouge Regime. Invading on Christmas Day 1978, Vietnam ousts the Khmer Rouge government on January 7, 1979, replacing it with a group of disaffected former Khmer Rouge leaders. The Khmer Rouge find support from China and Western countries, and civil war continues until the final collapse of the Khmer Rouge in 1998.
- 2001—Cambodia Adopts a Khmer Rouge Tribunal Law. Some 25 years after the Khmer Rouge were driven from power, Cambodia promulgated a law to establish a mixed national-international tribunal to judge the authors of the Cambodian genocide. More than a year after the law was passed, progress toward actually convening the tribunal remains stalled due to disputes with the United Nations over the content of the law and

As the Khmer Rouge consolidated power, the search for "enemies" intensified. Everyone was suspect. Everyone was required to produce elaborate biographies, which were then studied by the Khmer Rouge to identify enemies. Were you related to a policeman or a teacher or a government soldier? If yes, the person was deemed an Enemy. Had your mind been "poisoned" by too much education? If a person was educated, then he/she was deemed an Enemy. Did you have unacceptable attitudes, such as love of your parents, or love of god? If so, that person was deemed an Enemy. Were you born in a city? Enemy. Did you live too close to the hated Vietnam, or Thailand? Enemy. Did you object to working 12 hours a day, 7 days a week, without enough to food to maintain your strength? Enemy. Did you fail to cheer loudly enough last night at the political meeting, when two teenagers were publicly executed for the crime of holding hands? Enemy. Everyone was a potential enemy; and according to the Khmer Rouge, there was only one thing to do with enemies: kill them. Millions of people ended up being classified as enemies of the Khmer Rouge.

One unique aspect of the Cambodian genocide is that the victims and perpetrators were mostly the same people—ethnic Khmer peasants raised as Buddhists. But ethnic and religious minority groups in Cambodia suffered even higher levels of repression under the Khmer Rouge. For example, prior to 1970, 5% of the population was ethnic Vietnamese, many of whom followed the Catholic religion. By the end of 1978, there were no Vietnamese remaining in Cambodia; many fled abroad, and any who did not were killed. Cambodia also had a substantial Muslim minority, an ethnic group called the Cham. Under the Khmer Rouge, Islam was prohibited. Because some Cham resisted this policy, many—as many as half of them—were killed. Similarly, the ethnic Chinese minority also endured a higher than average death toll under the Khmer Rouge, though for different reasons. Chinese Cambodians had always been relatively well assimilated into Khmer culture, but they also tended to be primarily urban dwellers, often working as merchants. But the Khmer Rouge reserved a special hatred for people who lived in cities, as well as anyone involved in business or commerce. Deported to rural areas as forced laborers, many Chinese Cambodians found the sudden change deadly. Up to half of them perished. Many of those were killed outright because the Khmer Rouge believed that "capitalists" had no place in their new society. Though ethnic minorities in Cambodia suffered more killing than did the Khmer majority group, it is still a fact that most of the people who died were poor Khmer farmers—the very people in whose name the Khmer Rouge waged their revolution.

The Khmer Rouge were extremely suspicious of neighboring Vietnam, fearing it might annex Cambodian lands as it had done in the past. Vietnam was much larger than Cambodia, with 10 times the population, and 10 times the military forces. Using what might be called a "porcupine strategy," the Khmer Rouge launched attacks on Vietnam, as if to say, "Don't tread on our sovereignty! If you do, you will suffer the consequences." To inspire their troops, the Khmer Rouge told their soldiers that their mission was to reconquer the ancient lands of the Angkor Empire, lost to Vietnam over the centuries. Thus filled with ideological hatred and nationalistic fervor, the Khmer Rouge army attacked unguarded communities in Vietnam, killing many innocent civilians. Vietnam attempted to negotiate a solution, but when that failed, they carried out a reprisal attack in 1977. This convinced the Khmer Rouge that war was inevitable, and thus they escalated their attacks on Vietnam. In 1978, Vietnam concluded that the only way out of the situation was to oust the Khmer Rouge. Vietnam invaded on Christmas Day, and two weeks later, they captured the capital of Cambodia.

The Khmer Rouge retreated to the forest with the rump of their army and resumed guerrilla warfare. Alarmed at seeing their traditional Vietnamese rivals approaching their border, Thailand assisted the Khmer Rouge. China also supported the Khmer Rouge, and before long, the United States as well as many European and Asian countries were also providing diplomatic, economic, or military assistance to the Khmer Rouge. This widespread support for the Khmer Rouge was mainly a function of the Cold War struggle between the United States and the Soviet Union, but the result was a civil war that lasted for another 20 years.

In 1991, the Khmer Rouge, along with two other guerrilla armies, signed a peace treaty with the Cambodian government that had been installed by Vietnam, leading to a massive United Nations peacekeeping operation in Cambodia during 1992 and 1993. But the peace treaty and the peacekeeping mission did not immediately bring peace. In fact, the war did not end until Khmer Rouge military forces finally collapsed in 1998.

The reaction of the world to the Cambodian genocide illustrates a sad reality about international attitudes toward human rights: during the 1970s and 1980s—and to a significant extent it is still true today—geopolitical concerns had a greater impact in shaping responses to serious violations of human rights than did humanitarian concerns about the welfare of human beings. Between 1975 and 1979, Western governments condemned Khmer Rouge abuses, while communist nations like China, Russia, and Vietnam denied that any such abuses were occurring. After Vietnam invaded Cambodia and overthrew the Khmer Rouge, there was a near-complete role reversal in these policies. Western nations then downplayed reports of human rights abuses by the Khmer Rouge and instead condemned the Vietnamese occupation of Cambodia, while communist nations such as Vietnam and Russia now declared that the Khmer Rouge had committed genocide. The only thing that had changed was the geopolitical situation. The entire time, the international community did nothing to punish the authors of the Cambodian genocide.

Even after the end of the Cold War and the creation of mechanisms for prosecuting genocide such as the International Criminal Tribunal for the former Yugoslavia, political considerations continued to hamper efforts to achieve accountability for the Cambodian genocide. The Chinese government insisted that the Cambodian genocide was an internal affair of Cambodia and should not be subject to "interference" by the United Nations, while the policies of the United States toward the establishment of a genocide tribunal for the Khmer Rouge varied depending on which political party controlled the White House. Once again, political interests trumped human interests.

ESSENTIAL ANTECEDENTS AND KEY ISSUES

So, why did the Cambodian genocide happen? We can consider a variety of factors which played a greater or lesser role in bringing about these tragic events. One factor was the economic crisis that afflicted Cambodia during the 1960s. As a result of a growing population, a declining area of land under agricultural cultivation, and an increasing concentration in landholdings, Cambodians began to experience food shortages, rising food prices, and rising interest rates and rents in the 1960s. In the weakened economy, more and more people began to experience acute hunger. A growing number of farmers found themselves with no land to farm. This group of landless, destitute peasants—a phenomenon previously unknown in Cambodia—provided a ready pool of potential recruits for the revolution. Hunger has the power to radicalize people.

In the mid-1960s, the economic pressures also began to have a serious impact on cities. In response to the economic crisis, the government printed more currency and imposed new taxes. This contributed to rising inflation and declining living standards. As economic tensions increased, so did the tendency to look for someone to blame. That blame was often directed at Cambodia's minority groups, particularly ethnic Chinese and Vietnamese Cambodians.

Cambodia's economic crisis was also exacerbated by the war in neighboring Vietnam. Large purchases of rice and other commodities by the North Vietnamese military to support their troops' fighting in southern Vietnam contributed to rising prices in Cambodia. As the North Vietnamese army entrenched itself in Cambodia all along the Vietnam border, and the South Vietnamese and U.S. military increasingly began to attack those forces, more and more Cambodian land fell out of cultivation, and increasing numbers of farmers were driven from their homes and fields. In turn, this fueled increasing ethnic tensions in Cambodian society, igniting resentment against Vietnamese Cambodians, a resentment which always smoldered just beneath the surface.

The economic troubles and the resentment of Vietnamese occupation of Cambodian land finally led to a coup d'état against Cambodia's long-time leader, Prince Norodom Sihanouk, in 1970. This fateful decision plunged the country headlong into war. When Cambodia's new leaders abruptly demanded that North Vietnam remove its troops from Cambodia, the Vietnamese communists temporarily turned from their war in South Vietnam and pummeled the unprepared Cambodian military, driving the government out of many areas. To make matters worse, weeks later, U.S. and South Vietnamese forces launched a major attack across the border into Cambodia, striking against the Vietnamese communists. North Vietnamese forces retreated from their bases along the border deep into Cambodia, seizing territory as they went, and then turning it over to the previously weak Khmer Rouge. Thus, in the first few months of the Cambodian civil war, the government lost most of the countryside to the Khmer Rouge.

The Vietnam War then expanded dramatically, with the U.S. Air Force attacking all across Cambodia in strikes against communist forces—dropping more than a billion pounds of bombs on Cambodia. Because the communist troops often operated near Cambodian villages, there were many civilian casualties. Many farmers fled to Cambodia's cities to avoid the danger, quickly overwhelming social services and food supplies. Others responded to Khmer Rouge appeals, blaming their government for the U.S. bombing. As a result, the Khmer Rouge army grew rapidly.

Cambodia's leaders called the communist Khmer Rouge "unbelievers," encouraging the army to attack their followers without mercy. In turn, the Khmer Rogue called their enemies "parasites" and "microbes," thus dehumanizing them, and making it easier for their followers to kill without feeling any guilt. After they came to power, the Khmer Rouge continued their habit of describing their perceived "enemies" as less than human, and of being not worthy of existing.

The Khmer Rouge returned Cambodia to what it deemed "Year Zero," a preindustrial agrarian society, without modern conveniences like money and machines. But in another sense, the Khmer Rouge were also inspired by modern ideological ideas, such as the impulse to eradicate superstition. For the Khmer Rouge, this included religion. Because 90% of Cambodians were Buddhists, the elimination of Buddhism was a central part of the Khmer Rouge plan. Buddhist monks were ordered to leave their temples, remove their robes, and work in the fields with everyone else.

Any who refused were killed. Over time, more than 90% of Cambodia's monks died of disease or overwork, or were killed outright. Those who objected to the treatment of their religious leaders were also killed. Buddhist temples were turned into pig styes, and even torture chambers.

Other religions received the same treatment. The Muslim minority was severely repressed, and nearly all of the Muslim religious leaders were murdered. The Christian minority fared even worse; largely Vietnamese, if they were not able to escape the country, they were killed.

In Cambodia, villagers in remote areas have often been suspicious of outsiders, concerned that strangers would only bring bad things to their homes. Farmers were perhaps especially suspicious of city people, remembering bad experiences with the king's tax collectors, his governors, or his security forces. Thus, when the entire population of Cambodia's cities was deported to rural areas by the Khmer Rouge, well-entrenched prejudices were reinforced as the Khmer Rouge told the villagers to carefully watch the newcomers. Treated as a category of persons below the level of citizens, the city people succumbed in huge numbers to starvation, overwork, disease, and execution. In some cases, no doubt, villagers agreed with the Khmer Rouge saying, "To keep you is no gain, to lose you is no loss."

Adapting a twisted version of Marxism, the Khmer Rouge attempted to make a "super-great leap forward" from a peasant society into a utopian, futuristic classless society. They were willing to sacrifice any number of human beings in order to accomplish this goal. The old saying that "the road to hell is paved with good intentions" is amply illustrated by what the Khmer Rouge did in Cambodia. The country was destroyed, along with a third or more of its people.

Even so, everything we have learned about the Cambodian genocide suggests that people who carried out the actual killing were not monsters, or even necessarily particularly evil, as such. They were ordinary people. This is one of the most horrific realities about genocide. It is usually committed by ordinary people who find themselves in extraordinary circumstances.

A SPECIAL CONSIDERATION FOR U.S. EDUCATORS

Particularly for U.S. educators, the U.S. role in events leading up to and following the Cambodian genocide needs to be treated with care, because it raises a number of difficult issues. What part did U.S. military intervention in Southeast Asia play in the Khmer Rouge rise to power? Is it perverse, as former U.S. Secretary of State Henry Kissinger has argued, to allocate any responsibility to the United States, given that the United

States was fighting to prevent a Khmer Rouge victory? Or, as others have argued, does the United States bear some responsibility for the tactics employed by U.S. military forces, tactics which arguably proved helpful to the Khmer Rouge? What about U.S. support for the exiled Khmer Rouge coalition government in the 1980s? Can it be justified as a necessary element of the struggle against Soviet tyranny? What about the leading U.S. role in the peace process, ultimately resulting in a political settlement, which in turn created the conditions for the final destruction of the Khmer Rouge? All of these questions should be addressed during a course of study.

Recommended Readings on the Cambodian Genocide

Becker, E. (1986). *When the war was over: The voices of Cambodia's revolution and its people*. New York: Simon & Schuster.

This volume stands out as the single most well-written account of the catastrophe which swept over Cambodia. With haunting, lyrical prose, the author brings the instincts of a reporter and an eye for the telling anecdote to the task of trying to make sense of an apparently senseless terror.

Chandler, D. P. (1999). *Voices from S-21: Terror and history in Pol Pot's secret prison*. Berkeley: University of California Press.

This volume is a specialized study of Khmer Rouge secret police headquarters, analyzing the methods used to discover and purge alleged "enemies" of the revolution.

Etcheson, C. (in press). *Crimes of the Khmer Rouge: The search for peace and justice in Cambodia*. London: Mellen, forthcoming.

This volume presents the most recent scholarly research on the Khmer Rouge. It also covers the history of the Khmer Rouge after they were driven from power in 1979. Of particular note is the treatment of mass grave studies, and the failed negotiations to establish a Khmer Rouge tribunal.

Jackson, K. J. (Ed.). (1989). *Cambodia 1975-1978: Rendezvous with death*. Princeton, NJ: Princeton University Press.

This volume brings together contributions by a number of leading experts on Cambodia and the Khmer Rouge, covering economic, political, demographic, ideological, and organizational and social aspects of the revolution. The analytical tone of the book well serves the editor's goal.

Pran, D. (Ed.). (1997). *Children of Cambodia's killing fields: Memoirs by survivors*. New Haven, CT: Yale University Press.

This volume is a collection of first-person accounts by Cambodians who experienced the Khmer Rouge revolution as children. Appropriate for all ages, the book describes in excruciating detail what it was like for urban, middle-class youngsters to suffer and survive the Khmer Rouge.

CASE STUDY 8

THE 1988 ANFAL OPERATIONS
IN IRAQI KURDISTAN

Michiel Leezenberg

INTRODUCTION

The 1988 Anfal operations conducted by the Iraqi regime against part of
its Kurdish population are among the best-documented cases of genocide.
Ostensibly a counterinsurgency measure against Kurdish rebels, they in
fact involved the deliberate killing of large numbers of noncombatants.
They were characterized by an unusual degree of bureaucratic organization, centralized implementation, and secrecy. But because they are documented not only by eyewitness and survivor testimonies, but also by a vast
number of Iraqi government documents captured in the 1991 uprising,
they provide one of the strongest and most unambiguous legal cases for a
genocide tribunal. Nevertheless, the perpetrators have not yet been
brought to justice, and it is becoming increasingly uncertain that they
ever will be.

Teaching About Genocide: Issues, Approaches, and Resources, 181–191

OVERVIEW OF THE 1988 ANFAL (GENOCIDAL) OPERATIONS IN IRAQI KURDISTAN

In Spring 1987, Iraq's predicament was bleak. In the first Gulf War, its enemy Iran appeared to be regaining momentum; in March, Iran had reopened its Northern front, in collusion with Iraqi Kurdish guerrillas. Moreover, in the course of 1987, the major Iraqi Kurdish parties, including the Kurdistan Democratic Party headed by Massoud Barzani and the Patriotic Union of Kurdistan (PUK), headed by Jalal Talabani, decided to join forces, ending their long-standing differences and years of infighting. The Kurds' tactical alliance with Iran posed a new threat to the Iraqi regime, which reacted by implementing increasingly drastic counterinsurgency measures. On March 29, Saddam Husayn promulgated decree no. 160, making his cousin Ali Hasan al-Majid director of the Baath Party's Directorate of Northern Affairs, which was responsible for the autonomous Kurdish region in Northern Iraq. al-Majid, until then the director of General Security, was granted sweeping powers over all civilian, military, and security institutions of the region.

In April, he ordered the first attacks, including chemical bombardments, against the PUK headquarters but also against the Kurdish mountain villagers and villages that could provide them with shelter and supplies. In this campaign, alone, at least 703 Kurdish villages were destroyed. After a few months, however, these operations were discontinued, possibly because the Iraqi army was too preoccupied with Iranian offensives. But al-Majid's June 1987 directives give a clear indication of what was to come. His document 28/3650, dated June 3, imposed both a total blockade and a shoot-on-sight policy on the areas outside government control: "The armed forces must kill any human being or animal present within these areas"; document 28/4008 of June 20 provides a standing order for the summary execution of all (male) captives: "Those between the ages of 15 and 70 shall be executed after any useful information has been obtained from them."

The next organizational step toward Anfal was the nationwide census that was held on October 17, 1987. For the Kurdish North, this was less a registration of population data than a sweeping gov-

---

List of Essential Antecedents That Need to Be Addressed When Teaching About This Genocide

- Baathist ideology, especially the preoccupation with "betrayal of the national pact"
- Arabization of the Kirkuk region since the early 1970s
- Deportations and disappearances of Kurdish civilians since 1975
- Village destructions and shoot-on-sight zones since 1975;
- The 1983 disappearance of some 8,000 Barzani clansmen;
- The 1988 chemical attack against Halabja

ernment directive that not only identified the target population of the future operations, but also indiscriminately marginalized and criminalized it. All traffic to and from areas outside government control was forbidden, and relatives of alleged saboteurs were expelled from government-held areas. All individuals who consequently failed to participate in the census were stripped of their citizenship, and considered as deserters or saboteurs who deserved the death penalty.

The Anfal operations proper did not start until February, 1988; presumably, by then, the Iraqi regime felt that Iranian pressure had eased sufficiently to allow for the redeployment of large numbers of troops in the north. They were conducted on a much larger scale, and were of a much more systematic character than the Spring 1987 operations: several army divisions participated in them, together with personnel of general intelligence and the Baath party, along with Kurdish irregulars.

The first Anfal operations, starting February 23, were primarily directed against the PUK headquarters near the Iranian border, but also against the surrounding villages. Most villagers, however, appear to have escaped into Iran or to the larger cities of the Kurdish region in Iraq. The first Anfal, in other words, does not appear to have involved the large-scale disappearance of civilians. This was to change in the following operations.

In the following months, seven further operations were carried out, systematically targeting the different areas that had remained under Kurdish control. They typically involved the surrounding of the target area, which was then exposed to massive shelling and air attacks, including the use of chemical weapons; apparently, the latter were intended primarily to destroy the morale of the villagers and guerrillas (who had long become used to conventional bombardments). With the target population dislodged, government forces would gradually close the circle, and mount a massive ground attack by army troops and irregulars, or alternatively have the irregulars persuade the villagers to surrender.

The Kurdish captives were first brought to local collection points, mostly by Kurdish irregulars; subsequently, government personnel took them to centralized transit camps at military bases near Kirkuk, Tikrit, and Duhok. Here, they were divided by age and gender, and stripped of their remaining possessions. The vast majority of captured adult men were loaded onto windowless trucks, and taken to execution sites in central Iraq. Several men, however, survived these mass executions. They all report having seen rows of trenches dug by bulldozers, each holding hundreds of corpses. It is more than likely that tens of thousands of Kurdish men were massacred in this way, merely on account of their Kurdish ethnicity and of their living in an area declared out of bounds by the regime.

Unknown numbers of women are also suspected to have been massacred. More typically, however, women were left alive and relocated; there also is credible testimony that many younger women were sold off as brides, or rather into virtual slavery, to rich men elsewhere in Iraq, but also in Kuwait and Saudi Arabia. Many elderly captives were initially resettled in the Nugrat Salman concentration camp in southern Iraq; in the appalling living conditions existing there, up to 10% of the inmates may have died in the space of a few months. Often, the corpses were refused a proper burial, and were left exposed in the summer heat for several days.

On August 20, 1988, a cease-fire between Iran and Iraq came into effect. The Iraqi army now had its hands free to finish its campaign against the Kurdish insurgents. On August 25, it initiated the Final Anfal, directed against what remained of the traditional KDP strongholds in the Badinan region bordering on Turkey. This area was not entirely sealed off, however, and over 60,000 Kurds managed to escape to Turkey. Now, for the first time, substantial eyewitness reports about the Iraqi regime's chemical attacks against its Kurdish civilians reached the international community. Press coverage led to some minor and inconsequential protests by Western governments; in international forums like the U.N. Security Council, the Iraqis avoided condemnation by cleverly manipulating remaining Cold War cleavages and existing fears of Iran.

The violence against the civilian population did not end with the successful completion of the Final Anfal. A number of refugees were lured back by the September 6 announcement of a general amnesty for all Iraqi Kurds, but many of them were "disappeared" on returning. Especially gruesome was the fate of the returning members of minority groups like the Yezidis and Assyrian Christians. Unbeknownst to themselves, the government had excluded these groups, which it wished to consider as Arabs rather than Kurds, from the amnesty. On returning to Iraq, they were separated from the Muslim Kurds; many of them, including women, children and elderly, were taken to unknown destinations and never seen again.

After the amnesty, the surviving deportees were brought back to the north, and simply dumped on relocation sites near the main roads, surrounded by barbed-wire fence. Unlike the victims of most earlier deportations, they were not provided with any housing, construction materials, food or medicine (let alone financial compensation), but just left to their own devices.

In early 1989, the Anfal operations officially ended with a new decree revoking Ali Hasan al-Majid's sweeping powers. To all appearances, the Kurdish insurgency had been solved once and for all. The major Kurdish parties had been thoroughly demoralized, and indeed discredited, by the government's brutal actions, and faced fierce internal criticism because of their tactics which had left the civilian population exposed to the Iraqi onslaught. Virtually the entire surviving rural population had been cowed into submission and relocated in easily controlled resettlement camps.

In these operations alone, an estimated 1,200 Kurdish villages were destroyed. The number of civilian casualties has been variously estimated: Kurdish sources, basing themselves on extrapolations from the numbers of villages destroyed, at first spoke of some 182,000 people killed or missing. Human Rights Watch, an international human rights organization, makes a more conservative estimate of between 50,000 and 100,000 civilian dead. And during the spring 1991 negotiations between the Kurds and the government, Ali Hasan al-Majid himself at one point exclaimed: "What is this exaggerated figure of 182,000? It could not have been more than a hundred thousand!"

EVIDENCE

The Anfal operations formed the genocidal climax of the prolonged conflict between the successive Iraqi regimes and the Kurdish nationalist movement. Their full scale and bureaucratic nature, and indeed full horror, did not become widely known until the aftermath of the 1991 Gulf War. In the popular uprising against the Iraqi regime, literally tons of documents from various government institutions were captured that provided ample, if partly indirect and circumstantial, evidence for the 1988 genocide. Although many questions remain unanswered, they appear to contain sufficient material for a legal genocide case against the Iraqi regime.

The authenticity of these documents has been contested by the Iraqi government; but it is extremely unlikely that they are forgeries. They form an extremely complex network of interlocking texts of a highly bureaucratic nature; moreover, in many cases, they closely match the testimony provided by eyewitnesses and survivors. References to government actions are often quite indirect or opaque; thus, few documents openly refer to mass executions or chemical weapons. Even internal documents usually, though by no means without exceptions, euphemistically speak of "special attacks" and "special ammunition" when referring to chemical warfare, or of a "return to the national ranks" when talking about surrender to government forces.

Organization and Implementation

The documentary and other evidence also provides detailed insight into the chain of command and into the motives of the perpetrators. Among the personnel participating in them were the first, second, and fifth army divisions, General Security, and numerous members of the

Baath party, in particular those associated with the Northern Affairs Bureau, as well as irregular troops mostly provided by Kurdish tribal chieftains. The command was firmly in the hands of Ali Hasan al-Majid, acting as the head Baath party's Northern Bureau, and overruling all other authorities. It appears to have been the regional Baath party apparatus, rather than the intelligence services, the police or the army, that was at the heart of the operations; in all likelihood, the firing squads also consisted first and foremost of party members.

There are significant differences in the execution of the successive operations. In the first Anfal, few noncombatants were disappeared. While in the Final Anfal captured men were often executed on the spot, in later operations adult males were taken to mass execution sites far away from the Kurdish region. It is not clear whether such variations reflect an escalating logic of violence, a differentiated reaction to the degree of resistance encountered, or simply the whims of local field commanders.

Among the Kurdish population at large, and also among Arab civilians and even among some government officials, there were a few but important episodes of resistance or support for the victims. In the third Anfal, the local population of Chamchamal rose up in revolt against the deportation of villagers. During the fifth, *peshmerga* [which literally means, "those who face death," which is the name the Kurdish fighters adopted] resistance turned out to be so strong that two further operations against the same area were mounted, keeping government forces occupied for over three months.

Especially in Arbil, the local urban population, at times at great personal risk to themselves, made a prolonged effort to help the deported villagers. The documents not only show the high degree of secrecy surrounding the operations, but also the extreme concentration of power, and the bureaucratic structure that made them possible. There are indications, for example, that military intelligence did not know precisely what was going on; and that lower army officers were incredulous at the standing order to execute all captives. Unlike many other cases of genocide, then, the Anfal operations were made possible far less by mobilizing latent or open ethnic hatred against the target group among the population at large than by a highly efficient, centralized, and secret organization.

In the operations, the Kurdish irregular troops, or *jash* ("donkey foal") as they are disparagingly called among Kurds, played an important but ambiguous role. Formed in the early 1980s, as a means of relieving the Iraqi army in the northern countryside, numerous *jash* leaders in fact maintained contacts with the Kurdish insurgents. For many Kurds, enlisting as an irregular was a convenient means of escaping active front duty in the war with Iran (and of making a living). Other tribal leaders siding with the government, however, had their own accounts to settle with either the Kurdish parties, or with tribes and villages in nearby areas.

The Kurdish irregulars appear to have had a relatively low position among the personnel involved in the operations. They had a better knowledge of the mountainous territory than the regular security forces, and they could more easily persuade the population to surrender; but not all of them were wholly reliable in the implementation of al-Majid's orders.

It is unlikely that all irregular troops were equally well informed about the operations' true character. Apparently, most of them had merely been told to help in the rounding up of villagers for the purpose of relocation, and made a genuine effort to help the captives. Others, however, participated with glee in the rounding up of civilians and looting their possessions. In some cases, acts of clemency were simply bought by bribes. A better appreciation of the role of the *jash* is hampered to some extent by the fact that all (powerful) government collaborators were granted a general amnesty in 1991, and most of them continued to wield considerable power under the new Kurdish rulers.

PARALLEL CASES

There are two well-documented parallel cases, showing the Iraqi regime's readiness to resort to the killing of Kurds as such:

- the 1983 disappearance of Barzani clansmen and,
- the 1988 chemical attack against Halabja.

The background of both involves not only the armed Kurdish insurgency, but also the Iraqi war against Iran. After the 1975 collapse of the Kurdish front, hundreds of thousands of Kurdish villagers had already been deported to relocation camps or *mujamma'at*; their traditional dwellings had been destroyed and declared forbidden territory. Thousands of members of the Barzani clan had been deported to Southern Iraq in 1976; in 1981, they had been relocated in the Qushtepe *mujamma'a* just South of Arbil. Then, in 1983, after Iran had captured the border town of Haj Omran with the aid of KDP guerrillas, the Iraqi government took its revenge on the Barzani clan. Some five to eight thousand men were taken away from the Qushtepe camp and never seen again; the remaining women were reduced to a life of abject poverty. Thus, government policies not only aimed at the physical elimination of Kurds associated with disloyal elements, but also aimed at the symbolic destruction of the honor of both the male and female members of the proud Barzani tribe.

It has been the March 16, 1988 chemical attack against the town of Halabja, in which an estimated 5,000 Kurdish civilians died a gruesome death, rather than the Anfal operations, that has entered collective memory as a symbol of the Iraqi repression of the Kurds. It has been captured in the indelible image of a Kurdish father clutching his infant son, both killed by poison gas, made by the Turkish-Kurdish photographer Ramazan Öztürk. Although the Halabja attack was not part of the Anfal operations proper (which only targeted rural areas, not cities), it certainly follows the same destructive logic. Apparently, it was provoked by the Kurdish-Iranian occupation of the city as an attempt to ease the pressure on the PUK headquarters, which at the time was bearing the brunt of the first Anfal operation. After the attack, nothing happened to Halabja for several months. It was not until July 1988 that Iraqi troops reoccupied the city, which they then proceeded to demolish. The remaining population was relocated in "New Halabja"' mujamma'a, a few miles down the road.

Secrecy and Complicity

The Iraqi regime made a strong effort to keep the true nature of the Anfal operations entirely secret, or at least to maintain strict control over the flow of information. Throughout much of 1988, Iraqi radio proudly broadcast news of the "heroic Anfal campaigns," allegedly directed against saboteurs and collaborators of Iran; but these reports carefully avoided reference to the use of chemical weapons, the deportations and executions of civilians, and the razing of villages that accompanied the operations. On several occasions, victims of chemical attacks were dragged out of hospitals and disappeared; this may have been a form of collective punishment, but it is more likely that the regime tried to eliminate all witnesses at this stage.

Despite several substantial investigations by journalists, academics, and parliamentary committees, the extent of international knowledge of, and indeed complicity in, Iraq's crimes still awaits assessment. Various European companies continued to supply Iraq with ingredients for chemical weapons, even at a time when its use of such weapons against the Iranian army was well documented. In the United States, the Reagan and Bush Sr. administrations actively supported Iraq with military advisors and equipment, and blocked diplomatic initiatives against it. It is virtually certain that the U.S. government had detailed knowledge about the campaign of destruction, of its scale, and of Iraq's systematic use of chemical weapons against its own civilians. Meiselas (1997) reproduced a Joint Chiefs of Staffs document from the National Security Archives, dated August 4, 1987, which already speaks of a campaign coordinated by Ali Hasan al-

Majid, in which 300 villages had been destroyed, and of "the ruthless repression which also includes the use of chemical weapons" (pp. 312-313). Moreover, the U.S. troops that were stationed in Northern Iraq in 1991 had extremely detailed maps in their possession, which not only indicated name and location of every village in the area but also stated which of these had been destroyed, and when the destruction had taken place.

Motives

The Anfal operations cannot simply be explained as a drastic form of counterinsurgency; but characterizing the mindset that made them possible is no easy task. The question of whether, and how far, the Anfal operations were driven by racist animosity has not yet adequately been answered; but this question does not, of course, detract from their criminal character. Racism does not appear to be an obvious feature of Iraqi society, Baathist ideology, or the perpetrators' personalities. Although there have been, and are, occasional ethnic tensions among the different segments of the Iraqi population, there is no widespread racial hatred between Kurds and Arabs in Iraq. In official Baathist discourse, categories of loyalty, treason, and sabotage (which are of an ultimately Stalinist inspiration) are much more prominent than ethnic or racial terms; the latter appear to have been rather flexible items, given the Baath regime's at times rather arbitrary and voluntaristic way of creating and dissolving ethnic identities by bureaucratic fiat. And when overtly racist language was used, this typically concerned Iranians and, predictably, Jews, rather than Kurds. Baathist ideology is of an undeniably Arab nationalist character, but it has always been ambivalent as to the inclusion of Iraq's sizable Kurdish population. There are indications, however, that in the course of the 1980s, emphasizing one's Kurdish or other non-Arab ethnicity was in itself increasingly becoming a criminal offense, if not an act of treason. For example, smaller ethnic groups, like Yezidis and Christians, were forcibly registered as Arabs, and when they changed their ethnicity to "Kurdish" in the 1987 census, Ali Hasan al-Majid had them deported and their villages destroyed.

It is even questionable whether al-Majid himself can be simply labeled a racist. On tape recordings of meetings with senior party officials, he can be heard speaking in a coarse and derogatory manner of Kurds; but his remarks hardly betray any generic hatred of Kurds as an inferior race; rather, he speaks of saboteurs and of uneducated villagers who "live like donkeys." Whatever such personal motives and animosities, official discourse consistently proclaimed both Kurds and Arabs as equal parts of the Iraqi people or nation, on condition of their political loyalty.

Likewise, religious considerations do not appear to have been a prime motivating or legitimating factor. The name Anfal, or "spoils," which comes from the eighth sura of the *Koran*, has little specifically religious significance here; it appears to refer primarily to the right granted to the Kurdish irregulars involved in the operations to loot the possessions of the captured civilians. The Baath party, which has ruled Iraq since 1968, is largely secular, and was (and is) inspired more by twentieth-century ideologies and practices of Nazism and Stalinism than by any specifically Islamic tradition.

Of the violent and indeed murderous character of Baathist rule, however, there can be no doubt at all. After the conclusion of the Anfal operations, only 673 Kurdish villages still stood in the whole of Iraqi Kurdistan; over the years, the regime had demolished 4,049 villages. Now, state violence increasingly turned toward Kurdish cities. In June 1989, the city of Qala Diza, with a population of close to 100,000, was evacuated and destroyed. It is impossible to tell where this continuing process of repression and destruction would have led, if it had not been interrupted by the 1990 gulf crisis and the ensuing war and uprising.

CONCLUSION

Increasingly, the political, moral, and legal significance of the Anfal operations has tended to be overruled by subsequent political developments. Human Rights Watch has been pushing to have a genocide case against Iraq opened at the International Court of Justice, but it has not yet found any country or group of countries willing to initiate proceedings. In the United States, a campaign to have Saddam indicted for genocide and crimes against humanity, largely on the basis of the captured Anfal documents, was initiated in the late 1990s, but it was pursued erratically, and appeared to reflect changing U.S. policies toward Iraq (not to mention domestic political rivalries) rather than any concern for the victims.

As the years pass, successful legal action against the perpetrators of the Anfal is becoming more and more unlikely. Any attempt to bring the perpetrators to justice, and to counter propagandist suggestions of bias and partiality, will require the sustained and concerted efforts of one or several neutral countries.

Recommended Reading on the 1988 Anfal (Genocidal) Operations in Iraqi Kurdistan

Human Rights Watch/Middle East. (1995). *Iraq's crime of genocide: The Anfal campaign against the Kurds*. New Haven, CT: Yale University Press.
   The only detailed study thus far in a Western language; absolutely indispensable.

Human Rights Watch/Middle East. (1994). *Bureaucracy of repression: The Iraqi government in its own words*. New York: Author.
   A selection of the most telling Anfal documents captured in 1991, with translation and commentary. Parts of these were reproduced in *Iraq's Crime of Genocide: The Anfal Campaign Against the Kurds*. They are also available on the Internet: http://www.fas.harvard.edu/~irdp/

Makiya, K. (1993). *Cruelty and silence: War, tyranny, uprising, and the Arab world*. New York: Norton.
   This book, together with the same author's earlier *Republic of Fear: The Politics of Modern Iraq*, provides the best introduction to the mind-set and discourse that "legitimated" the Anfal operations.

Meiselas, S. (1997). *Kurdistan in the shadow of history*. New York: Random House.
   A solid history of Kurdistan.

McDowall, D. (1997). *A modern history of the Kurds* (2nd ed.). London: I.B. Tauris.
   Good on the general historical context of the Kurdish movement.

REFERENCES

Meiselas, S. (1997). *Kurdistan in the shadow of history*. New York: Random House.

CASE STUDY 9

GENOCIDE IN BOSNIA

Eric Markusen

INTRODUCTION

During the last decade of the twentieth century, the Federal Republic of Yugoslavia—which had been a relatively peaceful, multiethnic nation in southeastern Europe—descended into a series of three wars when the republics of Slovenia and Croatia declared independence in 1991 and Bosnia and Herzeogvina did the same in 1992. Before the wars began in 1991, Yugoslavia consisted of five semiautonomous republics: Serbia, Croatia, Bosnia and Herzegovina, Macedonia, and Montenegro. The largest of them, Serbia, contained two autonomous provinces: Vojvodina in the northern part of Serbia, and Kosovo in the south. Among the most ethnically diverse nations of Europe, Yugoslavia included Serbs (36%), Croats (20%), Muslims (9%), Albanians (nearly 8%), and numerous other, smaller groups.

The aforementioned conflicts claimed hundreds of thousands of lives, left many more hundreds of thousands physically and mentally trauma-tized, forced millions to leave their homes, and resulted in the wholesale destruction of countless cities and villages.

The war between Slovene forces and the Yugoslav national army lasted only 10 days (June 27-July 7, 1991) and caused relatively few casualties. Then came the war in Croatia (May 1991-January 1992), which pitted

Teaching About Genocide: Issues, Approaches, and Resources, 193–202
Copyright © 2003 by Information Age Publishing
All rights of reproduction in any form reserved.

Serbs living in Croatia who opposed independence against Croatian forces. It lasted more than seven months, cost as many as 10,000 lives, and resulted in the destruction of many cities, towns, and villages. It also included numerous massacres and other atrocities, including the practice of ethnic cleansing, that is, the forcible removal of an ethnic group from territory it inhabits by another ethnic group that desires the territory solely for itself. The majority of atrocities and war crimes were committed by the Yugoslav National Army and by Croatian Serb Separatist Forces that had been armed by Belgrade, although Croatian soldiers and para-military units also engaged in massacres and deliberate destruction of vil-lages and towns.

The war in Bosnia (April 1992-December 1995)—which took place in Yugoslavia's most ethnically-diverse republic with approximately 40% Muslims, 33% Serbs, and 18% Croats—lasted more than three years, included atrocities committed by all the major sides (Serbs, Croats, and Muslims), and caused many times the death and destruction than had occurred in Croatia. This war also involved ethnic cleansing, when Bos-nian Serbs attempted to remove Muslims and other non-Serbs from areas they claimed for themselves, and when Bosnian Croats did the same to Muslims. However, in the case of the Bosnian Serbs (and their allies from Serbia/Yugoslavia), the atrocities were done in such a systematic manner and on such a vast scale that a number of observers, including the United Nations ad hoc International Criminal Tribunal for Former Yugoslavia (ICTY), along with prominent genocide scholars, concluded that geno-

---

List of Essential Antecedents That Need to Be Addressed When Teaching About This Genocide

- The death of Josef Tito in 1980 deprived Yugoslavia of a leader com-mitted to ethnic coexistence under the slogan "brotherhood and unity."
- Tito's death was followed by a period of economic instability in Yugoslavia which made many people feel insecure and frightened.
- Nationalist political leaders—notably Slobodan Milosevic in the republic of Serbia and Franjo Tudjman in the republic of Croatia—assumed power and began fomenting fear and anger between ethnic groups.
- The media, particularly television, were used by both Milosevic and Tudjman to increase tensions between ethnic groups.
- Milosevic deliberately revived painful memories from World War II, when Croatia had perpetrated a genocide against Serbs living in Croatia and Bos-nia.
- Tudjman "rehabilitated" the genocidal World War II-Croatian regime by nam-ing streets after leaders of the geno-cide, resurrecting symbols from the genocidal regime, and in other ways.
- The war in Croatia contributed to the genocidal character of the war in Bosnia by spawning ruthless Serb paramilitary groups who engaged in many massacres and by aggravating fear and anger among Serbs living in Bosnia.

cide—the intentional destruction of a human group, in whole or in significant part—had been perpetrated by Serbs against Muslims. The ICTY issued indictments on charges of genocide against more than a dozen Bosnian Serbs and to one Serb from Yugoslavia, the latter being its former president, Slobodan Milosevic. The ICTY has convicted one Bosnian Serb, General Radislav Krstic, for genocide, due to his responsibility for the massacre of thousands of Bosnian boys and men in the so-called "safe area" of Srebrenica in July 1995. The Srebrenica massacre is discussed below.[1]

AN OVERVIEW OF THE GENOCIDE IN BOSNIA, 1992-1995

In March 1992, on the day after the Bosnian referendum for independence, Serbs in Sarajevo, the capital city of Bosnia and Herzegovina, erected barricades across city streets in what was to be the first step toward a division of the city into Serbian and non-Serbian sections. Bosnian Serb forces controlled most of the area around the city and used artillery bombardment and sniping to terrorize and murder the defenseless civilians. According to the ICTY, "The [Serb] shelling and sniping killed and wounded thousands of civilians of both sexes and all ages, including children and the elderly" (Indictment of Bosnian Serb General Stanislav Galic: Case No. IT-98-29-I, p, 2). Serb forces also prevented food, water, heating oil, and other necessities from reaching the besieged people in the non-Serb areas. They also targeted many of the mosques in Sarajevo, and other Muslim cultural monuments.

While the citizens of Sarajevo were beginning their long ordeal, Serbs began ruthless attacks against Muslims living in eastern and northern Bosnia. On April 1, 1992, the notorious Serb paramilitary leader, Arkan, crossed the river from Serbia into the city of Bijeljina and began a campaign of terror against its Muslim population. He and his men murdered dozens of defenseless civilians and drove the rest from their homes and businesses, which were then systematically looted, then destroyed. Aided by military forces from Yugoslavia, Arkan's "Tigers" and other paramilitary units conducted similar actions in numerous other towns and cities. One week later, they struck the city of Zvornik, separated by only a bridge from Serbia. Yugoslav army forces across the river in Serbia bombarded the city with artillery, and then the paramilitaries went on a rampage of killing, looting, and destruction. By chance, a high-level United Nations official, Jose Maria Mendiluce, passed through Zvornik at the time, en route to Sarajevo after meetings in Belgrade. He later described what he witnessed to journalists: "I could see trucks full of dead bodies. I could see militiamen taking more corpses of children, women and old people from their houses and putting them on trucks. I saw at least four or five trucks full of corpses" (Quoted in Silber & Little, 1997, p. 223). In this reign of terror, which took place throughout Bosnia, non-Serbs identified as intellectuals, professionals, and/or political leaders were generally singled out for summary execution.

During the spring and summer of 1992, Serbs also established a number of so-called detention camps in which as many as 10,000 people were killed (Power, 2001, p. 269). While there were many detention camps in Serb-controlled territory, several were exposed by journalists in July and

---

Key Events: Genocide in Bosnia

- The declarations of independence by the Yugoslav republics of Slovenia and Croatia in 1991 led to the outbreak of war.
- The war in Croatia included widespread destruction, ethnic cleansing, and massacres.
- The declaration of independence by the Yugoslav republic of Bosnia and Hercegovina led to a war that lasted more than three years and was far more lethal and atrocious than the war in Croatia.
- In March 1992, the city of Sarajevo was placed under siege by Bosnian Serb forces who used artillery and sniping to murder thousands of civilians.
- In April 1992, the Bosnian Serbs, aided by Serbs from the republic of Serbia, began a campaign of massacre and ethnic cleansing in eastern and northern Bosnia.
- During the summer of 1992, Serbs established so-called "detention camps" in which non-Serbs were tortured and killed. Other parties in the conflict also established brutal detention camps. Rape of women, particularly Muslims, was widespread.
- In 1993, Bosnian Croats fought against Bosnian Muslims; many atrocities were committed, particularly by Croatian forces.

- In July 1995, Bosnian Serb forces overran the so-called "safe area" of Srebrenica, expelled Muslim women and children, and slaughtered approximately 7,000 Muslim men and boys in the worst massacre on European soil since World War II.

    Also in July 1995, the International Criminal Tribunal for the Former Yugoslavia publicly indicted Radovan Karadzic, the political leader of the Bosnian Serbs, and Ratko Mladic, commander of the Bosnian Serb military, on charges of genocide, crimes against human, war crimes, and grave breaches of the Geneva Convention.
- After the Srebrenica massacre, the international community, including the United States, intervened with airstrikes against Serb forces.
- In November 1995, representatives of Croatian, Muslim, and Serb factions met in Dayton, Ohio, and reached an agreement to end the conflict.
- On August 2, 2001, the International Criminal Tribunal for the Former Yugoslavia found Bosnian Serb general Radislav Krstic guilty of genocide for his role in the Srebrenica massacre.
- On February 12, 2002, the trial of Slobodan Milosevic, former President of Yugoslavia, began at the ICTY. Milosevic has been charged with genocide, crimes against humanity, and war crimes in Bosnia, Croatia, and Kosovo.

August of 1992 and became notorious, including Omarska and Kereterm. In addition to overcrowding and lack of food, medical care, and sanitary facilities, the inmates suffered from horrific abuse. According to the ICTY, "The camp guards, and others who came to the camp and physically abused the prisoners, used all manner of weapons during these beatings, including wooden batons, metal rods and tools, lengths of thick industrial cable that had metal balls affixed to the end, rifle butts, and knives. Both female and male prisoners were beaten, tortured, raped, sexually assaulted, and humiliated" (Amended indictment against Zeljko Meakic and others. Case No. IT-95-4-I, p. 2). Inmates perceived as political, cultural, or military leaders were subjected to special abuse and, frequently, execution. It should be emphasized that not only Serbs operated camps in which atrocities were committed. Croatian forces and Muslim forces also killed, tortured, and otherwise abused captive Serbs and others in a number of camps. While the ICTY has convicted a number of Croatian and Muslim individuals for crimes against humanity and war crimes, it has yet to indict or convict any Croat or Muslim for the ultimate crime of genocide.

Widespread and systematic sexual abuse against women captured by Serb forces was used as a tool of genocide. According to Beverly Allen (1996), tens of thousands of rapes took place, on orders from Serb authorities, and in many cases with the explicit purpose of impregnating the women. While rapes were committed by all sides in the Bosnian conflict, there is a general consensus that the vast majority of the perpetrators were Serbs, and the vast majority of victims were Muslims.

By spring of 1993, tens of thousands of Muslims were confined in three enclaves in eastern Bosnia—Srebrenica, Gorazde, and Zepa—that were surrounded by Serb-controlled territory. Although they had been declared "safe areas" by the United Nations, their inhabitants remained at grave risk. In addition to frequent shelling of towns and villages in the enclaves, Serb authorities frequently refused the United Nations permission to deliver food, medical supplies, and other humanitarian necessities. The populations of the enclaves were swollen by desperate refugees from areas that had already been "cleansed" by Serb forces, thus aggravating the overcrowding, hunger, and inadequate medical resources.

In early 1994, the Bosnian Muslims and Croats, who had been fighting against each other during 1993, resumed cooperative efforts to defeat the Serbs and reclaim the vast tracts of territory that the Serbs had conquered and "cleansed." (It was during the Croat-Muslim conflict that Croatian forces engaged in ethnic cleansing against Muslims involving massacres of civilians and destruction of homes and villages.) In response, the Serbs intensified their bombardment of Sarajevo and blocked shipments of humanitarian supplies to Gorazde and other "safe areas" areas where large numbers of Muslims were trapped. When the United Nations authorized NATO to conduct airstrikes against Serb forces, the Serbs seized U.N. personnel as hostages and continued their onslaught.

Then, in July 1995, the largest genocidal massacre of the entire war took place when Serb forces overran the "safe area" of Srebrenica, where tens of thousands of refugees had eked out a bare existence since the spring of 1992. On July 13, women, children, and elderly people were put on buses and driven to the front line where they were forced to walk to Muslim-controlled territory. Between July 13 and 19, as many as 7,000 boys and men were systematically slaughtered in a carefully-planned operation. While some were killed individually or in small groups, the vast majority were carried on buses to execution sites where they were executed at the hands of men using automatic weapons and machine guns. After the slaughter, trucks were brought in to collect the bodies and haul them to mass graves.

When the world learned about the Srebrenica massacre, there were intensified efforts by the international community, including the United States, to stop the violence. With the support of the United States, NATO (the North Atlantic Treaty Organization) began launching air strikes against Serb military targets, seriously degrading Serb capabilities. At the same time, Muslim and Croatian offensives recaptured some of the territory that had been seized by the Serbs. Finally, in November 1995, after intense negotiations organized by the United States and held at a U.S. airforce base near Dayton, Ohio, the war—and the genocide—were brought to an end.

ANTECEDENTS TO THE GENOCIDE

As the wars in former Yugoslavia broke out in the early 1990s, some observers argued that "ancient hatreds" stemming from a long history of conflict and ethnic cleansing in the Balkans were somehow responsible for the shocking mass violence. This assumption, which implied that violence in the region was somehow inevitable and, therefore, unlikely to be curbed by outside pressure or intervention, helped justify inaction by the international community.

The reality, however, was quite different. Under Tito, the powerful leader of Yugoslavia from 1945 until his death in 1980, politics based on ethnicity was officially discouraged, under the slogan "brotherhood and unity." Moreover, there was a high rate of intermarriage among the different ethnic groups, particularly in urban areas. According to one source, based on the 1981 Yugoslav census, "if children of mixed marriages were included, over half the population of Bosnia had a close relative of a different nationality" (quoted in Burg & Shoup, 1999, p. 43).

Rather than "ancient hatreds," the genocidal violence of the Bosnian war reflected a lethal combination of economic and political instability, the rise of nationalistic leaders who manipulated the media before and during the war to anger and frighten ordinary citizens and reinforce the relative minority of extremists, and the war, itself, in Croatia.

Following Tito's death in May 1980, the Yugoslav economy faced many serious problems, including high unemployment that made the relatively high standard of living to which Yugoslavs had grown accustomed harder to attain. According to Sudetic (1998), "Young professional people could not find work and leapt at any chance to flee abroad. High school graduates who had managed to acquire occupational training were now almost unemployable except as physical laborers toiling for slave wages" (p. 76). During the same period, there were growing tensions both within republics, for example, between Albanians and Serbs in the Serbian province of Kosovo, and between republics, particularly between Slovenia and Croatia, on the one hand, and Serbia on the other.

Tito's death left a leadership vacuum in Yugoslavia. No leader emerged with sufficient power and motivation to unify the diverse peoples of Yugoslavia and suppress the growing tensions among them. Instead, ruthless, nationalistic leaders, notably Slobodan Milosevic in Serbia and Franjo Tudjman in Croatia, assumed power. Rising to power in the late 1980s, Milosevic repudiated the ideology of ethnic coexistence and asserted that he would protect the interests of Serbs both in Serbia and other republics. Tudjman, a deeply-committed advocate of a Greater Croatia inhabited mainly by Croats who had been imprisoned for such beliefs by Tito, assumed power in 1990 and quickly pushed through new laws that were openly discriminatory against Serbs.

Milosevic and Tudjman controlled and exploited the media, particularly television and radio, as effective propaganda tools that aggravated fears and tensions between Croats and Serbs and eventually led to open warfare. Both leaders deliberately revived memories of genocide during World War II, when Croatia, an ally of Hitler, slaughtered hundreds of thousands of Serbs, as well as Jews and Gypsies, living in Croatia and Bosnia. The genocidal Croatian regime, known as the Ustasha, was so barbaric that German military officers stationed in Croatia complained about the torture and massacres perpetrated by them.[2]

When Milosevic spoke before large crowds of Serbs, he frequently emphasized the theme of Serb vulnerability and past victimization. Documentaries on the Croatian genocide against Serbs were often aired on television in Serbia. Mass graves of murdered Serbs were exhumed and the bones reburied amidst pomp and ceremony. While the Milosevic regime was deliberately exploiting the World War II genocide to induce fear and anger among Serbs, the Tudjman regime was engaged in "rehabilitating" the genocidal regime. Streets in Croatia were renamed after leaders of the Ustasha regime; the Croatian parliament selected, as the new national coat of arms, a design very similar to that used by the Ustasha. According to one scholar from the region, "Tudjman's every new step seemed to lend credibility to Milosevic's outrageous warnings that unless due measures were taken, the Serbs could never feel safe again" (Stitkovac, 1997, p. 156).

The war in Croatia contributed to the viciousness and, ultimately, genocidal character of the war in Bosnia in several ways. As noted earlier, the practice of ethnic cleansing began in Croatia, establishing a pattern that was repeated many times in Bosnia, where it was so widespread and lethal, as to suggest genocidal intent, including the systematic murder of many leaders of non-Serb communities and the destruction of mosques and other cultural monuments. The war in Croatia also spawned some of the paramilitary units, including Arkan's Tigers, that later engaged in massacres and destruction. Atrocities were perpetrated by Croats, as well as Serbs, and Serb propaganda deliberately inflamed fear and anger among Bosnian Serbs, warning them that similar fates awaited them, should Bosnia become independent.

All of these forces combined to persuade a sufficient number of Serb leaders and followers that their own, Serb, safety and survival could be assured only by eliminating, and, eventually, destroying, non-Serb groups.

CONCLUSION:
WHAT HAVE WE LEARNED FROM THIS TRAGEDY?

The tragedy of war and genocide in Bosnia suggests a number of lessons. First, hateful propaganda, atrocities in war, and ethnic cleansing are often early warning signs of potential genocide. Second, mass media, particularly television, can be powerful means of creating hatred and fears that can lead to a genocidal mentality, that is, the willingness to consider the destruction of a group as a means of solving one's own group's problems. Third, once a regime has decided to engage in genocide as a social policy, only intervention from the outside world is likely to stop it.

Recommended Reading on the Genocide in Bosnia

Cigar, N. (1995). *Genocide in Bosnia: The policy of "ethnic cleansing."* College Station: Texas A&M University Press.
A detailed examination of the ideological and political preparations for Serbian genocide against Bosnian Muslims. Cigar also discusses Croatian genocidal violence against Bosnian Muslims.

Gutman, R. (1993). *A witness to genocide: The 1993 Pulitzer Prize-winning dispatches on the "ethnic cleansing" of Bosnia.* New York: Macmillan.
An account of the barbaric conditions in Serb detention camps, including the beatings, torture, and frequent murders.

Markusen, E., & Mirkovic, D. (1999). Understanding genocidal killing in the former Yugoslavia: Preliminary observations. In C. Summers & E. Markusen (Eds.), *Collective violence: Harmful behavior in groups and governments* (pp. 35-67). Lanham, MD: Rowman & Littlefield.
The authors examine a number of findings from the field of genocide studies (including the work of Leo Kuper and Robert Melson) and their relevance to the case of genocidal violence in the former Yugoslavia. They also summarize the Croatian genocide against Serbs in 1941-1945, and show how, prior to and during the wars in the 1990s, both the Serb leader, Slobodan Milosevic, and the Croat leader, Franjo Tudjman, deliberately revived memories from World War II.

Rohde, D. (1997). *Endgame: The betrayal and fall of Srebrenica, Europe's worst massacre since World War II.* New York: Farrar, Straus & Giroux.
Reporting for the *Christian Science Monitor*, Rohde went to Srebrenica shortly after the massacre and was arrested by Bosnian Serb officials while trying to investigate what had happened. This book, an expansion of his Pulitzer Prize-winning newspaper accounts, describes the massacre by detailing the experiences of seven people who experienced it: three Muslims from the Srebrenica area, two men involved in the killing, and two U.N. peacekeepers.

Silber, L., & Little, A. (1997). *Yugoslavia: Death of a nation.* New York: Penguin Books.
The best book for an overview of the background of the wars in former Yugoslavia, the atrocities and ethnic cleansing campaigns, and the efforts by the international community to stop the conflict.

NOTES

1. Genocide scholars who have concluded that the Serbs committed genocide in Bosnia include Helen Fein and Robert Melson. See Fein (1994); and Melson (1995).
2. For a summary of the Croatian Ustasha genocide, see Mirkovic (1993).

REFERENCES

Allen, B. (1996). *Rape warfare: The hidden genocide in Bosnia-Herzegovina and Croatia*. Minneapolis: University of Minnesota Press.

Burg, S. L., & Shoup, P. S. (1999). *The war in Bosnia-Herzegovina: Ethnic conflict and international intervention*. Armonk, NY: M.E. Sharpe.

Fein, H. (1994). Genocide, terror, life integrity, and war crimes: The case for discrimination. In G. Andreopolous (Ed.), *Genocide: Conceptual and historical dimensions* (pp. 95-107). Philadelphia: University of Pennsylvania Press.

Melson, R. (1995). Paradigms of genocide. *Annals of the American Academy of Political and Social Sciences, 548*, 156-168.

Mirkovic, D. (1993). Victims and perpetrators in the Yugoslav genocide of 1941-1945: Some preliminary observations. *Holocaust and Genocide Studies, 7(3)*, 317-332.

Power, Samantha. (2001). *"A problem from hell": America and the age of genocide*. New York: Basic Books.

Silber, L., & Little, A. (1997). *The death of Yugoslavia*. New York: Penguin Books.

Stitkovac, E. (1997). Croatia: The first war. In J. Udovicki & J. Ridgeway (Eds.), *Burn this house: The making and unmaking of Yugoslavia* (pp. 153-173). Durham, NC: Duke University Press.

Sudetic, C. (1998). *Blood and vengeance: One Family's story of the war in Bosnia*. New York: W.W. Norton.

CASE STUDY 10

THE RWANDA GENOCIDE

René Lemarchand

INTRODUCTION

In the spring of 1994, Rwanda burst into the headlines when half a million to 800,000 people, mostly Tutsi, were savagely butchered by Hutu extremists. In what will go down in history as one of the biggest genocides of the last century, about 10% of the population were wiped out in 100 days. The consequences have been felt far and wide, not only in Rwanda but throughout Central Africa and beyond. In Burundi, where the ethnic map is roughly the same as in Rwanda, the net result has been to greatly intensify the simmering conflict between Hutu and Tutsi. In the Congo the threats posed to its eastern neighbors by the huge outflow of Hutu refugees provided the basis for a temporary alliance of domestic and external forces that led to the overthrow of the "Mobutist" state in May 1997, and a year later to the virtual disintegration of the country under the impact of a civil war involving the intervention of neighboring states, most prominently Rwanda, Uganda, Zimbabwe and Angola. In the United States the astonishing passivity of the Clinton administration in the face of mass murder has led to bitter criticisms of U.S. policies by human rights groups, and in time to public apologies by President Clinton to the Tutsi-dominated government of strongman Paul Kagame. Last but not least, the appalling performance of the United Nations during the crisis has cast serious discredit on the international community. Though much of the blame was directed at the Secretary General, Boutros-Boutros Ghali, Kofi Annan, at the time in charge of the Department of Peace Keeping Operations, also came under fire. Measured by the sheer scale of the bloodletting, and the impotence or unwillingness of the international community to stop the killings, few human tragedies capture more cruelly the eerie quality of the phenomena that have ravaged the late twentieth century.

OVERVIEW

At 8:30 p.m. on April 6, 1994, as Rwandan President Habyalimana's plane was about to land in Kigali, two SA 16 missiles fired from near the airport scored direct hits, bringing the plane down in a matter of seconds.

List of Essential Antecedents That Need to Be Addressed When Teaching About This Genocide

- A number of large-scale killings took place in Rwanda before the 1994 bloodbath: in December 1963 as many as 10,000 Tutsi civilians were killed by Hutu mobs in response to a nearly successful invasion of Tutsi refugees from Burundi; immediately following the FPR invasion in October 1993 some 300 Tutsi were killed in the Kilibira communes; and in January 1991 an estimated 1,000 Tutsi-Bagogwe were massacred in the north, and some 300 in the Bugesera region in March 1992, and hundreds more in and around Ruhengeri and Gisenyi in January 1993.
- The Rwandan government responded to the FPR invasion by rounding up and arresting some 13,000 Tutsi suspected of supporting the FPR, thus initiating a process of polarization that went on unabated in the years preceding the genocide.
- As the FPR military campaign picked up momentum in the north, thousands of Hutu were killed while an estimated one million ended up in internally displaced persons (IDP) camps, many of whom later joined the genocidaires.

- Hutu-Tutsi polarization reached new levels of intensity during the Arusha peace process, with the creation of the militantly anti-Tutsi Coalition pour la Défense de la République (CDR) in March 1992, and after the assassination of President Melchior Ndadaye, a Hutu, in October 1993, when the Hutu Power movement became a synonym for anti-Tutsi extremism.
- The exodus of some 300,000 Hutu refugees from Burundi in the wake of Ndadaye's assassination, fleeing the repression of the army, was a major contributory factor to the growing radicalization of Hutu public opinion
- From 1992 onward, anti-Tutsi hysteria received a powerful boost from the Hutu-dominated media and repeated incitements to murder by some leading Hutu politicians, notably Leon Mugesera, who, in an infamous speech in November 1992, exhorted his audience to "rise up" and "exterminate this scum."
- The shooting down of the plane carrying Habyalimana, in Kigali on April 6, 1994, marked the "tipping point" in Hutu perceptions of their own security dilemma—to kill all Tutsi or being killed by them— thus precipitating the bloodbath.

Among the victims were President Habyalimana, his counterpart from Burundi, Cyprien Ntaryamira, and several high-ranking Rwandan officials, including General Nsabima, the army chief of staff.

Early the next morning, the exterminating mechanism was set in motion. Under the guidance of Colonel Theoneste Bagosora, the chief organizer and planner of the genocide, units of the Presidential Guard proceeded to arrest and kill every opposition leader, Hutu and Tutsi. Among the first to be killed were Agathe Uwilingiyimana, the Hutu Prime Minister and leading figure of the *Mouvement Democratique Republicain* (MDR), Lando Ndasingwa, the Tutsi president of the Liberal Party (PL), and Felicien Ngango, of the *Parti Social Democrate* (PSD). The military was quickly joined by the youth wing of the ruling *Mouvement National pour la Revolution et le Developpement* (MNRD), the so-called *interahamwe* ("those who stand together"), who formed the bulk of the grass-roots killers. In the countryside the prefects, burgomasters, and communal councilors played a key role in organizing the massacre of Tutsi civilians. Contributing in no small way to the carnage, day-after-day incitements to ethnic

---

Key Events: Rwanda Genocide

- Hutu revolution, 1959-1962. The monarchy is abolished; Rwanda becomes a republic under Hutu control; approximately 150,000 Tutsi flee the country to seek asylum in neighboring states.
- In July 1973 the army overthrows the Kayibanda government, and proclaims the Second Republic. Colonel Juvenal Habyalimana becomes president. Power is now in the hands of northern politicians and army men. Two years later Rwanda is officially declared a single-party state under the National Revolutionary Movement for Development (MRND).
- On October 1, 1990, some 6,000 guerrillas of the Rwanda Patriotic Front (RPF) invade the country from Uganda, with substantial military and financial support from President Museveni of Uganda, plunging the country into the throes of a bitter civil war.
- In mid-October troops from France, Belgium, and Zaire help contain the invasion. Belgium and Zaire withdraw their military assistance shortly thereafter, leaving only the French military

assistance mission on the ground to train the Rwandan army.
- In June 1991 a new constitution is proclaimed which formally allowed a multi-party system to replace the single-party state.
- The Arusha Accords are signed on August 4, 1993, formalizing a power sharing agreement among parties and the installation of a broadly-based transitional government (BBTG). The extremist *Convention pour la Defense de la Republique* (CDR) is excluded from both the negotiations and the BBTG.
- The first of 2,548 troops assigned to the United Nations Mission in Rwanda (UNAMIR) arrive in Kigali under the command of Brigadier General Romeo Dallaire.
- On April 6, 1994, the plane carrying Presidents Juvenal Habyalimana of Rwanda and Cyprien Ntaramira of Burundi is shot down above Kigali on its return flight from Dar-es-Salaam. Early the next day begins the mass slaughter of Tutsi and moderate Hutu.

hatred and murder were disseminated on the airwaves of Radio Mille Collines, a private radio station that served as the mouthpiece of Hutu extremism.

As the genocidal fury picked up momentum, it became transparently clear that the 2,500 strong United Nations Mission to Rwanda (UNAMIR), under the command of General Romeo Dallaire, would do little to stop the killings. Not only did the UNAMIR mandate specifically prohibit the use of force against the genocidaires; more serious still were the divergences of interests among the permanent members of the U.N. Security Council, most notably France and the United States. After 10 Belgian paratroopers were killed by the Presidential Guard, on April 7, Belgium felt it had no choice but to withdraw its contingent (428 troops), and on April 21 the U.N. Security Council took the fateful step of reducing the size of UNAMIR to 270 men, a move immediately interpreted by the killers as a sign of capitulation. While the United States, still reeling under the blow of a debacle in Somalia where it lost military personnel in a particularly vicious manner, refused to get involved, and stubbornly resisted the use of the term genocide to describe the slaughter, the French, who never ceased to support the Habyalimana government against the FPR invader, were understandably reluctant to endorse policies whose consequences would no longer serve the interests of their Rwandan protégés.

To this day a complete mystery surrounds the identity of the actors responsible for the downing of the presidential plane. While some observers detect a conspiracy of the hard-liners in the entourage of Habyalimana—the so-called *akazu* ("the little hut" in *Kinyarwanda*)—aimed at eliminating Habyalimana for being far too willing to make concessions to their domestic and external enemies, others firmly believe that the blame lies with the *Front Patriotique Rwandais* (FPR), the Tutsi-dominated party associated with the group of Uganda-based "refugee warriors" who fought their way into Rwanda on October 1, 1990, before they finally captured the capital city on July 4, 1994.

Although the truth may never be known, several things are reasonably clear: (a) to this day, for most politically conscious Hutu, there is no doubt whatsoever that the FPR was behind the shooting down of the plane; (b) this catastrophic event occurred in a climate already saturated with fear and anxieties, which provided Hutu extremists with an ideal environment for manipulating such fears to their advantage; and (c) although the tragedy of April 6 must be seen as the precipitating factor behind the genocide, its roots lie much deeper in the history of Rwanda.

From the standpoint of its ethnic map, traditional Rwanda was the example par excellence of a highly stratified, pluralistic society: through much of their recorded history, Tutsi, Hutu, and Twa lived side by side, the first two sometimes linked by intermarriage, while sharing the same culture, the same language (*Kinyarwanda*) and the same monarchical institutions. At the top of the social pyramid stood the Tutsi minority, a predominantly pastoralist group, accounting for approximately 15% of the population, while the Hutu majority represented the "lower orders," most of them agriculturalists; the Pygmoid Twa, representing 1% of the population, were a marginal group in every sense of the word. Adding cohesion to the social system was a "normative charter," or body of traditions, which gave legitimacy to both the monarchy and the claims of the Tutsi to represent a higher order of humanity.

The potential for conflict inherent in this kind of "ranked" society was greatly magnified by the impact of colonial rule. Not only did the European colonizer help reinforce the "premise of inequality" built into the traditional order in the name of indirect rule, but after World War II the spread of egalitarian ideas sowed the seeds of a conflict that reached its climax in the Hutu revolution of 1959-1962. Tens of thousands of Tutsi elements were forced to flee their homeland and seek refuge in neighboring states. Few would have imagined that 30 years later the sons of the refugee diaspora in Uganda would form the nucleus of a Tutsi-dominated politico-military organization, the Rwandan Patriotic Front (RPF), that would successfully fight its way into the capital city and defeat an army three times its size.

No sooner did Rwanda cross the threshold of independence, in 1962, than portents of disaster began to loom on the horizon. Thousands of innocent Tutsi civilians were massacred in the wake of armed raids by Tutsi refugees, the so-called *inyenzi* ("cockroaches"). The worst of such raids occurred in 1963, when a group of armed refugees from Burundi fought their way into the country and nearly seized the capital, triggering a massacre of an estimated 5,000 Tutsi civilians in the prefecture of Gikongoro. Not until October 1990 would the country face an even greater threat, when thousands of determined refugee-warriors from Uganda crossed the border into northern Rwanda under the banner of the FPR.

The nature of the threat and the context in which it occurred were quite different from what could be observed in the early 1960s. For one thing, the invading force was far better organized, better equipped, and more numerous than the *inyenzi* in the early 1960s. Numbering about 6,000, many had acquired valuable experience while fighting in the ranks of Museveni's National Resistance Army (NRA) against the Obote regime in Uganda. After the NRA took Kampala by storm, in January 1986, Paul Kagame became head of military security, while the late Fred Rwigema served for a while as army commander-in-chief and minister of defense.

Even more importantly, they had the full support of President Museveni of Uganda. Rwanda in the 1990s differed from the Rwanda of the early 1960s in several important respects. First, since the 1973 coup and the rise of Major Juvenal Habyalimana to the presidency, power rested firmly in the hands of northerners. It is well to emphasize that to this day the northerners form a distinctive subculture; they were incorporated into the fold of the monarchy at a comparatively late date; unlike what could be observed in the south, few married Tutsi women. Compared to their kinsmen from the south-central region, their attitude toward the Tutsi is notoriously more distant and distrustful. Little wonder if by 1990 very few southerners held positions of influence in the government or the army. It was this critically important regional dimension that inspired in the minds of some northerners the nightmarish vision of a possible alliance of Hutu politicians from the south with the FPR.

Secondly, by 1991 Habyalimana had formally proclaimed Rwanda's vocation to be a multiparty democracy—a move designed in part to take the wind out of the FPR sails—and since the bases of support for the leading opposition party, the MDR, were in the south, it is easy to see why the southerners were now seen by the ruling MRND as doubly threatening. Meanwhile, the proliferation of parties introduced a climate of intense competition among rival Hutu elites. The dominant trend has been toward an ever deeper split between extremists and moderates, the former identified with the rabidly anti-Tutsi *Convention pour la Defense de la Republique* (CRD) and the MRND, and the latter with the MDR, PSD, and the ethnically mixed PL. By late 1993, however, the extremist versus moderates polarity had contaminated every single party, including those which seemed least likely to succumb to the appeals of Hutu extremism.

Thirdly, and most importantly, Rwanda was now a country in the throes of a bitter civil war. As the FPR began consolidating its positions in the north, the enemies from within were increasingly identified with the enemies from without by Hutu extremists. The resident Tutsi population and the invading Tutsi were seen as two faces of the same coin. Indeed, the threats posed to Tutsi civilians became all too clear after the massacre of as many as a thousand Tutsi/Bagogwe in the north in January 1991, and hundreds in the Bugesera region in March 1992.

Three decisive events prepared the ground for the apocalypse. The first was the year-long Arusha talks, from August 1992 to August 1993, leading to an agreement on a power-sharing formula during the transition period preceding general elections to be held three years down the road. In what became known as the Broadly-Based Transition Government (BBTG), the FPR and the MRND would each hold an equal number of cabinet seats (four each), the rest distributed among the MDR, PSD and PL. Another key provision concerned the restructuring of the army. It was agreed that the officer corps would be divided evenly between the RPF and the *Forces Armee Rwandaises* (FAR), and the troops on a 40-60 basis in favor of the MRND. Of crucial importance to an understanding of the eruption of anti-Tutsi violence in January 1993 was the exclusion of the CDR from the BBTG, and the strong objection raised by both the CDR and the extremist fringe of the MRND, within and outside the FAR, to the provisions of the Arusha accords concerning the future integrated national army. The killing of scores of Tutsi civilians in the Gisenyi prefecture in January 1993 was clearly intended to scuttle the accords. Even though they failed to accomplish their objective, they nevertheless succeeded in dramatically ratcheting up tensions between Hutu and Tutsi. For many Tutsi youth common sense dictated that they might as well join the ranks of the FPR rather than be the innocent victims of Hutu extremists.

The second critical event took place in neighboring Burundi: the assassination of Melchior Ndadaye on October 21, 1993, had immediate and devastating repercussions in Rwanda, because the newly elected Hutu president of Burundi stood as the embodiment of the hopes of millions of Hutu (and not a few Tutsi) and because his assassination at the hands of an all-Tutsi army appeared to make one thing crystal clear: "You simply cannot trust the Tutsi!" Relayed by Radio Mille Collines, and projected in the most graphic and symbolically evocative fashion by the cartoonists of the newspaper *Kangura*, the message did not go unheeded. In the days following Ndadaye's death, a more virulent form of extremism emerged, in the form of Hutu Power. Widely used to designate a variety of extremist fringe groups, Hutu Power drove a deep wedge within the MDR, PSD, and PL leaderships, pitting extremists against moderates, and, in effect, greatly increasing the receptivity of the Hutu masses to the former's appeals.

The third, and cataclysmic, event was the shooting down of the presidential plane. Fear is a powerful tool of mobilization, and in the hours following the crash an intense, diffuse, uncontrollable fear suddenly seized the population. It was at this tipping moment that Hutu extremists proceeded to rationally manipulate the surge of irrational fears. Largely through the air waves of Radio Mille Collines diffuse anxieties were displaced on a substitute target: the Tutsi was the embodiment of evil and should be dealt with accordingly.

CRITICAL ISSUES

It is one thing to recognize the atrocities committed by the genocidaires, and quite another to explain their roots. What were the conditioning circumstances of the carnage, as distinct from the precipitating factor(s)? To what extent was the 1994 genocide inscribed in the structural characteristics of Rwanda's plural society? What is the part of responsibility assumed by the Belgian colonizer in "racializing" the Hutu-Tutsi divide? What is the significance of war and revolution as predisposing circumstances? What kind of logic, if any, can one detect in the behavior of the genocidaires? *These are crucial issues in any attempt to make sense of the senseless violence that swept across Rwanda in the spring of 1994.*

The case of Rwanda seems like a perfect illustration of genocide scholar Leo Kuper's (1981) argument that pluralism—meaning "divided societies, communally fragmented societies, multiethnic or multiple societies, segmented societies and internally colonized societies" (p. 57)—provides the necessary condition for domestic genocide. Few societies anywhere in Africa were as rigidly stratified as Rwanda before and during the colonial era. Nor did social exclusion disappear with the end of colonial rule. Independence profoundly altered the structure of society but it did not eliminate the domination of one ethnoclass over another; from a subordinate ethnoclass, the Hutu, in effect, became the dominant one. Clearly, the persistence of the plural nature of Rwanda society—notwithstanding major variations over time as to the identity of the top-dogs, and the severity of "differential incorporation"—must be seen as a major predisposing factor, and the fact that precolonial Rwanda experienced "ethnic cleansing" on a substantial scale (Vansina, 2001) would seem to add further weight to the Kuper argument.

While pluralism enhanced the vulnerability of Rwanda society to ethnic violence, some might argue with Mahmood Mamdani (2001) that the really critical factor was the transformation of an ethnic fault line into a racial divide under the impact of colonial rule. It is worth remembering in this connection that early European missionaries and administrators were almost unanimous in their endorsement of the Hamitic hypothesis: the notion that the Tutsi as Hamites belonged to a different "race," intrinsically superior to the Bantu, was rarely called into question. But is this a sufficient reason to argue that only racial divisions, as distinct from ethnic ones, correlate with genocide? What is beyond doubt is that colonial rule significantly increased the vulnerability of Rwanda society to genocidal violence by reinforcing inequalities between Hutu and Tutsi and thereby strengthening their social awareness as distinct groups. Furthermore, by assisting the Hutu elites in their revolutionary seizure of power, the departing colonizer unwittingly sowed the seeds of future genocide. One wonders, however, whether one can one seriously believe that, had Rwanda acceded to independence under Tutsi rule, ethnic violence on a genocidal scale could have been avoided.

War and revolution, as Robert Melson (1992) convincingly shows, are critically important factors in the dynamics of conflict leading to genocide. Revolution is a deadly serious business. Not only because of the human losses it involves, but because it seeks to redefine the boundaries of the community. The threat posed to Rwanda by the FPR was not just military, but social and political. It threatened to undo everything that had been accomplished since 1962 in the name of the "*revolution sociale.*" As the prospect of a return of Tutsi hegemony came to haunt the Hutu elites, the civil war instilled new fears in their minds that the resident Tutsi population might join hands with the invaders. The threat came from both the Tutsi "fifth column" inside the country, and from the FPR aggressors. In Nazi Germany, too, Jews were seen as a threat to non-Jews, but whereas Nazi suspicions of a Jewish plot clearly belonged to the realm of phantasms, in Rwanda such threats, at least as far as the external enemy was concerned, were only too real.

If so, there is reason to view the Rwanda genocide as a "retributive" genocide, to use Helen Fein's (1990) phrase; as such, it differs in its underlying motivations from "ideological" genocides, of which the Holocaust is a prime example. A retributive genocide occurs in reaction to a perceived threat, real or imaginary, rather than as a response to racist propaganda. This is not to deny the fact that anti-Tutsi propaganda played a major role in mobilizing the genocidaires, only to underscore the importance of the receptivity of the masses to such propaganda.

The image of the Tutsi as the embodiment of a mortal threat has deep roots in the history of the region. Long before the assassination of President Ndadaye, in 1993, Burundi had been the scene of a major genocide, when in 1972, in the wake of an abortive Hutu insurrection, anywhere from 100,000 to 200,000 Hutu men and children were massacred at the hands of the predominantly Tutsi army. Though almost forgotten, it is difficult to believe that this huge bloodletting did not have a crucial conditioning impact on Hutu attitudes. But this in itself does not explain the logic behind the 1994 Rwanda genocide.

If there was any rationality at work in the minds of the genocidaires, it is not to be found in the "Ordinary Men" argument set by Christopher Browning (1992) about the killing of Jews in occupied Poland: peer-pressure, careerism, self-serving ambitions were not significant factors in Rwanda. The logic behind the slaughter of Tutsi took on different dimensions depending on the circumstances. At the first, in the weeks following the RPF invasion, the killing of Tutsi in the north was intended to deprive the FPR of the potential support of Tutsi civilians and make the invaders aware of the price to be exacted by their recklessness. During the Arusha talks, a different logic emerged: planned and engineered by the CDR and the extremist groups in the MNRD, the massacre of Tutsi in the Gisenyi prefecture was meant to create bitter dissensions among the parties to the talks in hopes that the whole peace process might collapse. In the hours following the crash of the presidential plane, yet another set of motivations comes into view: the urge to kill stemmed from the straight rational choice proposition inscribed in the "security dilemma" created by the crash—"either we kill them first, or else we'll be killed." Thus framed, and forcefully articulated by the media, the logic of the security dilemma left no alternative but to annihilate the enemies of the nation.

CONCLUSION

Each genocide is different, yet the Rwanda genocide claims more distinctive characteristics than most. This is where the parallel with the Holocaust calls for considerable caution. For one thing, the Rwanda bloodletting is the epitome of a retributive genocide, occurring as a response to an invasion from a neighboring state. What gave anti-Tutsi propaganda its genocidal thrust was not the conjuring up of an imaginary Tutsi bogeyman, but the very real threat posed by the invading RPF. No other group was more cruelly aware of this threat than the million or so internally displaced persons (IDPs) who, by 1994, had fled the advance of Tutsi troops. Little wonder if the IDPs formed the bulk of the *interahamwe* militias. The regional dimension of the Rwanda tragedy also deserves special emphasis: its roots lie as much in the domestic arena as in the projection of external events into the collective consciousness of Hutu elites: the 1972 genocide of Hutu in Burundi, Ndadaye's assassination, the rising tension between Tutsi refugees and autochtons in eastern Congo and in Uganda, all of these and more were critical factors in shaping Hutu perceptions of a Tutsi threat. Hutu extremists played a crucial role in manipulating popular perceptions, but their task was greatly facilitated by the tendency of Hutu masses to heed binding orders from above. Perhaps more than in any other case of mass murder has this culturally engrained disposition contributed to the killings. Yet it is only fair to add that there were many exceptions to the rule. Countless examples have been cited to this writer of Hutu taking grave risks in trying to save the lives of their Tutsi neighbors. That these acts of individual courage are never mentioned in the media of postgenocide Rwanda is not an oversight; it is a commentary on the depth of the wounds left by one of the most appalling butcheries of the last century.

Recommended Reading on the Rwandan Genocide:

Des Forges, A. (1999). *Leave none to tell the story: Genocide in Rwanda*. New York and Paris: Human Rights Watch and International Federation of Human Rights.

This heavy tome (789 pages) is the definitive work on the Rwanda tragedy. It combines exhaustive field research with a wealth of hard to get documentary sources.

Jones, B. (2001). *Peace-making in Rwanda: The dynamics of failure*. Boulder, CO: Lynne Rienner.
An outstanding contribution to the complexities of the peace-process before and during the Arusha peace talks vis-à-vis the conflict in Rwanda.

Lemarchand, R. (2000, December). Disconnecting the threads: Rwanda and the Holocaust reconsidered. *The Journal of Genocide Research*, 4(4), 499-518.
A brief inquest into the logics of mass murder in Rwanda as distinct from the Holocaust.

Lemarchand, R. (1997). The Rwandan genocide. In S. Totten, W. Parsons, & I. W. Charny (Eds.), *Century of genocide: Eyewitness accounts and critical views* (pp. 408-423). New York: Garland.
Examines the causes and impact of the Rwandan genocide. Includes a set of first-person accounts by survivors of the genocide.

Prunier, G. (1995). *The Rwanda crisis: History of a genocide*. New York: Columbia University Press.
The most readable book on the Rwanda genocide, and by far the most informative on France's shameful role during the slaughter.

Smith, R. (Ed.). (1999). *Genocide: Essays toward understanding, early-warning and prevention*. Williamsburg, VA: Association of Genocide Scholars.
Includes four pieces on the Rwandan genocide: René Lemarchand's "Rwanda: The Rationality of Genocide"; Mark Levene's "Connecting Threads: Rwanda, the Holocaust, and the Pattern of Contemporary Genocide"; Howard Adelman's "Preventing Genocide: "The Case of Rwanda"; and Frank Chalk's "Radio Broadcasting in the Incitement and Interdiction of Gross Violations of Human Rights Including Genocide."

REFERENCES

Browning, C. (1992). *Ordinary men: Reserve Police Battalion 101 and the Final Solution in Poland*. New York: HarperCollins.

Fein, H. (1990, Spring). Genocide: A sociological perspective. *Current Sociology*, 38(1).

Kuper, L. (1981). *Genocide: Its political use in the twentieth century*. New Haven, CT: Yale University Press.

Mamdani, M. (2001). *When victims become killers: Colonialism, nativism and genocide in Rwanda.* Princeton, NJ: Princeton University Press.

Melson, R. (1992). *Revolution and genocide: On the origins of the Armenian genocide and the Holocaust.* Chicago, IL: University of Chicago Press.

Vansina, J. (2001). *Le Rwanda ancien: La monarchie Nyiginya* (Ancient Rwanda: The Nyiginya monarchy). Paris: Karthala.

CHAPTER 7

INSTRUCTIONAL STRATEGIES AND LEARNING ACTIVITIES

Teaching About Genocide

Samuel Totten

INTRODUCTION

Teaching about genocide is a complex and difficult task. Not only is the subject matter often horrific, but the issues are often extremely technical and/or thorny.

Different instructors, of course, approach the subject matter in different ways and use vastly different techniques and resources to assist their students to gain an understanding of key theories, concepts, issues, policies and actions. That said, the pedagogy used in such a study should be one that is student-centered—one in which the students are not passive but rather actively engaged in the study. It should be a study that, in the best sense of the word, "complicates" the students' thinking, engages them in critical and creative thought, and involves in-depth versus superficial coverage of information. It should also involve students in reading and examining primary sources (e.g., contemporaneous documents issued during the genocide under study, diaries and letters written by the

Teaching About Genocide: Issues, Approaches, and Resources, 215–237
Copyright © 2004 by Information Age Publishing
215

victims and others, trial records, etc.). Using such resources, in conjunction with first-person accounts by survivors and other witnesses, can help students examine the broader historical themes in powerful and personal ways. Any and all genocides are difficult to comprehend, therefore it is critical that educators choose resources that not only facilitate the achievement of course goals, but help students grapple with the choices, decisions, actions and inaction made by perpetrators, collaborators, bystanders, victims, and others. Secondary sources, as well, should not be overlooked—particularly those that were issued/published during the period of the genocide, such as newspaper articles and editorials, newsreels, and political cartoons. Certain feature films (such as "The Killing Fields" about the Cambodia genocide) and literature by the victims, survivors, and others (e.g., novels, shorts stories, poetry, drama) are also valuable sources for use in the classroom.

The rest of this chapter highlights some of the more powerful learning activities and teaching strategies that this author, his colleagues, and others have used in their courses on genocide.

STRATEGIES AND ACTIVITIES USEFUL FOR INTRODUCING A STUDY OF GENOCIDE

At the outset of any study, it is important to ascertain the students' knowledge base about the topic (e.g., what they know and do not know). To neglect to do so may result in a waste of time (e.g., possibly addressing information the students already know), not to mention potential frustration on both the students' part (e.g., their attempting to learn something that does not mesh with their present understanding) as well as the teacher's (e.g., not accomplishing one's goals and objectives). It is also important for students to have an opportunity to posit any questions or issues they want addressed during the course of the study.

Developing a Cluster/Mind-Map

An engaging and effective strategy for ascertaining the depth of student knowledge vis-à-vis any aspect of genocide (or for that matter, any topic) is to have them develop a cluster (alternatively referred to as a mind-map, web or conceptual map) around a "target" word or concept (e.g., "genocide") or event (e.g., a specific genocide such as the Armenian genocide). A cluster has been defined as "a nonlinear brainstorming process that generates ideas, images, and feelings around a stimulus word until a pattern becomes discernible" (Rico, 1987, p. 17).

To develop a cluster around the concept of genocide, for example, have students write the term "genocide" in the center of a piece of paper (a minimum of 8-1/2" by 11"), circle it, and then draw spokes out from the circle on which to place related terms or ideas. Each time a term is added, it should be circled and new spokes should be drawn out from it in order to delineate related terms and concepts. Each new or related idea may thus lead to a new clustering of ideas. As Rico (1987) points out: "A cluster is an expanding universe, and each word is a potential galaxy; each galaxy, in turn, may throw out its own universes. As students cluster around a stimulus word, the encircled words rapidly radiate outward until a sudden shift takes place, a sort of 'Aha!' that signals a sudden awareness of that tentative whole" (p. 17). Furthermore, "[s]ince a cluster draws on primary impressions—yet simultaneously on a sense of the overall design—clustering actually generates structure, shaping one thought into a starburst of other thoughts, each somehow related to the whole" (Rico, 1987, p. 18.)

Clustering is a more graphic, generally easier and more engaging way to delineate what one knows about a topic than by outlining it. That said, some students may prefer to use outlining rather than clustering, and that option should be available to them.[1]

To help students understand both the purpose for and the method of clustering, the instructor should choose a topic and demonstrate the development of a cluster, progressing from simple to more complex stages (see Figure 7.1). More specifically, the teacher should first create a simple, almost perfunctory, cluster and then a second, more complex,

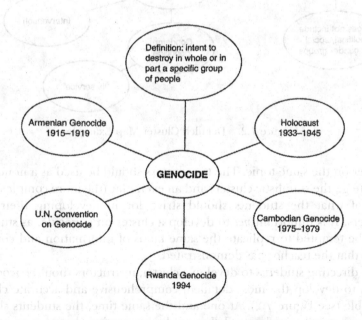

Figure 7.1. Simple Cluster Map Example

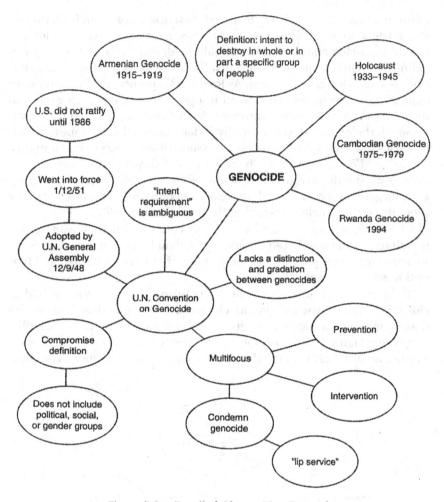

Figure 7.2.   Detailed Cluster Map Example

cluster on the same topic. The two clusters should be used as a nonexemplar (e.g., the simplistic cluster) and an exemplar (the more complex cluster) of what the students should strive for in developing their own clusters. It is important *not* to develop a cluster on genocide, as students may be tempted to replicate the same kinds of information and connections that the teacher has demonstrated.

In directing students to develop a cluster, instructors should encourage them to develop the most detailed, comprehensive and accurate cluster possible (see Figure 7.2). At one and the same time, the students should be encouraged to strive to delineate the connections between or among key items, concepts, events, and ideas. If such directions are not given *and emphasized*, then many students are likely to develop very simple and perfunctory, if not simplistic, clusters.

Once each student has completed a cluster, groups comprised of three to four students should meet in order to share and discuss their individual clusters. Each student should be given time (two to three minutes) to explain his/her cluster by noting the key points, issues, and connections. Instructors will need to emphasize that the students need not—and, indeed, should not—go over each and every word/phrase on his/her cluster as that would be extremely time-consuming and boring for the other group members to sit through.

As individual students present their clusters, others in the group may add items to their own clusters—in a color other than the one used originally in order to indicate the number and type of ideas gleaned/borrowed from their peers. At the end of this session, all of the clusters may be taped to the classroom wall or stored for revisiting at various points during the course of study.

Developing clusters serves a number of key purposes. First, it may assist a student to recognize what he/she does and does not know about the subject. Second, the instructor gains a vivid illustration of a student's depth of knowledge (or lack thereof) of the subject, as well as the sophistication of his/her conceptual framework. Third, the instructor is able to pinpoint specific inaccuracies, misconceptions, and/or myths that students hold about the concept of genocide and/or the facts of a specific genocide. In other words, the clustering activity can and does serve as a powerful preassessment exercise.

Clustering also provides students with a unique method to express their ideas; and in doing so, it allows them to tap into an "intelligence" (e.g., spatial) other than the typical one of writing ("linguistic").

At the conclusion of the study, students could be asked to complete another cluster, and required to compare and contrast, in writing, the information in each. This, of course, serves as a powerful way of conducting a postassessment exercise.

Using the Cluster to Develop a Working Definition of "Genocide"

Next, using the information (facts, concepts, connections) they have included in their clusters, students can develop a working definition of the term/concept "genocide." The students should be instructed to look carefully at all of the components of their cluster and then make every effort to develop the most comprehensive and accurate definition they possibly can based on the information they have delineated in their cluster. The students should also be informed that if they discover, as they develop their definition, they have left out key facts, concepts or connections in their clusters, then they should add such information to their clusters.

Once everyone has developed a definition, students should be placed in groups of three to four (either the original or new groups) to share their definitions. A recorder in each group should take down the salient points of the discussion that ensues. At the conclusion of the small group discussions, a general class discussion can be held during which any questions or points the students still have can be placed on a large sheet of paper with the heading "Genocide: Issues to Resolve and/or Examine in More Detail." As the class proceeds with its study it can return to such questions and concerns and attempt to answer them.

Again, just as the clusters do, the development of student definitions of genocide provides the instructor with valuable insights into his/her students' basic understanding of the concept of genocide (or the genocidal event under study), including their depth of knowledge, misconceptions, and so forth. This, of course, serves as another powerful preassessment exercise.

Positing Burning Questions About the Concept of Genocide or the Genocidal Event Under Study

Another powerful strategy is to have students write down (anonymously) three to five "burning questions" they have about the concept of genocide or the particular genocide under study. They should be informed that their questions can be about any and all facets of the concept of genocide or the genocidal event, and that throughout the study a real attempt will be made to locate answers to the questions.

Soliciting students' questions and concerns helps to make the study of genocide more focused and personal. Moreover, it encourages students to become active researchers versus passive participants in a class. By raising their own questions, students can actively join in in seeking answers to their questions.

Among some of the many questions (all quoted exactly as they were written) that my students have posited over the years are as follows:

Questions Posited in Regard to the Definition of Genocide
- What is the exact difference between a massacre and genocide?
- What is the point of the term genocide? Why not just use mass killing?
- Who coined the term genocide, and when and why?

Questions Posited in Regard to the Perpetration of Genocide
- When do historians think the first genocide was committed? By whom and why?
- What are the main causes of genocide?
- Who (on the outside) decides when a genocide is taking place and how?
- Has a genocide ever been prevented?
- Other than outright war is there any way to stave off genocide?

Questions Posited in Regard to The Holocaust
- Why exactly did the Holocaust happen?
- What was the main factor that caused this genocide?
- Why were Jews blamed in the first place?
- How did Hitler get into office?
- Why did Hitler have so much power?
- What was Hitler's main motive for killing all the Jews?
- Why didn't all the people in the camps rebel at once, so that their (sic) may have been a (sic) hope for freedom?
- Was there any dissention (sic) among the German ranks regarding following orders which lead to the mass extermination of Jews?
- Why didn't the United States try and step in sooner?
- What are the warning signs of something like the Holocaust, and how can we stop it from happening again?
- Why???

Questions Posited in Regard to the Rwandan Genocide
- Where was the rest of the world?
- Why didn't the United States step in to stop it?
- Why was the genocide committed?
- What has happened to the murderers?

Even a "misinformed" question is valuable as it provides the instructor with additional insights to his/her students' knowledge base.

At the conclusion of the study, it is wise to ascertain whether all of the students have had their questions answered; and if they have not, the instructor should try to set aside time to address them with the class.

Three Unique Strategies for Introducing a Course of Genocide: A Mathematical Approach, A Philosophical Approach, An Existential/ Autobiographical Approach

Herewith are three strategies that Dr. Henry Huttenbach, Professor of History at the City University of New York, has devised on how to introduce a course on genocide.

- *The Arithmetical Approach or The Mathematics of Genocide*:   This approach is designed to concretize mass killing. It consists of several exercises: (a) one and a half million Armenians died between 1915-1916; six million Jews died between 1941-1945, and half a million Roma in the same time span; 800,000 Tutsis in four months in 1994. What is the killing rate per day, per week, per month, per year for each group? (b) If a gas chamber holds 150 people and it takes one hour to load and empty, how long will it take to kill one million Jews, who perished in Auschwitz as against, a gas-truck which holds 50 persons and requires 35 minutes to kill them? (c) In Riga, in a 12-hour shift, 14,000 Jews were shot. What is the average rate of killing per hour, per minute? (d) One and half million Jewish children were killed: how many classrooms of 25 students does this come to?

- *Philosophical Approach*:   After showing extracts from *The Triumph of the Will* and newsreels of the British liberation of Bergen-Belsen, the question is posed; "What did you see?" "What is the relationship of the Hitlerian 1935 utopian vision of a race-perfect society to the piles of bodies filmed in 1945?" "What lessons do you draw from this?"

- *The Existential Autobiographical Approach*: I announce I should not be. I was labeled for death. I then show a family tree and illustrate how my life/death existentially affects my five children, 12 grandchildren and two great grandchildren. All of them would not be if certain circumstances had not saved me. Was it luck? Chance? Clever thinking by my parents? God's protecting me? and so forth (H. Huttenbach, personal communication, April 10, 2003).

STRATEGIES AND ACTIVITIES IDEAL FOR USE THROUGHOUT THE STUDY OF GENOCIDE

A Method to Assist Students to Develop a Deep Understanding of the Definition of Genocide

Clive Foss, Professor of History at the University of Massachusetts, Boston, engages his students in the following activity in order to assist them to gain a solid understanding of the definition of genocide:

I ask my students at the beginning of the seminar (on genocide) what they think genocide is. They think all sorts of things. Then I ask them to read the press continually and to bring in all examples of the term they can find. The results are amazing: Genocide is used to mean almost anything (usually something the writer doesn't like) from the mistreatment of cats and dogs to any number of real and imagined criminal activities.

Throughout the course I ask them to work out a definition on the basis of case studies.... [In relation to each case of genocide, the following questions are asked: Who did it? Why did they do it? How did they do it? Who were the victims? How did they react? What was the result? In each case, too, the central question—"Was this a genocide?"—will be asked, so that a suitable definition can be worked out during the course of the discussion].

At the end, I ask them to write a definition on the basis of what they have learned. (Foss, 1992a/b, pp. 3, 27)

A Comparative Study of Genocide

A comparative study of genocide provides students with an opportunity to examine the similarities and differences between different cases of genocide. During the course of such a study, students should, ideally, glean unique insights into a wealth of issues, including but not limited to the following: the preconditions of genocide; the motives and reasons for the genocide (in other words, the "why" behind the policies and actions); and the different types (or typologies) of genocide. Ultimately, such a study should assists students to glean insights into "the patterns and causes of genocide" (Fein, 1992, p. 11).[2]

Processing Information, Concepts, Ideas, and Thorny Issues

Eric Markusen, Professor of Sociology at Southwest State University in Marshall, Minnesota, believes, and rightly so, that it is imperative to provide ample time for students in classes on genocide and human rights to engage in small group discussions. More specifically,

Given the emotionally as well as intellectually challenging subject matter of such courses, students need time to process what they're confronting in readings and lectures. My class meets once a week for two and a half hours, and I generally have between three and five discussion sessions during the semester.

Generally, there are about 20 to 25 students in the class. I randomly divide them into groups of about eight students. Students remain in the same group for the duration of the term, thus allowing them to get to know each other [fairly well]. For each session, one student serves as "recorder" and lists the names of students who participate and records responses to the question under discussion.

All groups are given the same question. Examples of the kinds of questions I ask are: "If you were to return to your high school to give a talk to the senior social studies class about what you have learned about genocide [or on a particular genocide], what would be the five most important facts/issues that you would focus on, and why?"; "What are the five most surprising aspects you have learned about genocide [or a particular genocide] thus far in our class, and why?; "Identify at least five reasons why you feel it is important to learn about genocide"; and "Based on what you have learned in class, identify five approaches to reducing the threat of future genocides."

After the groups have had about a half hour to discuss the question and list their responses, all the groups come together and the recorder writes each group's list on the board. This enables each group to see what the other groups have come up with. I then lead the discussion by examining the points listed by all of the groups, broaching other relevant points and concerns, and probing students for the reasons for their choices. The entire exercise takes about 50 minutes. (E. Markusen, personal communication, April 5, 2003)

A Microcosmic Case Study-within-a-Case Study

Paul Bartrop, a historian based in Melbourne, Australia, notes that his

approach to teaching about the Bosnian war and the accompanying genocide of 1992-1995 begins with a chronological outline of the events leading up to Bosnia's secession from the Yugoslav federation in April 1992. That is followed by a discussion of the complex, and often confusing, ethnic and religious environment, along with the principal players. The latter, I have found, must also be thoroughly explained and understood before attempting to move on to any meaningful analysis. Once the students and I are speaking the same contextual language, we delve into such matters as expulsion, rape warfare, killing, concentration camps, and the notion of "ethnic cleansing."

It is via the case study approach, however, that the true drama and horror of the war and the ethnic cleansing and genocide come through, for both students and teacher. Accordingly, I make use of a video entitled *Srebrenica: A Cry from the Grave* (Woodhead, 2000). With accounts related in a very personal way by many of the participants (though not from the Serbs' side), and interspersed with contemporary news footage and film taken of the most intimate—and heartrending—negotiations, the film relates the story of the nine days during which the city fell to the Bosnian Serbs when the Dutch U.N. peacekeeping force stood aside and allowed the ethnic cleansing to take place. Without a single exception during the years I have shown this video, students have been thoroughly transfixed by the drama unfolding before them.

The real work for the teacher begins, of course, once the viewing has finished. My educational approach is largely one of wide-ranging discussion based on a full appreciation of historical detail. I supplement the students' viewing with readings, lectures (and hopefully, a corresponding degree of note taking!), and the constant revisiting of key points via short quizzes.

Broader assessment takes the form of an essay in which students can choose one of the major players in the Srebrenica story—for example, the United Nations, the Dutch peacekeepers, the Bosnian Serb forces, the Bosnian Muslim population—and engage in detailed research on the fall of the town and the subsequent massacre from the perspective of the group or organization they have chosen. By the time of the final examination, there is barely a stone that has been left unturned on the topic of Srebrenica specifically, and the Bosnian conflict more generally.

Why do I light upon this classroom topic in particular? Many of us use videos in our teaching, and a single documentary about Srebrenica is far from the only (or last) word on the war in Bosnia. Yet I find something particularly alluring about the use of a microcosmic case study-within-a-case study; it not only reduces the great blooming, buzzing confusion of a near-contemporary event to manageable proportions, it also serves to humanize the conflict by introducing a localized dimension which students can understand and with which they can often readily identify. The viewing of one video does not a genocide scholar make, to be sure; but by utilizing the video as the centerpiece of a wider topic of study in which an array of educational devices is employed, students can tease out an array of issues and truly begin to "own" the learning process. (P. Bartrop, personal communication, April 14, 2003)

Homework Assignments and Research Projects/Papers

Succinctly highlighted below are several homework assignments, research projects and papers that various educators have assigned or required in their courses on genocide. Two excellent sources for additional information along this line are the two books of course syllabi that Joyce Apsel and Helen Fein edited in 1992 and 2002, respectively. The complete bibliographical information for both books is located in the references section of this chapter and a description of each is included in the annotated bibliography at the end of this book.

Homework Assignments:
- "Each student should prepare a typed journal entry of at least 1,000 words for weeks 2 through 10. It should contain her/his reactions to and reflections on the readings and topics of discussion for the previous week…. Students are free and encouraged to balance scholarly reflection with personal/emotional, ethical, and political reflections" (Theriault, 2002, p. 23).
- "Please answer these questions [which are listed on the class schedule and unique and germane to the week's reading, lecture and discussion] as you read each weekly assignment and come to class prepared to discuss your answer to them and to present your own

analyses ... [Example from Week One, which is entitled Part One: The Conceptual Framework]: 1. Why is it important to have a precise and rigorous definition of genocide?; 2. What is a typology?; ... 4. What were the major components of Raphael Lemkin's definition of genocide?; 5. How did the General Assembly of the United Nations characterize genocide in 1946? How did the General Assembly modify its earlier definition of genocide in 1948 when it adopted the United Nations Genocide Convention" (Chalk, 2002, p. 59).

- "A two to three page written critique of the readings [is due each week. It need not] be a long and detailed summary; a brief review of the major strengths and weaknesses of the readings [shall suffice].... These critiques will provide the basis for much classroom discussion.... At the end of the semester, I will review all the critiques in order to determine a grade for them" (Alvarez, 2002, p. 173).

Research Papers/Projects

- "Each student will research and write a paper on some aspect of genocide such as: comparative approaches to the study of genocide; analyses of causes and types of genocide—religious, colonization, decolonization, ideological; the aftermath and consequences of genocide in general or a particular genocide, including the problem of denial; attempts to prevent the crimes of genocide; and proposals relating to early warning-signals and intervention. The papers should be approximately 25-30 pages" (Hovannisian, 1992, p. 40).

- "The research paper in this course will be based on your examination and analysis of original sources.... This assignment should contribute to your ability to find evidence, to evaluate sources, to assess contradictory evidence, and to write a coherent presentation of your findings accompanied by your reasoned conclusions, with emphasis on your argument in favor of those conclusions" (Chalk & Jonassohn, 1992, p. 50).

- "I [William Fernekes] integrate the issues of genocide and massive human rights violations into U. S. History classes by helping the students examine how the world responded to genocide, specifically national governments and international organizations such as the United Nations. The central question posed is: How have and how should individual governments and international organizations respond[ed] to the problem of genocide?

  "I introduce the unit by utilizing an activity I designed dealing with multiple definitions of genocide ('Defining Genocide: A

Model Unit,' *Social Education*, 1991). After students have grappled with the difficulty of establishing a single, all-encompassing definition of genocide, we examine specific cases of genocide where national governments planned and implemented genocides, paying particular attention to the stages of implementation. During the Armenian genocide, the Holocaust and the Khmer Rouge-perpetrated genocide in Cambodia, each national government in its role as perpetrator identified their targeted group(s), isolated them, removed their material basis for existence, and then sought to annihilate them. Such stages are detailed quite clearly in Hilberg's monumental study of the Holocaust, *The Destruction of the European Jews*; and on a broad scale, they are equally applicable to each case noted above. Incorporated into our study is content about the response(s) of targeted victim groups, relying heavily on survivor testimonies, eyewitness accounts by observers, and survivor memoirs. I believe that it is critical to provide students with diverse sources, not only from the victims, but from the perpetrators and bystanders as well, as this places the complexity of each case in high profile.

"Students regularly ask 'so what did the U.S./world do' about each case? Rather than give them the answer, I provide them with materials that illustrate the policies of governments and organizations that responded to each genocidal case, either aggressively (very few) or timidly (most). For example, we employ the Harbord Report, written by a U.S. military expert about whether the U.S. should accept a League of Nations mandate for Armenia in the post World War I era, along with dispatches by Ambassador Morgenthau and news reports from U.S. newspapers that detailed the suffering of the Armenians during the genocide. Regarding the Holocaust, the international response to the refugee crisis of the 1930s and the failure of the U.S. and other Allied powers to intervene and attempt to rescue Jews in Occupied Europe can be, given the huge amount of material available, thoroughly studied. As for Cambodia, the growing literature on that genocide, both in survivor accounts and in the study of the almost nonexistent world response to the killings of the Khmer Rouge, provides ample resource material.

"The culminating assessment for the unit is posed as a scenario where each student is invited to write a memo recommending how the U.S. government should respond to a potential genocide that may be imminent in one area of the world. Specific conditions evident in the three cases prior to the onset of mass killing are included in the description of the potential genocide, and the student is asked to recommend how the U.S. government should respond—alone, or in concert with other world states and international organizations. The student must draw upon historical examples and frame a recommendation that reflects their understanding of U.S. foreign policy priorities during the twentieth century (for example, the tension between isolation and internationalism) and the continuing problem of genocide and massive human rights violations (W. Fernekes, personal communication, April 15, 2003).

STRATEGIES AND ACTIVITIES FOR CLOSING THE UNIT OF STUDY

All aspects of a study of genocide or a genocidal act should be carefully crafted to encourage, guide, and even prod students to think about the concept and/or history in an in-depth and reflective manner. Nothing in such a study should be perfunctory. This is as true for the close of a lesson or unit as it is for the introduction or the body of the lesson or unit.

In closing lessons and units on genocide, some teachers may be satisfied to conclude the study by giving a traditional quiz or examination. Examinations can, of course, be designed in a way that challenge students to truly synthesize as well as ponder and wrestle with what they have learned. However, as Wiggins (1989) notes and then asks: "[C]ompetence [or mastery of a subject] can be shown in various, sometime idiosyncratic ways. Why must all students show what they know and can do in the same standardized way?" (p. 48). It seems as if this is a question that all educators should contemplate.

There are, in fact, numerous concluding activities that can be used to complete a study of genocide. Some are ideal for use prior to and/or in conjunction with traditional or authentic assessments, while others are capable of standing on their own. Instructors need to use their own judgment in selecting the type of closing activities that will guide their students in accomplishing one or more of the following three goals: (1) synthesize what they have learned, (2) reflect on what they have learned, especially as it pertains to their own lives and the world in which they reside, and (3) plant seeds for on-going rumination about their newfound knowledge.

Various closing activities that have been effective in completing a study of genocide are discussed below.

Prelude to the Conclusion of a Study of Genocide

An engaging and valuable way to set the stage for a final discussion of genocide (which could last one or more class sessions) and/or student-designed extension projects is to have the students address any and/or all of the following issues in writing: Is there anything you are still perplexed about in regard to the specific aspect of genocide we have studied; and if so, what is it and why? What issues and concerns still elude you in your effort to gain a clear understanding of the whats, whys, hows, wheres, and whens of the genocide under study? What concepts, issues, events, and concerns do you still feel the need to learn more about, and why? What are the most significant insights, concepts or pieces of information you have gleaned from your study of genocide and why?

Closing Discussion Based on Probing/Philosophical Questions

In addition to addressing the questions broached in the "Prelude to the Conclusion of a Study," a final discussion around such questions as the following can be useful: (1) Can any lessons at all be learned from the genocide we have studied? If so, what are they and why? If not, why not? (2) Can it be said that the history of humanity has been a history of progress in human relations? Why or why not? (3) Why should those far from (via great expanses of time, land, ocean, and/or culture) the perpetration of genocide even care about genocide? (4) Now that you have a deeper understanding of genocide, do you think or feel you have a responsibility to be more aware and/or concerned about human rights abuses and genocide perpetrated in your own life time? Why or why not? If you do, what will that look like in regard to your own actions? These and other open-ended questions can serve as a means for the students to reflect on what they have learned from their study of genocide, as well as how the study has impacted them as human beings.

Having Students Note What They *Never* Want to Forget About a Particular Genocide

A simple but powerful closing activity is to have the students, individually, write down those facts, concepts, events, issues, and images that they *never want to forget* in regard to what they have learned during the course of their study of genocide. Taking part in such an activity prods students to articulate that which is most meaningful to them as a result of their study. It also has the potential of planting the seeds for on-going concern about the ubiquitous deprivation of human rights across the globe.

This activity also provides the instructor with a sense of those facts, ideas, concepts, issues, events, discussions, films, and so forth, that most powerfully impacted students. Such information, obviously, can provide invaluable insights for him or her as he/she revises and hones the focus of the lessons, unit, or course.

A Letter Assignment

A thought-provoking exercise involves each student in writing a letter to a larger audience about what he/she has gleaned as a result of studying genocide. At the outset of this activity, the teacher might wish to share the following letter with his/her students:

> Dear Teacher, I am a survivor of a concentration camp. My eyes saw what no man should witness. Gas chambers built by learned engineers. Children poisoned by educated physicians. Infants killed by trained nurses. Women and babies shot and burned by high school and college graduates. So, I am suspicious of education. My request is: Help your students become human. Your efforts must never produce learned monsters, skilled psychopaths, educated Eichmanns. Reading, writing, arithmetic are important only if they serve to make our children more humane.

Once the above letter has been read and discussed, students should reflect on what *they* ardently wish to share with others in regard to what they gleaned from their study. After reflecting on what they have learned, the students should jot down key words, phrases, and thoughts that come to mind. Next, students should pair up and discuss their insights. Then, each student should write a letter to whomever they want (a letter to the editor of a local, regional, or national newspaper; the congregation of a church, synagogue, or mosque; the local school board; their parents; students in other classes within the school) in which they succinctly but powerfully convey thoughts, ideas and/or warnings that they would like others to ponder and heed.[3]

Developing an Encyclopedia Entry/Article

An excellent synthesizing activity is to form groups of two to three students for the purpose of having them develop an encyclopedia-like entry/

article that thoroughly and accurately summarizes what they have learned about genocide. Prior to assigning the task, the instructor should select and copy a rather lengthy entry from a respected academic encyclopedia (that is, one that addresses a single subject such as the Holocaust), and assist the students in examining how a solid encyclopedic entry is constructed. During the examination of the entry, the instructor should also direct the students' attention to the following: the succinctness of the writing; how the entry/article is comprised of key—*versus* superfluous—information; how the role of key personages is delineated; and how the chronology of events is interwoven into the fabric of the entry/article.[4]

After the students complete their entries, any number of things could be done with them: they could be exchanged with other groups for the purpose of having them critiqued, and subsequently the critiques could be used for the purposes of revision; the instructor could read each, make suggestions for revisions, and upon revision, each could be placed in a booklet for use by future classes; or the pieces generated by the students could be compared and contrasted with entries on the same topic found in various encyclopedias.

Addressing a Key Quote for the Purpose of Synthesizing *and* Reflecting on One's New Knowledge of the Holocaust

Both Steve Feinberg, a noted Holocaust educator, and this author are strong advocates of concluding exercises that are reflective in nature. Two ways in which Feinberg has prompted students to both synthesize their new knowledge about the Holocaust and to reflect on its meaning for them is through reflective journal entries and final essays. More specifically, he states that:

Concluding activities should encourage students to reflect upon the history and/or literature studied in the unit. Students need to be encouraged to combine the various elements of their study of the Holocaust into a coherent whole. Activities that can assist students in this synthesis of information is a reflective journal-writing assignment or a reflective essay. This activity permits students to blend and unify their thoughts about this particular history into an integrated whole.

Providing students with an evocative quote and asking them to respond to the quote is a solid way to accomplish the above. For example, Gerda Lerner has said that "It is not the function of history to drum ethical lessons into our brains. The only thing one can learn from history is that actions have consequences and that certain choices once made are irretrievable" (quoted in Lerner, 1997). Students can be asked to respond to this quote (or others like it) in either essay or journal form, using the information they have examined in their Holocaust unit. Hopefully, the general historical nature of such quotes will serve as a catalyst for thinking reflectively about the history of the Holocaust. (S. Feinberg, personal communication, March, 2000)

Again, *with some adaptation*, this is an ideal assignment/task for the conclusion of any study of genocide.

Student-Developed Questions for the Final Examination

For those teachers who wish to use a final essay examination at the completion of a study of genocide, an engaging activity is to have each student develop a minimum of two essay questions, with the understanding that *they might be used on the final exam*. Students need to be given directions as to what constitutes a sound question: a question that truly addresses what the class has studied and not something so far afield that a fellow student would not have the knowledge base to answer the question; one that is not so narrow or so broad that an individual would have a torturous time addressing it; one that is thought-provoking and requires the writer to bring to bear both a broad and deep knowledge base about the subject; a question that does not call for rote recitation of facts but an analysis and/or synthesis of facts, concepts, ideas, issues; and one that is crystal clear in regard to what the respondent needs to address in his/her paper. *If such criteria is not provided then many students are likely to write questions that are of little value or use.*

Students should be informed that the instructor will select those questions that are well written, comprehensive, and most thought-provoking. They should also be told that the instructor reserves the right to revise and/or combine questions for use on the exam. When selecting the final set of questions (a total of six to eight from which the students select a single question to answer), an instructor needs to make sure that he/she includes a wide variety of questions so that students have ample choice in regard to what they choose to write about.

Having the students design their own questions is, in and of itself, an excellent synthesizing activity. The act of creating solid questions forces students to wrestle with what they have studied. In order to emphasize the seriousness of developing these questions, a grade can be given for the students' efforts.

Designing such questions also provides students with an opportunity to have a real say about the exam. Most students find this refreshing.

Finally, nothing, of course, precludes a student from answering his/her own question on the exam. This, too, is often enticing to students.

A Challenging Essay Examination

Well thought-out and carefully crafted, open-book essay examinations (whether take-home or in-class) are also an excellent means for students to synthesize their new-found knowledge. William Fernekes, a high school social studies supervisor and teacher at Hunterdon Central High School in New Jersey, finds that such essays are extremely useful for assessing his students' knowledge in an elective course entitled "The Holocaust and Human Behavior." Speaking of such an exam, he states that

> Students are assigned a take-home essay constituting 50% of their final exam grade in the course. The essay topic integrates learning from the entire course while permitting flexibility in the choice of sources to support the student's arguments. A critical requirement for the essay is the application of key course concepts regarding human behavior: prejudice, stereotyping, discrimination, in group/out group relationships, psychological distancing and compensating behaviors, and the creation of "the other" (dehumanization). Students must employ evidence from their two outside readings (survivor memoirs) as well as a selection of three or more additional eyewitness accounts by participants in the Holocaust (perpetrators, victims, rescuers, or bystanders).

The following question is one that Fernekes has required his students to answer:

> The Holocaust can be viewed as the outgrowth of choices made by individuals and groups in a wide variety of situations. Drawing primarily upon personal eyewitness accounts, explain what you consider to be (1) the key factors which significantly influenced the choices made by perpetrators and victims groups, and (2) the most important insight for understanding human behavior in today's world based upon your analysis of these factors and the choices that were made. (W. Ferenekes, personal communication, October, 2002).

Yet again, such an assignment/task could be adapted for use with any study of genocide.

Performance-based Assessment

Noted educator Theodore Sizer observes that performances and exhibitions "serve at once as evaluative agent and expressive tool: We expect people to show us and explain to us how they use content—it's more than mere memory. It's the first real step towards coming up with some ideas of their own.... In its original form, the exhibition was the public expression by a student of real command over what she'd learned" (Coalition of Essential Schools, 1990, p. 1). And, as Grant Wiggins (1989) has pointed out in a thought-provoking article entitled "The Futility of Trying to Teach Everything of Importance," a good way of assisting students to ultimately develop such demonstrations of mastery is to frame the study of a subject and the assessment of the study around key or "essential" questions such as: "'What must my students actually demonstrate to reveal whether they have a thoughtful as opposed to thoughtless grasp of the essentials?' and 'What will 'successful' student understanding (with limited experience and background) actually look like?'" (p. 208).[5]

Such performances or exhibitions may take many different forms (e.g., an individual project, a group project, a preparation of a portfolio, or an oral presentation and "defense"). No matter what form it takes, though, "the performance must engage the student in real intellectual work, not just memorization or recall. The 'content' the students master in the process is the means to an end, not the end itself" (Coalition of Essentials Schools, 1990, pp. 3-4).

Even an abbreviated list of some of the Coalition's "qualities of 'authentic performances'" provides a good sense of the rigor factored into well-structured performances and exhibitions of mastery. More specifically, each should: "require some collaboration with others; [be] constructed to point the student toward more sophisticated use of his/her skills or knowledge; [consist of] contextualized, complex intellectual challenges, not 'atomized' tasks corresponding to isolated 'outcomes'; involve the student's own research or use of knowledge, for which 'content' is a means; assess student habits and repertoires, not mere recall or plug-in skills; [constitute] representative challenges—designed to emphasize depth more than breadth; [be] engaging and educational; involve criteria that assesses essentials, not easily counted but relatively unimportant errors; [be] graded not on a curve but in reference to performance standards (criterion-referenced, not norm-referenced); make self-assessment a part of the assessment; use a multifaceted scoring system instead of one aggregate grade; ferret out and identify (perhaps hidden) strengths; minimize needless, unfair, and demoralizing comparisons; allow appropriate room for student learning styles, aptitudes, and interests; [be] attempted by all students, with the test 'scaffolded up,' not 'dumbed down,' as necessary; and reverse typical test-design procedures. A model task is first specified; then, a fair and reliable plan for scoring is devised" (Coalition of Essential Schools, 1990, p. 2).

CONCLUSION

Again, the pedagogical strategies and learning activities described in this chapter are only a few of the many ways in which an instructor can effectively engage his or her students in rich and meaningful interactions during a study of genocide. Ultimately, it is hoped that through the use of one or more of them, *students will gain valuable insights into the issue of genocide and glean powerful insights into the human condition—and themselves*. If they do, then such a unit or course will be worth all of the time, effort, and stress involved in teaching and studying such complex concepts and horrific events.

NOTES

1. For excellent and thought-provoking discussions by classroom teachers regarding the clustering method, see Olson (1987).

2. For a detailed discussion as to how to undertake a comparative study of genocide, see Chapter 8 (this volume), "Conducting a Comparative Study of Genocide: Rationale and Methodology," by Henry Huttenbach.

3. The "danger" in sharing the above letter with the students is that they may latch on to the format and thoughts presented therein, and then simply present a rough facsimile as their own. Thus, instructors who share the letter with their students must urge them to create their own format and incorporate their own thoughts and voice into their respective letters.

4. *It is important that the students have a rubric as a guide of sorts for developing their entry.* In developing the rubric, the following should be taken into consideration: the historical trends that conjoined and resulted in the genocide; the chronology of the genocide; the various groups involved and/or impacted by the events, including: the victims, perpetrators, collaborators and bystanders; the different stages of the genocidal period; the world's response to the mass killings; and the aftermath of the genocide. Ultimately, each and every entry needs to address the "whys, hows, whens, wheres, and whos" of the genocide, otherwise the entry is bound to be incomplete and inaccurate.

5. Teachers who are interested in developing a project- or performance-based assessment may wish to contact The Coalition of Essential Schools project at Brown University, Box 1938, Providence, Rhode Island 02912, and request their materials on "demonstrations (performances and exhibitions) of mastery." The coalition has developed numerous outstanding and detailed models for the development of rigorous performances that, when implemented with care and thought, can truly tap students' critical and creative faculties.

REFERENCES

Alvarez, A. (2002). Genocide, war crimes, and human rights violations. In J. Apsel & H. Fein (Eds.), *Teaching about genocide: An interdisciplinary guidebook with syllabi for college and university teachers* (2nd ed., pp. 172-175). Washington, DC: American Sociological Association.

Apsel, J., & Fein, H. (Eds.). (2002). *Teaching about genocide: An interdisciplinary guidebook with syllabi for college and university teachers* (2nd ed.). Washington, DC: American Sociological Association.

Chalk, F. (2002). The history and sociology of genocide from ancient times to 1920. In J. Apsel & H. Fein (Eds.), *Teaching about genocide: An interdisciplinary guidebook with syllabi for college and university teachers* (2nd ed., pp. 55-109). Washington, DC: American Sociological Association.

Chalk, F., & Jonassohn, K. (1992). The history and sociology of genocide. In J. Freedman-Apsel & H. Fein (Eds.), *Teaching about genocide: A guidebook for college and university teachers: Critical essays, syllabi and assignments* (pp. 45-53). New York: The Institute for the Study of Genocide.

Coalition of Essential Schools. (1990, March). Performances and exhibitions: The demonstration of mastery. *Horace, 6*(3), 1-12.

Fein, H. (1992). Teaching about genocide in an age of genocides. In J. Freedman-Apsel & H. Fein (Eds.), *Teaching about genocide: A guidebook for college and university teachers: Critical essays, syllabi and assignments* (pp. 9-11). New York: Institute for the Study of Genocide.

Foss, C. (1992a). Introduction. In J. Freedman-Apsel & H. Fein (Eds.), *Teaching about genocide: A guidebook for college and university teachers: Critical essays, syllabi and assignments* (pp. 1-5). New York: The Institute for the Study of Genocide.

Foss, C. (1992b). Genocide in history. In J. Freedman-Apsel & H. Fein (Eds.) *Teaching about genocide: A guidebook for college and university teachers: Critical essays, syllabi and assignments* (pp. 27-30). New York: Institute for the Study of Genocide.

Freedman-Apsel, J., & Fein, H. (Eds.). (1992). *Teaching about genocide: A guidebook for college and university teachers: Critical essays, syllabi and assignments.* New York: The Institute for the Study of Genocide.

Hovannisian, R. (1992). Comparative study of genocide. In J. Freedman-Apsel & H. Fein (Eds.), *Teaching about genocide: A guidebook for college and university teachers: Critical essays, syllabi and assignments* (pp. 40-44). New York: The Institute for the Study of Genocide.

Lerner, G. (1997). *Why history matters: Life and thought.* New York: Oxford University Press.

Olson, C. B. (1987). *Practical ideas for teaching writing as a process.* Sacramento: California State Department of Education.

Rico, G. (1987). Clustering: A prewriting process. In C. B. Olson (Ed.), *Practical ideas for teaching writing as a process* (pp. 17-20). Sacramento: California State Department of Education.

Theriault, H. (2002). The Armenian genocide. In J. Apsel & H. Fein (Eds.), *Teaching about genocide: An interdisciplinary guidebook with syllabi for college and univer-*

sity teachers (2nd ed., pp. 23-27). Washington, DC: American Sociological Association.

Wiggins, G. (1989). The futility of trying to teach everything of importance. *Educational Leadership, 47*(3), 44-48, 57-59.

Woodhead, L. (Producer/Director). (2000, January) *Srebrenica: A cry from the grave.* New York: WNET.

CHAPTER 8

CONDUCTING A COMPARATIVE STUDY OF GENOCIDE

Rationale and Methodology

Henry R. Huttenbach

INTRODUCTION

An earlier chapter ("Defining Genocide: Issues and Resolutions") by this author focused on issues of defining genocide, ways of determining whether an event qualifies as being categorized as a genocide or not. Therein, it was shown that once one is in possession of a fundamental definition, it will minimize disputes as to whether an event is or is not a genocide. The key is conceptual precision and an absence of bias for or against a particular event in order to avoid the "politics" of inclusion or exclusion.

In the process of determining an event's status as genocidal or non-genocidal, considerable research is required on the empirical level. Theories and hypotheses and generalizations serve no purpose until a satisfactory database has been established. The designation "genocide" must fit

Teaching About Genocide: Issues, Approaches, and Resources, 239–247
Copyright © 2004 by Information Age Publishing
239

the facts and not vice versa. The immediate result is a reliable body of specialized articles and monographs that lay the groundwork for concluding one way or the other whether an event indeed belongs to the category defined as genocide. Certainly, areas of disagreement will always exist as scholars probe the specific identity of a single case of suspected genocide, especially as new evidence is uncovered. In genocide studies, as in other fields of academic investigation, there is always room for additional interpretation. Yet, in most cases of bona fide genocide, differences of opinion will be less about the status of an event's being a genocide than about the *kind* of genocide it is. It is precisely here that the comparative study of genocide enters into the picture.

Comparison is the sine qua non of genocide studies. It lies at the heart of the methodological way of understanding genocide per se and not just on the basis of familiarity with one or perhaps two instances of genocide. In order to break out of the parochial limits posed by knowledge of but one or maybe just two genocides, scholars and educators must survey the field from a broad comparative approach. To do so, researchers must set up categories or clusters of genocide with common elements to set them apart from other sets of genocide having other commonalties. To stress the obvious, comparison is as much about similarities as about dissimilarities; it is a necessary tool with which to stress simultaneously the specific singular identity of a genocidal event that sets it apart from others, and also to highlight common features that individual genocides share with others.

Clearly not all similarities and/or differences are equi-important or insightful. There are degrees of significance, from the trivial to those opening up ground-breaking new vistas. How can one be assured of avoiding the former and compare the more pertinent aspects of genocide shared or not by two or more incidents of genocide?

TOWARD AN ANATOMY OF GENOCIDE

Having already fixed the conceptual epicenter of genocide—*nullification* or a variant thereof—one must have a skeletal grasp of what genocide consists of. That is to say, what is the basic anatomy of genocide? What is its fundamental structure? What are its component parts? To determine this, one must move from the conceptual definition to the descriptive, that is, to the "anatomical parts" of genocide. Like any anatomical scheme, one begins with fundamental and essential segments: the equivalent of muscles, arteries, neural systems, the organs, all the way to the least important and most superficial. In the case of setting up an "anatomical" schemata for genocide, one should not carry the analogy of a real anatomy too far and take it too literally. Nevertheless, as the following schemata illustrates, it is a useful approach in preparation for a probing comparative approach.

The Anatomy of Genocide

I.   Pregenocide: 1.   General Background (economic, cultural, political)
                     2.   Specific Antecedents (massacres, propaganda)
                     3.   Immediate Circumstances (emerging crises)

II. The Event: 1.   Dramatis Personae
                          i.   The Genocidaires and Collaborators
                          ii.   The Victims
                          iii. Rescuers and Resistance
                          iv.   The Bystanders and Neutrals
                     2.   The Blueprint of Genocide:
                          i.   The plan
                          ii.   The means
                          iii. The results

III. Postgenocide: 1.   The Survivors and Restitution
                     2.   Trials, Tribunals, and Punishment
                     3.   Social Reconstruction and Reconciliation
                     4.   Denial
                     5.   Long-range Repercussions

Essentially, this schemata is quite simple. Basically chronological, it breaks genocide into three major sequential parts: before, during, and after. Each of these is broken into subtopics. The goal is to achieve both temporal and topical contextualization. Obviously, additional topics can be appended; nevertheless, the fundamental anatomy remains the same, affording a workable way of selecting central aspects of genocide to be compared, thereby insuring a symmetrical approach to comparison. This, in turn, will permit systematic grouping for each instance of genocide, alongside those with which it shares common features, that is, a typology of genocide.

TOWARD A METHODOLOGY OF SYMMETRICAL COMPARISON OF GENOCIDE

Without a systematic comparison of genocide, genocide studies will remain a fragmented field, a hodge-podge of uneven monographs, each devoted to one particular genocide. Yet this problem cannot be overcome unless genocides, properly identified as such by a governing definition, are also systematically and symmetrically compared according to a set of rational guidelines. These must originate from an accepted anatomy or skeletal structure, serving as a reference point, a source of key aspects of genocide which need to be compared, as a means of distinguishing and, equally importantly, relating genocides to one another as types.

If common agreement of a conceptual definition of genocide can be mustered, then identifying individual cases becomes easier. And once individual events have been categorized as genocides they can then be more readily classified into clusters, into types sharing common characteristics, whether primary or secondary.

Establishing a typology is still in its infancy for two reasons: (1) the lack of agreement as to what is an objective conceptual definition; and (2) the absence of agreement as to what and how to compare. The task of what to compare is made difficult, if not impossible, due to an absence of a workable anatomy of genocide. There is yet no consensus as to how to break genocide down into its essential primary and secondary component parts. Once these have been established, only then can systematic and consistent comparison begin. As to the how to compare genocides, this is a methodological problem still to be developed. Once the *what* and the *how* of genocide comparison have been resolved, valid types or groupings can be set up.

Which leads to the last point: namely, what is legitimate comparison and what is not? There are two broad approaches, one academic (functional) and one political (biased). The latter, basically impressionistic, strives to set up a highly subjective, *vertical* hierarchy of genocides according to nonacademic criteria; ranking genocides according to such fuzzy (essentially subjective) concepts as uniqueness, primacy, significance, importance, impact, et cetera. This approach, ascribing degrees of prominence to each genocide—such as a scale of suffering—is inherently skewed and is employed by those promoting a favored genocide. Intellectually, this kind of pseudo-comparativism is an intellectual cul-de-sac. The former, in contrast, seeks to look at genocides *horizontally* and, therefore, in clusters (types) according to objective common criteria such as those suggested in the "anatomy." For example, a useful rubric is the colonial and postcolonial context in which there have been several genocides.

This comparative approach needs considerable development to overcome monographic parallelism or isolation, and its twin, monographic parochialism. The former leads to single genocides studied by experts independently of each other, one rarely referring comparatively to another genocide, largely because expert knowledge in most cases is limited exclusively to *one* case of genocide, primary knowledge of a second being fragmentary. The latter, monographic parochialism, is also the result of single-case specialization from which, in this case, one draws unsubstantiated broad conclusions about genocide in general from but one instance, thereby committing the academic sin of leaping injudiciously from the particular to the general, a common temptation to be sure and one encouraged by the Katz (1994) model.

TYPOLOGIES OF GENOCIDE

Typology of genocide depends on two factors: sufficient knowledge about two or more instances of genocide and the creative imagination of the individual scholars in search of further elucidation. An example is the work of Robert Melson (1992). In his pioneering monograph, Melson (1992) studies the Holocaust and the Armenian genocide in the broad context of societies in revolution. Unfortunately, neither Melson (1992) nor other scholars have expanded on the possible central links between genocidal violence and societies in great flux to see whether the factor of revolution is a necessary or an incidental one in the tendency toward genocidal behavior. A broader approach would have raised the question: why do some (or most?) revolutions not culminate in genocidal violence. The origins of violence has puzzled countless scholars. For example, the dissolution of the Soviet Union in 1991 did not lead to wide-scale slaughter—though the collapse of the Tsarist Empire 70 years earlier did descend into multidimensional violence; the dismemberment of Yugoslavia in the early 1990s degenerated into several instances of genocidal violence, but in the divorce between Czechia and Slovakia (1990) there erupted no internecine war. This leads to classifying genocides according to those that were the products of revolution or civil war and those that were not. Why that inconsistency is so is the ongoing task of scholars engaged in comparing genocides according to common features and categories.

In search of comparative elements, the common denominator needs to be carefully defined. For example, as a result of decades of Holocaust study, the concepts "neutral" and "collaborator" were coined; they have been included in the sample anatomy outlined above. For several years these were rather simplistically accepted at face value till more recently, when several scholars reexamined the two categories and expressed serious skepticism about their utility as presently defined. Paul Levine (2002) broke down the status of "neutral" and concluded that in many instances, neutrality belonged more to the category of collaboration or to complicity than to that of a truly disinterested party. The present author took that one step further, declaring neutrality, in matters genocide, to be tantamount to conspiring with the forces of the genocidaires (Huttenbach, 1998). Both legally and morally, neutrality is unacceptable in a genocidal context: to do nothing for the victims is in fact to lend aid to the criminals engaged in genocide. Studies of events in Bosnia and Rwanda bear out this view. What this means is that the previously accepted (in Holocaust studies) broad status of neutrality needs to be narrowed severely, and the previously narrowly conceived concept of collaborator must be broadened to meet the exigencies of reality, of the facts, drawn both from the data of the Holocaust itself, but, more importantly here, from facts associated with other genocides. This is a typical example of the value of the comparative approach in seeking to unlock the problems of genocide study.

The comparative approach is especially useful in avoiding extracting a paradigm for genocide from a single genocide. For years, Holocaust scholars have held the Holocaust up as *the* genocide of all genocides, the event from which everything substantial associated with genocide can be gleaned. This is not only intellectual nonsense, but delays, if not prevents, opportunities to learn about the Holocaust from other cases of genocide.

The role of religion is a case in point. To be sure, Holocaust scholars have been aware of the involvement of Christian churches in the Nazi regime, and, in particular, of the Vatican's spurious policies during World War II. This raised many questions; some but not all have been answered satisfactorily, partly because of the failure to compare the role of churches, both Protestant and Catholic, in other genocidal settings. Again, Bosnia and Rwanda spring to mind. Unlike after World War II, a few clerics are today being brought to trial and held accountable for their murderous participation in recent genocides. If the Vatican provided safe passage to Nazis after the end of the war, then its policy of granting safe havens for criminal priests was even more flagrant and criminal in the case of Rwanda. Here is a rubric or category that cries out for future in-depth comparative exploration. The overall participation of clerics—Catholic, Orthodox, Muslim, and now Hindu—in genocides is a chapter yet to be fully written and assessed. Far from being benignly neutral, clergymen and their churches have been, are, and will continue to be collaborators in the crime of genocide. Only the comparative approach can permit this kind of broad interpretation resting on a solid base of incontestable evidence.

The phenomenon of denial is another concept (included in the anatomy) that suffers from the noncomparative and the lack of more sophisticated or nuanced understanding. The concept is taken far too one-dimensionally and lacks preparatory theoretical analysis, above all the philosophical implications of both the ideas and acts of denial. In this regard, Israel Charny (1991, 1997, 1999) has done fruitful work on the breadth and depth of denial and on its scope and many facets, despite his tendency to reach to the point of exaggeration. Nevertheless, not all genocides are denied (one category) while others are denied (another category): if so, why and why not? Apropos nuances: distinctions should be made between, for example, the denials by genocidists and the denial of postgenocide nonparticipants. These are at once related and unrelated categories that should not be lumped indiscriminately into one unmodified term. Then, of course, there is the denial by those who, for academic reasons, do not recognize—perceive—an event to be a full-scale genocide: they should not be accused of the same kind of denial as that expressed by the criminals and their postgenocide sympathizers.

The entire first rubric—pregenocide—needs careful comparative investigation. To locate the beginnings or roots of genocide in the near and more distant pasts is a much desired skill. What prompts this is the hope that by uncovering the causality of an event one can project—by extension—or anticipate genocidal incidents in the future. If one could fix a common past to most genocides, then one might come closer to anticipating and, thereby, preventing a future genocidal crisis. The search for early warning systems is a skill many scholars and politicians concerned with genocide hope to acquire. After all, a primary reason for genocide research is to see whether one or more genocides could have been avoided. Thus, hindsight, one hopes, can be translated into foresight if a pregenocide pattern or patterns can be ascertained.

To come close to this ideal one must first of all search for clues on a comparative level. Each genocide may have had its own distinct past; yet there may also be a common denominator. Only the comparative method will yield results to this problem. If, for example, one defines genocide as a particular mode of human rights violation, then by means of careful monitoring of the violation and/or collective human rights violators, observers of the present could be helped in pointing to a crisis that might mature into genocide. But again, this comparative monitoring of the present with an eye to the past and future relies on (1) the methodology of comparing of the past and (2) on the application of reasoning by analogy to anticipate the future.

CAVEAT EMPTOR

Some words of caution, however. The study of the "genocide past" is not a science; nor is projection into the future. The comparative approach—by definition multidisciplinary—is not infallible, anymore than our view into the as yet nonexistent future. There is nothing automatic or deterministic here. One past may culminate in genocide; another will not. No matter how refined our academic tools and how skillfully we apply them, there are always the elements of the incomplete past and the unknowable future. The comparative is necessary but it is not the last word in genocide research.

All goes back to accuracy of definition: of genocide and of all terms associated with it in the anatomy. Conceptual and descriptive terminology must lie at the heart of all attempts to understand a particular genocide and genocide in general. Comparison is a way of escaping broad generalities based on insufficient information. It is also a means to another crucial end, namely, contextualization.

Contextualization is one of the byproducts of the comparative approach. Clues to what a genocide was emerge both from comparing contemporaneous events and those preceding and following it. Thus, for example, the Holocaust, whatever is known about it, must also include the full context of when it took place: not just World War II, but specifically, for example, in terms of the *Generalplan Ost*, a much too little known Nazi scheme which, if properly understood, "threatens" to transform our grasp of the Final Solution from an event purely in itself to a part of a greater whole. Similarly, the Holocaust must be contemporalized beyond German history. As the Holocaust raged, Stalin was practicing variants of genocide against a dozen Soviet ethnic minorities. Even as the Warsaw Ghetto Uprising flared up, five million people died of manmade (!) starvation in British Colonial India. And even further afield, in the context of the Japanese invasion of China, the genocidal slaughter of hundreds of thousand of citizens of Nanjing took place. And—to come full circle—Fascist Croatia set up Auschwitz-like death camps such as the Jasenovac death camp in which to kill all its citizens of Serbian ethnicity.

Given this contemporaneous contextualization, certain unavoidable questions spring to mind. In full knowledge of these "other" events, some of them indisputably genocidal in character, can one really teach the Holocaust or any other genocide as an event totally apart? Can any one genocide really be meaningfully singled out as "more" significant to human history? Would it not make more sense to make students aware of the several genocidal global violences involved, of which the Holocaust, as each and every other incident, was but one, however tragic? Seen in this fashion, can one not conclude that the Holocaust was an extreme example of political violence, but by no means an out of the ordinary one? The answers—given the comparative approach on a contemporary level—is by no means simple, raising more issues than ones resolved.

There is a second dimension to contextualization, namely, the chronological. It is one thing to focus on one genocide; it is quite another to place it in the context of genocides preceding and following it. Once again, a comparative approach will help one assess in large measure the import of a genocide when measured against the chronological context of another.

It is one thing to view the Holocaust, for example, as unprecedented and, therefore, "unique" because one ignores what preceded and followed, and quite another when seen in sequence, as part of a genocidal trend, as one among many and not as the one (and *only*) in the twentieth century. This approach by no means diminishes a genocide; on the contrary, it enhances one's ability to understand and comprehend it in the all-encompassing context of human history.

Finally, the comparative approach raises a series of fascinating questions. Is there such a topic as "the history" of genocide? Is there, other than a chronological sequence, a causal one? Do later genocides take some of their cues from earlier ones? Those who teach other histories, such as, the history of revolution, for example, tend to extrapolate linkages and "evolutions." There is no simple answer to these questions. There are many simplistic ones, monosyllabic "yes" and "no," but very few reasoned ones based on a broad perspective. And that is as it should be.

REFERENCES

Charny, I. W. (Ed.). (1991). The psychology of denial of known genocides. In *Genocide: A critical bibliographic review* (Vol. 2, pp. 3-37). New York: Facts on File.

Charny, I. W. (1997). Commonality in denial: Classifying the final stage of the genocide process. *International Network on Holocaust and Genocide, 11*(5), 4-7.

Charny, I. W. (1999). *Encyclopedia of genocide* (Vol. 1). Santa Barbara, CA: ABC-CLIO Press.

Huttenbach, H. R. (1988). Locating the Holocaust in the genocide spectrum: Towards a methodology of categorization. *Holocaust and Genocide Studies, 3*(3), 287-303.

Katz, S. (1994). *The Holocaust in historical context: Vol. 1. The Holocaust and mass death before the modern age.* New York: Oxford University Press.

Levine, P. (2002). Swedish neutrality during the Second World War: Tactical success or moral compromise? In N. Wylie (Ed.), *European neutrals and non-belligerents during the Second World War* (pp. 41-56). New York: Cambridge University Press.

Melson, R. (1992). *Revolution and genocide.* Chicago, IL: University of Chicago Press.

CHAPTER 9

HUMAN RIGHTS, GENOCIDE, AND SOCIAL RESPONSIBILITY

William R. Fernekes and Samuel Totten

INTRODUCTION

In the latter half of the twentieth century, the field of human rights and, to a lesser extent, genocide studies blossomed from the seeds planted by the notable work of individuals and groups to protect human rights across the globe. While solid headway has been made in both fields, much remains to be done to meet the norms and goals established in the U.N. Universal Declaration of Human Rights and the U.N. Convention on the Prevention and Punishment of Genocide (UNCG). Similarly, human rights education and genocide education (the latter, a component of genocide studies), which are relatively new fields, require more intensive development in both theory and practice.

Many seem to consider the study of genocide and contemporary human rights violations to be mutually exclusive. The fields, though, are inextricably linked. First, genocide (often referred to as "the crime of crimes") is a human rights violation. Indeed, it is one of the worst violations in the long list of wrongs that governments and others perpetrate. Second, genocide does not develop in a vacuum. Genocide is typically preceded by a host of human rights violations against targeted groups,

Teaching About Genocide: Issues, Approaches, and Resources, 249–273
Copyright © 2004 by Information Age Publishing
249

ranging from deprivation of free speech to discrimination of all types, and from torture to extrajudicial executions.

While efforts around the globe have been successful in reducing infant mortality and the prevalence of major diseases such as polio and small-pox, expanding access to education for millions of people, and develop-ing a set of international norms for the protection of a wide array of international human rights, it is deplorable that so little effort has been made to stanch genocide. Indeed, despite the creation of the Universal Declaration of Human Rights and the United Nations Convention on the Prevention and Punishment of Genocide in 1948, state-sponsored vio-lence of genocidal proportions has remained constant over the years. The long, sordid catalogue of well-planned, systematic efforts to destroy human groups, in part or whole, is a tragic commentary on humankind's capacity for murderous violence—*as well as its unwillingness to make serious and all-out attempts to prevent genocide from being perpetrated.* Whether the location is Europe (e.g., The Holocaust and the former Yugoslavia), Asia (e.g., Bangladesh, East Timor, China, Cambodia/Kampuchea), Latin America (e.g., Guatemala, Paraguay, and Brazil), the Middle East (e.g., the former Ottoman Empire and Iraq), or Africa (e.g., Rwanda, Burundi, and Nigeria), genocide is a phenomenon crossing cultural and national boundaries. It is also a phenomenon whose prevention presents world states, international organizations, and nongovernmental organizations with a profound challenge to rethink their priorities in the twenty-first century.

To more effectively address the prevention of genocide, we believe that scholars, policymakers, and activists need to focus on how the deprivation of human rights often serve as the catalyst for genocide. Concomitantly, we are also in favor of the development of a strong human rights educa-tion process within genocidal studies. The latter could take many forms, including curriculum design, classroom instruction, and community-based social participation activities.

HUMAN RIGHTS AND GENOCIDE: HISTORICAL CONNECTIONS

Despite the continuing colonial domination of much of the globe by European states, the early twentieth century (1900-1939) was a period of rapid and often violent social change. During this same period various groups—including those opposed to colonialism, and those advocating the rights of women—assumed leadership roles in pressuring govern-ments to limit human rights violations via legislative enactment and through participation in international treaties.

Of great importance was *the momentum* (and momentum is the operative term here) generated by U.S. President Woodrow Wilson's advocacy of a "league of nations" where international conflicts could be peacefully resolved and human rights serve as a central focus of international diplomacy. The expectations which many delegates brought to the Paris Peace Conference of 1919 are summarized as follows:

> The experience of World War I, after all, had demonstrated the international aspects of shared life and death, and thus reinforced the importance of responsibilities beyond one's own national borders. Millions of people had sacrificed and suffered during the course of the war and many solemn promises about rights had been made that were now due to be fulfilled. In addition, they thought that the presence of new participants with different kinds of voices heretofore excluded from the inner sanctum of previous Great Power diplomacy would guarantee the realization of these visions of human rights. One of the most striking differences, for example, was the arrival of non-European states, a sign that the age had passed when it could be claimed that Europe represented the lever that moved the world. (Lauren, 1998, p. 93)

Some gains, in fact, were made. Most notable among them was the expansion of the rights of minority groups as embodied in the Minorities Treaties, in which members of the international community agreed to assume responsibility for the protection of minority rights not only within their own borders, but beyond them. Such international concern was to be guaranteed under the umbrella of the League of Nations (Buergenthal, 1988, pp. 8-11).

Be that as it may, the principle of self-determination, one of the cardinal principles of Wilson's 14-point plan for peace, was denied to peoples living in colonial possessions. Most striking was the failure of the conference to recognize the overriding need for a clause in the League of Nations Charter that would guarantee racial equality for all peoples. The Japanese delegation, representing one of the victorious powers in World War I, submitted a proposal to that end, and despite Wilson's strenuous efforts to block its consideration by the Commission on the League of Nations at the Peace Conference, the proposal was debated and voted on. Eleven of the 17 states represented on the commission voted in favor of including the proposal, but Wilson abruptly informed the assembled delegates that the proposal had failed.

Wilson's arbitrary decision highlighted the hypocrisy of the victorious European and North American powers in addressing universal human rights guarantees. The hopes of colonial and indigenous peoples, as well as people of color, for the creation of a new world order based on principles of equality and self-determination were frustrated, and at the end of the Paris Peace Conference, the right to self-determination had been limited to those of white skin in Eastern and Southern Europe. A coalition of European and North American powers had made their point—human rights were for some, not all.

Following 1919, it became painfully evident that compromises made at the Paris Peace Conference and embodied in the Covenant of the League of Nations hampered both the advocacy and enforcement of international human rights guarantees. While the League was successful in some initiatives—including the coordination of massive refugee relief for those displaced by World War I, the Bolshevik Revolution, and the Ottoman Turk genocide of the Armenians; the creation of the Geneva Protocol of 1925, which set standards for the treatment of prisoners of war in armed conflicts; and the development of the first truly universal declaration in regard to the rights of children, the Declaration of the Rights of the Child of 1924—the unwillingness of member-states to restrict the priorities of national sovereignty in the interest of enforcing international human rights guarantees seriously hampered the League's effectiveness. Compounding the problem was the failure of the United States to join the League, which weakened the latter's influence, as did the decision by such colonial powers as Great Britain and France to deny their colonial possessions' participation in the League's deliberations (Lauren, 1998, p. 124). But at the heart of the League's difficulties was a fundamental dilemma: how would member states respond if and when the League of Nations acted to advance international guarantees of human rights that conflicted with the policies and practices of a member state?

The U.S. position on this issue was clearly stated by Senator Henry Cabot Lodge, one of the architects of the U.S. rejection of the League of Nations in the U.S. Senate: "We [the U. S.] do not want a narrow alley of escape from the jurisdiction of the League. We want to prevent any jurisdiction whatever" (quoted in Lauren, 1998, p. 125).

The League also had structural problems which hampered its capacity for effective enforcement of human rights guarantees. More specifically, the two major operational bodies of the League, the Council and the representative Assembly, were hampered by the requirement for unanimous consent as a condition for action. Additionally, the Council was comprised primarily, though not exclusively, of the major European and North American powers, and they exercised considerable influence over the operations of the League. To make the situation even more difficult, exemptions were provided in the League's governing document, permitting specific issues to remain with the domestic jurisdiction of member states. Racial segregation in the United States was therefore a subject of "domestic jurisdiction" and the daily violations of human rights perpetrated under Jim Crow laws thus could not be addressed by the League.

In light of the deficiencies of the League's governance structure along with the resistance to supporting international human rights guarantees when they clashed with member state practices and policies, it is understandable why those individuals and groups who petitioned the League for relief from 1920-1939 experienced continuous frustration as they sought protections for what they believed were their fundamental human rights.

By the mid-1930s, totalitarian governments of the left and the right had come to power in a number of European states: Italy, Germany, and the Soviet Union. These governments' ideologies, policies, and practices clashed directly with international human rights guarantees, whether it was using violence and terror to crush internal dissent, systematically discriminating against minority groups, or implementing aggressive war and crimes of violence against selected population groups, both within their national borders and in other parts of Europe and Africa. In case after case—ranging from the individual petition of the German Jew Franz Bernheim to insure his rights as a minority in Germany under the terms of the German-Polish Convention for Upper Silesia to the invasion of Ethiopia by Mussolini's Italy and to the imposition by Stalin of a forced famine in the Ukraine against Ukrainian peasants—the League failed to intervene on behalf of individuals and groups being victimized by member-states. Minus an effective enforcement mechanism, and severely constrained by League members' unwillingness to put aside the priorities of national sovereignty when individuals and groups were being victimized across the globe, the League was marginalized as an international peacemaking and problem-solving organization. The organization's impotence in the face of totalitarian aggression sent a clear message to totalitarian leaders and their followers: minus an effective international response to state-sponsored violence, including mass murder, governments could implement their policies with impunity.

The Holocaust and Its Consequences vis-à-vis International Human Rights

World War II was the most disastrous war in the history of humankind. The scope and intensity of the suffering endured by people across the globe was unparalleled, with civilian populations experiencing horrors that were scarcely imaginable in earlier conflicts. Not only were civilians now likely targets of military strategists, but the annihilation of entire groups became state policy, as Nazi Germany systematically implemented its genocidal policies against Jews, the Sinti and Roma (Gypsies), and the physically and mentally disabled. Earlier in the century (1915-1918), the

major world powers had failed to intervene as the Young Turk government had sought to eliminate, through extermination, its Christian
Armenian population during World War I, and now another genocide was
being implemented at will and bereft of international interference. The
failure of the Great Powers to intervene against the Young Turks as they
committed genocide against hundreds of thousands of innocent Armenians was no secret to Adolf Hitler, who stated on the eve of the German
invasion of Poland in 1939, "Who after all is today speaking of the
destruction of the Armenians?" (quoted in Dadrian, 1997, p. 403). By the
end of World War II, an estimated five to six million Jews, between 300-
500,000 Sinti and Roma, and over 100,000 disabled had been murdered
by Nazi Germany and its allies and collaborators. Additionally, millions of
other European civilians had been enslaved as forced laborers and/or
killed outright, and an estimated three million Soviet prisoners of war
were killed or died from mistreatment in German captivity.

Confronted with the overwhelming evidence of Nazi crimes as presented at postwar trials, the leaders of the victorious Allied powers could
only with great difficulty turn their backs on pledges they had made during the war "to preserve human rights in their own lands as well as in
other lands" (Declaration of the United Nations from 1942). The principles of the Atlantic Charter, as well as those enunciated in Franklin
Roosevelt's Four Freedoms speech—freedom of speech and expression,
freedom of worship, freedom from want, and freedom from fear—had
been widely disseminated as core elements of the Allied crusade against
fascism. The blatant violation of these rights and the implementation of
mass murder as state policy by Nazi Germany clashed so directly with
such core principles that world public opinion demanded not only
accountability for those responsible, but also real commitment to the
development of an international peace and security system that placed
human rights its center—not, as it had been, on the periphery. As Lauren
(1998) states, "It was World War II that demonstrated as never before in
history the extreme consequences of the doctrine of national sovereignty
and ideologies of superiority. This, in turn, forced those engaged in the
war to look in a mirror and see their own reflections, and consider
whether they had any responsibility to change their ways" (p. 145).

With the creation of the United Nations Charter in April 1945, human
rights emerged as a core issue for international relations in the postwar
world. Following intense debate between the major powers (which sought
to severely limit human rights as part of the U.N. charter and structure)
and small states and nongovernmental organizations (which advocated a
much expanded presence for human rights in the new organization's
charter and structure), the final draft of the Charter inextricably linked
protection of human rights with the achievement of peace and security.

Never before had a formal international treaty placed such an emphasis on human rights, and never before had the traditional conceptions of national sovereignty been challenged so directly. Despite the inclusion of Article 2 (7), which limited U.N. intervention in matters "which are essentially within the domestic jurisdiction of any state ..." (U.N. Charter), the stage was set for a direct clash between advocates of international human rights norms and protections, and opponents who wanted to perpetuate traditional conceptions of national sovereignty and limit any efforts by international organizations to protect the human rights of individuals or groups in member-states.

Ironically, the victorious Allied powers' initiative to bring Nazi and Japanese war criminals to justice following the end of World War II undermined their own efforts to justify the notion that what occurs within the boundaries of a nation-state is beyond the concern of the world community. The indictment issued by the International Military Tribunal (IMT) against Nazi defendants included a revolutionary new category: crimes against humanity, which challenged the centuries-old prohibition against holding governments accountable for crimes committed against civilians, not only in areas they had conquered and occupied, but within their own borders. Despite claims by various defendants that not only did they bear no responsibility for the crimes their nation had committed but that which occurred within the boundaries of their own nation-state should not be the subject of international law, the deliberations at Nuremberg enshrined the principle in international law that states could no longer act with impunity against civilian populations, within or beyond their borders, without being held accountable by the world community.

The momentum generated by the creation of the United Nations, the postwar trials in Nuremberg and Tokyo, and the mounting challenges by indigenous peoples and other minority groups to policies and practices of colonialism and discrimination on the basis of ethnicity, race, and gender around the world resulted in the demand for further action. The United Nations responded by creating its Commission on Human Rights, which became the drafting body for the most widely disseminated international document on human rights, the Universal Declaration of Human Rights (UDHR). Approved by the U.N. General Assembly in 1948, the UDHR has served as the bedrock document for the over 20 subsequent U.N. declarations and conventions (binding international treaties) dealing with human rights. At one and the same time, worldwide pressure mounted on member states to ratify the U.N. Convention on the Prevention and Punishment of the Crime of Genocide, a treaty whose content was directly inspired by the horrors of the

Holocaust, but which fit well with the broad vision of international human rights norms stated in the UDHR.[1]

As Mills (1998) notes in *Human Rights in the Emerging Global Order: A New Sovereignty*, the Universal Declaration of Human Rights (UDHR)

> describes itself as "a common standard of achievement for all peoples and nations" and even though it is not binding in the same manner as formal treaties all members of the UN supposedly accept its contents. W. Michael Reisman argues that it is "now accepted as declaratory of customary international law." [That is,] the UDHR is perceived as having "considerable authority" and "lays down rules, which, irrespective of whether they are embodied in a binding document or not, are binding as customary international law." (p. 39)

It is worth noting that on December 16, 1966, the Commission on Human Rights adopted, by unanimous vote, two separate Covenants—the Covenant on Economic, Social, and Cultural Rights and the Covenant on Civil and Political Rights, which "aimed at achieving [a] certain, if extremely, limited degree of compliance" (Korey, 2001, p. 73). On a different but related note, the establishment of the International Criminal Court (ICC) in the late 1990s (see below) will, hopefully, and finally, put teeth into the Genocide Convention.

While great strides have been made in recognizing the need to protect the basic civil and human rights of individual citizens and groups, horrifying human rights violations (e.g., torture, mass rape, extrajudicial killings, and genocide) have been—and continue to be—perpetrated on a regular basis across the globe. Even slavery, which some had thought to have been eradicated, has reared its ugly face again in such parts of the world as the Sudan. With only certain exceptions—most notably, human rights activists, concerned journalists, and a small number of scholars—most of this activity goes unnoticed, much less decried. Shamefully, the perpetrators are not the only culprits; indeed, those nations' governments that have political ties to the perpetrators, that provide foreign aid and/or weaponry to the perpetrators, train, and arm the perpetrators' security forces, and trade with perpetrator states are complicit in such violations. In a world of realpolitik, the hands of many are stained with the blood of millions.

Convention on Genocide

Public revulsion at the devastating horror of the Holocaust "provided the impetus for the formal recognition of genocide as a crime in international law, thus laying the basis for intervention by judicial process" (Kuper, 1981, p. 20). While true, both the acceptance of the term "genocide" and the extensive groundwork for the establishment of a convention for the prevention of genocide resulted from the indefatigable efforts of Raphael Lemkin, a refugee Polish-Jewish jurist, and the individual who coined the word "genocide." Lemkin dedicated a great amount of time and energy in an effort to convince the member states of the new United Nations about the dire need to ratify the U.N. Convention on the Prevention and Punishment of Genocide (UNCG).[2]

Ultimately, the "declared purpose of the [UN] Genocide Convention, in terms of the original resolution of the General Assembly of the United Nations, was *to prevent and punish the crime of genocide*" (Kuper, 1981, p. 36, italics added).

In the UNCG, genocide is defined as follows: "genocide means any of the following acts committed with intent to destroy, in whole or in part, a national, ethnical, racial or religious group, as such: (a) Killing members of the group; (b) Causing serious bodily or mental harm to members of the group; (c) Deliberately inflicting on the group conditions of life calculated to bring about its physical destruction in whole or in part; (d) Imposing measures intended to prevent births within the group; and (e) Forcibly transferring children of the group to another group." Despite its extremely broad and compromise nature (for example, after much debate, "political groups" and "social groups" were excluded from its wording—and ostensibly, from protection under the Convention), the development, passage, and ratification of the Convention was hailed as a major milestone in the protection of basic human rights.[3]

Thus far, though, the effectiveness of the UNCG in preventing and prosecuting genocide has been extremely limited. This is evidenced by the fact that the post-Holocaust world has witnessed the perpetration of one genocide after another since 1945 in which millions have been brutally murdered. Among the many places where genocide has been perpetrated since 1945 are: Indonesia (1965-1966), East Timor (1975-1979), Bangladesh (1971), Burundi (1972), Cambodia (1975-1979), Iraq (the late 1980s), Rwanda (1994), and Bosnia-Herzegovina (early to mid-1990s). This does not include the many indigenous groups that have been subjected to both genocide and ethnocide (e.g., the attempt to destroy in substantial part or totally a group's culture, such as the destruction of the Tibetan way of life by the Chinese, which has been ongoing since the late 1950s). The international community has rarely intervened to prevent genocide from being perpetrated nor have more than a handful of the perpetrators been held accountable for their crimes.

Trying Perpetrators of Genocide

In addition to the Nuremberg trials, numerous other trials were conducted throughout Europe to hold Nazi war criminals and collaborators accountable, and war crimes trials were also held in the Far East at the conclusion of World War II. Still, most of the lower echelon perpetrators were not prosecuted. Among the most notable trials in the later years was that of Adolf Eichmann, who was captured in Argentina by Israeli agents in May 1960 and transported to Israel. He was tried, found guilty, and sentenced to death in 1961 in Jerusalem.

From the conclusion of the postwar trials to the early 1990s, the international community's record of bringing perpetrators of other genocides to justice was disgraceful. The genocide in Cambodia provides a classic example of such inaction. For many years following the Cambodian genocide, which resulted in the estimated deaths of between one to three million people, all of the major perpetrators—Pol Pot, Ieng Sary, Khieu Samphan, Non Chea—remained at large, untouched by the United Nations or any individual nation that had ratified the U.N. Convention on Genocide.[4] Instead, the perpetrators were allowed to roam the jungles of Cambodia and the towns on both sides of the Cambodian/Thailand border, where they carried on their ragtag but deadly insurgency. Pol Pot eventually died a natural death on April 15, 1998, and in early 1999 Khieu Samphan and Nuon Chea voluntarily gave up the resistance movement they had continued to lead. In 1999, Seth Mydans of the *New York Times* reported that

> The Khmer Rouge leaders, Khieu Samphan and Nuon Chea, came in from the cold on Christmas day after three decades of revolution that included brutal rule from 1975 to 1978. They were welcomed with embraces from both Mr. Hun Sen [the new leader of Cambodia) and the former United Nations Secretary General, Boutros Boutros Ghali, who was visiting town to promote the French language but took time out to pay his respects.
>
> Mr. Boutros-Ghali praised Mr. Hun Sen for his policy of "national reconciliation." He said the mass killings by the Khmer Rouge were the internal affair of a sovereign state, immune from the "interference" of outsiders. He said Cambodians must find their own route to resolving their rights issues.
>
> Not everybody sees it this way. The United Nations itself is investigating evidence and possible procedures to bring the Khmer Rouge killers to trial if either the [U.N.] Security Council or the Cambodians themselves will agree to move forward. (p. 11)

Up until the appearance of Khieu Samphan and Nuon Chea, most Cambodians quietly, if painfully, attempted to salvage what was left of their damaged lives. But the sight of the two perpetrators being treated like royalty, along with the prospect of them going free after having caused such widespread grief, has proved too much for them and many are demanding that the perpetrators face trial. Time will tell.

Two of the main architects behind the genocide in Bosnia-Herzegovina, Radovan Karadzic and Ratko Mladic, wartime politicians and military leaders of the Bosnian Serbs, have managed to elude capture by U.N. peacekeepers in Bosnia, despite their indictment in 1995 by an international criminal tribunal on charges of genocide, crimes against humanity, and war crimes. Both have flaunted their freedom, and have been seen in public places dining and socializing with friends, often in plain view of U.N. peacekeeping forces.

In September 1998, General Augusto Pinochet was arrested in London on a Spanish warrant for flagrant human rights abuses (including genocide, during which 2,000 people "disappeared") during his reign of terror in Chile from 1973 through 1990. Ultimately, Pinochet's lawyers extricated him from his legal entanglements and he was allowed to return to Chile where there was talk of placing him on trial. Those plans vanished when he was allowed to go free, reportedly because of illness and frailty. Still, the warrant and his arrest served notice to other perpetrators that their days of impunity may be numbered.

Until the international community decides once and for all that all perpetrators of all genocidal acts will—without fail—stand trial, the world will continue to see perpetrators committing their horrific deeds and then walking away. That said, the trials currently being conducted at the Hague and in Tanzania and Rwanda are a start in the right direction.

ICTY and ICTR

In 1993 the U.N. Security Council established a war crimes tribunal in The Hague to prosecute individuals responsible for serious violations of international humanitarian law in the territory of the former Yugoslavia. As Neier (1998) has observed, this decision by the Security Council "set a precedent for the world body: It was the first time in its forty-eight year history that it tried to bring anyone to justice for committing human rights abuses" (p. 21). Equally significant, "The charter for the ex-Yugoslavia war crimes tribunal ... uses language from the UN's 1948 Genocide Convention" (Neier, 1998, p. 21). Still, as previously mentioned, both Mladic and Karadzic remain at large. Many, including senior United States diplomats, have argued "that no lasting peace is possible in Bosnia until Dr. Karadzic and General Mladic are brought to justice" (Weiner, 1998, p. 8). And Karadzic and Mladic are not the only ones at large; in fact, "less than half of the 62 men indicted on war-crimes charges by the international tribunal are now in custody" (Weiner, 1998, p. 8).

In the aftermath of the Rwandan genocide in 1994, the U.N. Security Council established an international ad hoc tribunal in Arusha, Tanzania, to prosecute those who committed genocide and crimes against humanity. A separate tribunal was also established by Rwanda in Kigali.[5]

Though rife with major limitations, the establishment of the two international tribunals constitutes a major step toward holding perpetrators of genocide and related crimes against humanity accountable for their actions.

International Criminal Court (ICC)

Although the United Nations General Assembly authorized the U.N. International Law Commission (ILC) to initiate planning for an international criminal court (ICC) in 1951—it was not until the late 1980s and early 1990s that substantial momentum was generated to bring that vision into reality. Korey (2001) notes that

> In 1992, "ethnic cleansing" in Yugoslavia was shocking the international community. The Assembly again requested that the ILC prepare a draft statute for a permanent ICC. While the ILC was deliberating about a permanent court, the Security Council established two ad hoc Tribunals, one for the Former Yugoslavia and the other for Rwanda. These statutes helped expedite the ILC deliberation while, at the same time, they generated favorable international sentiments regarding the ICC. (p. 525)

In 1994, the ILC presented its final draft of recommendations for the creation of the International Criminal Court (ICC). Between 1994 and the 1998 Rome Conference, where the final version of the ICC Treaty was negotiated, nongovernmental organizations and U.N. member-states debated, compromised, and eventually created a framework for a court that will prosecute acts of genocide, war crimes, crimes against humanity, and aggression (as quoted by Axel in Cooper, 1999, pp. 313-322).

The most serious issue faced by the ICC Statute focused on when and how prosecutions would take place. The United States and certain other U.N. member-states supported a provision that would have necessitated a unanimous decision by the U.N. Security Council for the ICC to initiate a prosecution. This option was rejected, and the final version included three ways for prosecutions to be pursued: through referrals by the U.N. Security Council, from state-parties, and by the initiative of the prosecutor him/herself. The U.N. Security Council, by unanimous agreement, can block an ICC prosecution, but only for one year at a time.

Another important element concerning the court's jurisdiction is the requirement that the ICC only take on a case "if the state in which the crime occurred or the state of the accused's nationality has ratified the statute or agreed to the court's jurisdiction, except in cases that have been referred by the Security Council, in which case its jurisdiction is universal" (Cooper, 1999, p. 325). This provision caused consternation for both supporters and opponents of the ICC Treaty, as it posed the potential problem of making nonratification states' nationals subject to prosecution if apprehended in countries who are parties to the treaty, while simultaneously making it difficult for suspects "accused of crimes in their own countries in a situation that is still ongoing" (Cooper, 1999, p. 326) to be prosecuted, such as Saddam Hussein in regard to the genocide of the Kurdish population in northern Iraq.

Similar to the scenario following the end of World War I, and which has characterized the U.S. response to many U.N. human rights treaties, the United States has been a vehement opponent of the ICC, claiming it would undercut national sovereignty. Although President Clinton signed the ICC Treaty prior to his departure from office in 2000, President George W. Bush took the unprecedented step of having the U.S. "unsign" the ICC Treaty in the fall of 2001. To further undermine the ICC, the United States, in 2002, initiated discussions with ICC signatory states to develop bilateral agreements with those states exempting United States' nationals from being prosecuted under the ICC Treaty (Amnesty International, 2002). This retreat from responsibility has not deterred the ICC Treaty from coming into force as a binding U.N. treaty, but it calls into serious question the commitment of the United States in regard to creating an effective mechanism for the prosecution of the most serious human rights violations.

CURRICULAR FRAGMENTATION IN THE STUDY OF HUMAN RIGHTS AND GENOCIDES

Emergence of Holocaust/Genocide Studies in the Curriculum

For many years following the end of World War II, there was little to no discussion/study of the Holocaust in U.S. public schools. Among the many reasons for the latter were: (1) a lack of knowledge and/or interest by teachers about the Holocaust; (2) a lack of attention to the Holocaust in school textbooks; (3) the absence of the mention of the Holocaust in school, district, county, and state curriculum guidelines; and (4) a dearth of curricular resources. If the Holocaust was taught at all, it was by the individual teacher who perceived the need to do so and/or had the interest to do so.[6]

Certainly one factor in generating great interest in the subject of the Holocaust by the general public and teachers in public schools in the United States in the late 1970s was the televised production of the miniseries "Holocaust." Although widely criticized by survivors and scholars as vulgar and inaccurate, it was effective in raising awareness of the Holocaust among the American public. Later, in the early 1990s, the feature film "Schindler's List" and the opening of the United States Holocaust Memorial Museum (USHMM) in Washington, D.C. served similar purposes and spawned similar results.

According to the Education Department at the USHMM, as of 2003, five states (California, Florida, Illinois, New Jersey, and New York) mandate the teaching of the Holocaust in their public schools. Ten other states (Connecticut, Georgia, Indiana, North Carolina, Ohio, Pennsylvania, South Carolina, Tennessee, Virginia, and Washington) either recommend or encourage their public school personnel to teach about the Holocaust. The state of Mississippi is in the process of deciding whether to recommend or mandate Holocaust education.

In addition to the Holocaust, the New York, California, and Connecticut curricular guidelines address other genocides (e.g., The Armenian genocide, the manmade famine in Ukraine, and the Cambodian genocide), but the coverage in all three is bereft of adequate depth and accuracy.

While it is admirable that the Holocaust is no longer a subject that is either ignored by teachers and professors, it is disturbing that it is frequently taught with minimal reference to the history of human rights violations in the twentieth century. It is also disturbing that the ongoing problem of genocide and its direct connection to human rights violations is given little attention. This oversight, if in fact that is what it is, constitutes what Stanford Professor of Education Elliot Eisner (1979) refers to as the "null curriculum":

> It is my thesis that what schools do not teach may be as important as what they do teach. I argue this position because ignorance is not simply a void; it has important effects on the kinds of options one is able to consider; the alternatives one can examine, and the perspectives with which one can view a situation or problem. (p. 83)

Human Rights Education

State University of New York at Oneonta professor Dennis Banks conducted a comprehensive national survey of human rights education in 2000, and the results clearly demonstrated that although sporadic efforts have been made in certain states to include human rights content within state-mandated or recommended social studies curricula, human rights literacy in the United States is not widespread (Banks 2000, p. 1). He also found that although 20 states included some aspect of human rights education in state-mandated curricula, there was wide variability regarding (1) the scope of human rights knowledge required for study, (2) the use of curriculum models, which varied between focusing on knowledge and/or values and attitudes, and (3) the level of integration within actual school district curricula and classroom instruction of required or recommended human rights content (Banks, 2000, pp. 13-15).

Interestingly, Banks found that when asked to specify curriculum topics within human rights education at the state level, among those most often cited by respondents were the Holocaust and genocide. He also noted, however, that in many states, specific topics were not delineated for the study of human rights, leaving content selection to local discretion. One of many questions that arises is: when the issue of genocide is taught, how are human rights concepts and issues addressed?

It is illuminating to examine Holocaust/genocide curriculum guides from two states where the linkages between human rights and genocide studies content are clearly articulated: New Jersey and New York. Tellingly, Holocaust/genocide state level curriculum documents from these states *do not* elaborate on the potential linkages between human rights and genocides. Indeed, not only do they lack clear explanations of conceptual relationships (e.g., the protection of individual rights against abuses of state power), but few detailed examples are provided regarding how human rights advocacy and activism have informed the development of genocide prevention efforts.

In the New Jersey Commission on Holocaust Education (NJCHE) Grades 7-12 curriculum guide, for example, the use of the phrase "human rights" is confined to Chapters 2 and 6 of the guide. In Chapter 2, "Views of Prejudice and Genocide," it is included in the poem "Our Human Rights," which discusses issues of safety, self-expression and tolerance. What human rights are on the broader scale, meaning the full range of human rights included in the UDHR, is not introduced. In the introduction to Chapter 6, entitled "Genocide," the authors assert that "this unit will urge students to support and be knowledgeable of groups that monitor violations of human rights and promote resolution of violent social conflict through peaceful negotiation" (NJCHE, 1995). The United Nations, nongovernmental organizations (NGOs), and local cultural, religious, and civic groups are mentioned as groups that engage in human rights monitoring and advocacy. The Universal Declaration of Human Rights is included as a document for teacher and student use in this chapter of the resource guide, and is linked to the following learning objective:

"students will be encouraged and exposed to the proactive skills necessary to prevent genocide, atrocity, or random acts of violence in the future" (NJCHE, 1995). While this may appear to be adequate coverage of the UDHR for developing student awareness of human rights violations, no learning objectives are provided that elaborate on definitions or categories of human rights, nor is there any effort to have students examine the direct and compelling connections between historic human rights violations and the origins of the UDHR and the UNCG. Furthermore, despite the stated emphasis on the roles of nongovernmental organizations in the introduction to Chapter 6, no historical content is included about the roles of nongovernmental organizations in the defense of human rights.

In the New York State curriculum guide for the study of the Holocaust and genocides entitled *The Human Rights Series: Volumes I, II and III*, the authors state in the Foreword to the three volume series that in addition to assisting secondary school teachers in their teaching about the Holocaust, "this guide serves as an introduction to the concept of human rights" (New York State Department of Education, 1985, p. iii). On the same page, the authors claim that "Education represents the most effective weapon in guarding against future violations of human rights" (New York State Department of Education, 1985, p. iii). Examination of the content of the three volume series, however, reveals that few direct linkages are established between definitions and concepts of human rights and the history of the Holocaust. Human rights issues are confined, almost without exception, to the sections of Volume III dealing with case studies of the Soviet manmade famine in the Ukraine and the Khmer Rouge-perpetrated genocide in Cambodia. Within each of these units, the human rights content is contained within subsections entitled "Human Rights Violations in Ukraine" and "Human Rights Violations in Cambodia." To the authors' credit, there is a comprehensive attempt to have students compare and contrast standards of human rights as articulated in the UDHR and the Soviet constitution with the actual policies and practices of the Union of Soviet Socialist Republics in dealing with dissenters and prisoners of conscience from the Ukraine. The unit on Cambodia is much less comprehensive. The resulting compartmentalization reinforces the divisions evident in classroom study of human rights and genocides, rather than seeking to bridge the gap. Many opportunities for structured comparisons between historic and contemporary violations of human rights are available within the content included in the three volumes, but no objectives are listed that utilize the language and concepts of human rights to bring such comparisons to light. *The precursors to mass murder, particularly the denial of civil rights and all sorts of discriminatory actions and restrictions of targeted groups, were evident in all three cases, and by the time of the Cambodian genocide, statements of rights guarantees concerning these issues had been incorporated within the UDHR for over 30 years.* But the fact that the authors only introduced the concept of state-sponsored violations of human rights in the cases of Ukraine and Cambodia clearly demonstrates that they viewed human rights as having little applicability to study of the Holocaust, despite the title of the three volume curriculum guide. The omission of detailed explanations about the relationships between these two major content topics results in a simplistic understanding of extremely critical issues, and is little to no help to teachers who should be highlighting such linkages.

During the early 1990s, both authors of this chapter served as consultants to the Education Department of the United State Holocaust Memorial Museum (USHMM). A major project in which we participated (Totten as coauthor, and Fernekes as a contributor and a manuscript reviewer) was the development of the USHMM's *Guidelines for Teaching About the Holocaust* (initially published in 1993, and updated in 2000). Analyzing the contents of the *Guidelines* today, it is regrettable that connections between human rights content and concepts are largely absent. There is only one mention of the term "human rights" and virtually no elaboration of any relationships between the study of human rights and study of the Holocaust. Although the museum takes no official position on the preferred design and content of classroom curricula for the study of the Holocaust, the virtual absence of human rights in their most widely disseminated document for educators speaks volumes. The sole mention of the phrase "human rights" is on page 13 in the section "Incorporating Study of the Holocaust into Existing Courses," where students in government courses can "recognize that among the legacies of the Holocaust have been the creation of the United Nations in 1945 and its ongoing efforts to develop and adopt numerous, significant human rights bills (e.g., the U.N. Declaration of Human Rights and the U.N. Convention on Genocide)" (USHMM, 2000, p. 13). While the concept of genocide is incorporated into suggestions for curriculum integration in world history, world cultures, government and contemporary world problems, human rights violations and efforts to expand human rights protections merit only one mention. *Similar to the NJ and NY curriculum* guides, there is no examination or discussion of the relationships between *the history of the Holocaust and emerging conceptions of human rights, which is particularly striking given the vast scope and intensity of Nazi Germany's assault on human rights, whether they be civil/political or economic, social and cultural. Indeed, as Berenbaum (1993) has noted, Nazi Germany passed "over four hundred separate pieces of legislation enacted between 1933 and 1939 that defined, isolated, excluded, segregated and impoverished German Jews."* (p. 22, italics added).

Reasons for Curricular Fragmentation

The following reasons offer potential explanations for such curricular fragmentation:

- The Western world's dichotomous way of thinking about issues has no doubt been a factor—"deriving, perhaps, from the Platonic constructs of ideal versus real and intellect versus emotion" (Barash, 2001, p. 14)—and thus, out of such thinking, two separate conventions (UNDHR AND UNCG) and two separate fields have emerged: human rights studies and genocide studies.
- Early on (e.g., the late 1940s), the issues of human rights and genocide were perceived as distinctive enough issues that the United Nations developed a separate convention on genocide and a declaration about human rights. In turn, this seems to have compelled scholars, activists and even many governmental officials to perceive the two issues, both of which are obviously human rights-oriented, as being distinct and different.
- To this day, many mainstream and major international human rights organizations either do not focus on the issue of genocide or only do so in a peripheral manner.
- In the 1980s, the field of genocide studies was born, in part, out of a perceived need that the human rights community was not addressing the issue of genocide, and this reinforced the sense that the issues/concerns were distinct and different.
- Most scholars who study human rights issues only touch on the issue of genocide in peripheral ways and many in the field of genocide studies, while focusing on the antecedents to genocide, do not seem to invest as much time and thought into the critical need to avert human rights violations—other than massacres and genocide—as need be.
- Most educators—including those associated with human rights organizations as well as curriculum developers in state departments of education—have basically followed the lead of the international community, scholars, and human rights activists in their division of the two related areas.
- Few classroom educators have been eductaed about or engaged in systematic study of human rights content during either their pre-service or in-service education.
- Although human rights activities in educational settings are increasing due to the efforts of nongovernmental organizations (i.e., presentations at national, regional, and state professional conferences), human rights education is not a priority in teacher edu-

cation programs nor does it regularly appear on the agendas of in-service programs in school districts.

• The rapid development of Holocaust and genocide education in the United States since the early 1970s and mid-1980s, respectively, has emphasized study of a relatively confined set of curriculum topics, primarily the Holocaust and select other genocides of the twentieth century, such as occurred in the former Ottoman Empire against the Armenian population, or in Cambodia under the Pol Pot regime. These curricula and related pre-/in-service training programs tend to present genocide as an international problem requiring a solution, but are disconnected from broader considerations of international human rights protections.

The language of human rights is often missing from discussions and explanations of how state power was used to violate guarantees of human rights for individuals and groups in societies such as Nazi Germany, the Soviet Union, or Cambodia. For example, how often is the constitution of the Weimar Republic examined in relation to the actions taken by the Nazi state to violate human rights in the early years of the Hitler chancellorship (1933-1934)? The Nazi regime showed nothing but contempt for the concept of universal human rights, ranging from denying people their very livelihoods to their citizenship, and from forcibly and brutally limiting the reproduction of "racially unfit" offspring to the implementation the "final solution" of the Jewish question. Viewed through the lens of human rights, the actions of the Nazi state and their allies and collaborators retain both unique and universal characteristics, which makes all the more compelling the claims at the Nuremberg trials that the crimes of the Nazi state had universal implications. Justice Robert Jackson's statement that "The wrongs which we seek to condemn and punish have been so calculated, so malignant, and so devastating, that civilization cannot tolerate their being ignored, because it cannot survive their being repeated" (quoted in Marrus, 1997, p. 79) says a great deal about the need to protect fundamental human rights, yet this quote is often discussed only in relation to the prosecution of the major Nazi criminals, and not its significance for developing a culture of universal human rights protections.

POSSIBLE SOLUTIONS IN POLICY AND PRACTICE

Prior to deciding on what and how to teach a topic, it is wise to ask oneself the following: Why am I even teaching this topic, lesson or unit? Why is it important? Why, out of all of the other topics that are available to teach about, is this one so critical that I am going to dedicate time to it and ask my students to commit their time and energy and thought to it? What relevance does it hold for my students? Will the topic, lesson, unit engage my students in a thoughtful and valuable manner or will it simply be engaging for engagement's sake? What will time spent on this topic likely push out of the curriculum, and which of the two or three topics is more significant? Will this topic or unit of study raise profound questions in their minds about what it means to be human, and/or a member of this society and the world? If I teach this particular topic/unit, is it, in and of itself, complete or does it behoove me to segue into another topic/unit that will illuminate and/or extend the focus of the initial topic/unit?

What follows are two specific suggestions for establishing links between human rights and genocide issues and genocide education:

1.   Curriculum Integration. Using five overlapping dimensions from the emerging field of human rights education, as well as content drawn from genocide education, it is possible to develop core themes that unify the study of human rights and genocide. Flowers and Shiman (1997) have, for example, concisely articulated the five overlapping dimensions that define human rights education in the following way: "1. inform and instruct about human rights and the responsibilities which accompany them; 2. empower individuals and groups with the knowledge, skills and attitudes necessary to realize a more just global society; 3. mobilize individuals and groups to work on behalf of those needing support; 4. protect against future human rights abuses; and 5. reconstruct society in accordance with the principles of justice, caring and human dignity" (p. 174).

Building on these dimensions, Shiman and Fernekes (1999) describe a series of curricular themes that provide opportunities to integrate study of the Holocaust and other genocides within a human rights perspective. One of the key themes deals with "constructing the other," otherwise known as the devaluation and dehumanization by perpetrators of targeted groups, a process characteristic of virtually all genocides (Shiman & Fernekes, 1999, pp. 53-62). Through stereotyping, labeling, and eventually the concrete denial of human rights, perpetrators marginalize targeted groups by reinforcing negative images that are often disseminated through the mass media and the state's educational system. The human rights concepts which run counter to "constructing the other" deal not only with knowledge (combating stereotypes with accurate information) but also with attitudes and values (justice, equity) and skills (sustaining inclusive bonds with a wide universe of people). As students examine historical examples such as, for example, the prevalence of racism in Nazi ideology, the ever-increasing denial of civil rights in the early years of Nazi rule, and the use of strategies of dehumanization in the camp system and the euthanasia program, the language of human rights can and should be applied to these acts. The Nazi state worked diligently to promote inequality and foster the exclusion of targeted groups, while a fundamental tenet of democratic practice is to expand opportunities for civic participation and guarantee equal justice under the law. Embedded in daily violations of human rights, these processes of dehumanization erode human dignity and place targeted groups in danger. *These "preconditions" are found in many historic cases of genocide which strengthen the argument in favor of the integrated study of human rights and genocide.*

Instead of dichotomizing the study of these two important curricular topics, the use of integrative themes can facilitate a deeper understanding of how human rights and genocide are closely connected both theoretically and practically. *The lesson is clear—when human rights violations are systematic and pervasive, the seeds of a potential genocide are possibly being sown.* Thus, classroom instruction about human rights and genocide should not be conducted in isolation, but rather as an integrated enterprise.

2. **Influencing Policy and Curriculum Standards.** Major educational organizations (such as the National Council for the Social Studies, the Council of Chief State School Officers, the National Association of Secondary School Principals, the National Council of Teachers of English) need to make the integration of human rights education and genocide education a higher priority in state and national curriculum standards. Although selectively mentioned within such documents, human rights and genocide education require a more comprehensive presence in curriculum standards, and they should not be presented as unrelated, separate topics.

Social Participation and Citizen Responsibility

The historical record demonstrates that citizen action has been essential in (1) the development of international awareness about human rights abuses and genocidal acts and (2) the mobilizing of expertise, public opinion, and other resources to influence the creation of the international human rights legal system, which today functions on both regional and global levels. For students to understand how public policy can be influenced, they need to participate in activities, under teacher guidance, which assist them in the development of active citizenship skills. The following opportunities provide for such skill development—which, again, are tied to content about human rights and genocide:

1. When teaching about the Holocaust and other genocides, instructors need to establish clear linkages to core human rights concepts and themes. Human rights content (knowledge, values, attitudes and social participation behaviors) should be introduced early in a student's education, and integrated with the study of genocide-related issues at developmentally appropriate points.

2. Teachers and students can assist with the creation of libraries (both in school libraries and public libraries) of materials on human rights and genocide (including testimonies from victims and perpetrators) to educate the public about the issues.

3. A network of educators and students (the scope could be international) could be developed to advocate the inclusion of human rights concepts and themes in genocide education. This network should be inclusive of formal (schooling) and nonformal (families, workplaces, religious institutions) settings, and be informed by the work of various countries as reflected in their national plans for the U.N. Decade of Human Rights Education.

4. Teachers and students can cofound a student-led Amnesty International Adoption group. In such groups, students work on the behalf of prisoners of conscience across the globe. Although the main focus of such groups is a wide-range of human rights violations and not genocide per se, such work provides students with unique insights into problems faced by nations and individuals across the globe, some of which periodically result in genocidal massacres, if not outright genocide.[7]

5. Students could work with local, state, and national media to expand awareness of potential and actual genocides and their connections to human rights violations. Letters to the editor, petitions, sponsorship of public affairs announcements, and advocacy of campaigns to influence legislation are tactics which can be used.

6. Interested educators, students, and policymakers could form coalitions with human rights organizations to monitor the voting records of elected officials on issues directly related to genocide prevention and intervention. Such information could be disseminated widely and form the basis for discussions regarding the role of the U.S. government and its leaders vis-à-vis the prevention of genocide.

CONCLUSION

When all is said and done, the key to education about human rights and genocide is to educate citizens to care about the deprivation of human rights of others and to help them develop the means and methods to act upon their concerns. In a world rife with horrific daily acts of injustice—including torture, forced starvation, episodes of mass killings, and genocide—it behooves the global community to prepare its citizens to act on the behalf of others who are the brunt of such miscarriages of justice and horror. To focus, however, solely on human rights violations and not genocide or, conversely, on genocide but not the human rights violations that lead up to and can and often do culminate in genocide, is counterproductive. In order for the world community to combat both phenomena effectively, an effort must be made to prepare individuals—both the average citizen as well as policymakers and activists—to recognize the significance of defending human rights, and how the erosion of human rights' guarantees often serves as a precursor to genocide. An integrated approach to the study of human rights and genocide can promote citizen action to enforce human rights standards and to take proactive steps supporting genocide intervention and prevention.

NOTES

1. The UNCG was actually adopted—as opposed to ratified—prior to the approval of the Universal Declaration of Human Rights; the UNCG on December 9, 1948, and the U.N. Declaration of Human Rights on December 10, 1948.

2. For a fuller discussion of Lemkin's efforts, see Lemkin (2002.)

3. For a detailed discussion of the many problematic aspects of the definition settled on by the United Nations, see Chapter 2, "The Genocide Convention," in Kuper (1981).

4. In 1979 Pol Pot and his former Foreign Minister, Ieng Sary, were tried in absentia and convicted of genocide in 1979 by the Vietnamese-backed Government that replaced the Khmer Rouge, but that trial was never accepted abroad as legally sound (Mydans, 1997, p. 4).

5. For a detailed and highly readable discussion of the war crime tribunals and related events, see Neier (1998).

6. For a detailed history of the genesis and evolution of Holocaust education in the United States, see Totten (2001.)

7. For information about student A.I. Adoption Groups, contact Amnesty International USA at aimember@aiusa.org or 322 8th Avenue, New York, NY 10001; 212-807-8400.

REFERENCES

Amnesty International. (2002). *International Criminal Court: U.S. efforts to obtain impunity for genocide, crimes against humanity and war crimes*. London: Author.

Axel, D. K. (1999). Toward a Permanent International Criminal Court. In B. Cooper (Ed.), *War crimes: The legacy of Nuremberg* (pp. 313-322). New York: TV Books.

Banks, D. (2000). Promises to Meet: The National Survey of Human Rights Education [On-line]. Available http://hrusa.org/education/promisestomeet.htm

Barash, D. P. (2001, February 23). Buddhism and the "subversive science." *The Chronicle of Higher Education*, pp. B-13-B15.

Berenbaum, M. (1993). *The world must know: The history of the Holocaust as told in the United States Holocaust Memorial Museum*. Boston, MA: Little, Brown.

Buergenthal, T. (1988). *International human rights in a nutshell*. St. Paul, MN: West.

Cooper, B. (Ed.). (1999). Epilogue. In *War crimes: The legacy of Nuremberg* (pp. 323-327). New York: TV Books.

Dadrian, V. (1997). *The history of the Armenian genocide*. Providence, RI: Berghahn.

Eisner, E. (1979). *The educational imagination: On the design and evaluation of school programs*. New York: Macmillan.

Flowers, N., & Shiman, D. A. (1997). Teacher education and the human rights vision. In G. J. Andreopoulos & R. P. Claude (Eds.), *Human rights education for the twenty-first century* (pp. 161-175). Philadelphia: University of Pennsylvania Press.

Jackson, R. H. (1997). Opening address for the United States, November 21, 1945. In M. R. Marrus (Ed.), *The Nuremberg war crimes trial 1945-1946: A documentary history* (pp. 79-85). New York: Bedford Books.

Korey, W. (2001). *NGOs and the universal declaration of human rights: "A curious grapevine."* New York: Palgrave.

Kuper, L. (1981). *Genocide: Its political use in the twentieth century*. New Haven, CT: Yale University Press.

Lauren, P. G. (1998). *The evolution of international human rights: Visions seen*. Philadelphia: University of Pennsylvania Press.

Lemkin, R. (2002). Totally unofficial man. In S. Totten & S. Jacobs (Eds.), *Pioneers of genocide studies* (pp. 365-399). New Brunswick, NJ: Transaction.

Marrus, M. (1997). *The Nuremberg War Crimes Trial 1945-46: A documented history*. Boston: Bedford Books.

Mills, K. (1998). *Human rights in the emerging global order: A new sovereignty?* New York: St. Martin's Press.

Mydans, S. (1997, July 27). Khmer Rouge say Pol Pot is tried and sentenced. *The New York Times*, p. 4.

Mydans, S. (1999, January 3). The long, erratic arm of justice. *The New York Times*, p. 11.

Neier, A. (1998). *War crimes: Brutality, genocide, terror, and the struggle for justice*. New York: Times Books.

New Jersey Commission on Holocaust Education. (1995). *The Holocaust: The betrayal of mankind. curriculum guide for grades 7-12*. Trenton: NJ Commission on Holocaust Education.

New York State Department of Education. (1985). *Teaching about the Holocaust and genocide: Introduction. The human rights series, Vol. I*. Albany: University of the State of New York.

Shiman, D. A., & Fernekes, W. R. (1999, March/April). The Holocaust, human rights and democratic citizenship education. *The Social Studies, 90*(2), 53-62.

Totten, S. (2001). Holocaust education. In W. Laqueur (Ed.), *The Holocaust encyclopedia* (pp. 305-312). New Haven, CT: Yale University Press.

United States Holocaust Memorial Museum. (2000). *Teaching about the Holocaust.* Washington DC: United States Holocaust Memorial Council.

Weiner, T. (1998, July 26). U.S. drops plan to raid bosnia to get 2 Serbs. *The New York Times*, pp. 1, 8.

CHAPTER 10

THE INTERVENTION AND PREVENTION OF GENOCIDE

Where There is the Political Will, There is a Way

Samuel Totten

INTRODUCTION

As is readily apparent from the mass of killing fields that stained the globe throughout and late into the twentieth century, there is a dire need for an efficacious system of intervention and prevention of genocide. Although far from accomplishing the latter goal, progress has been made and there is hope on the horizon. Such hope, though will only become reality if there is the will to act early and effectively when genocide early warning signals are first detected.

CONFRONTING GENOCIDE

Over the course of the past century, many perpetrators of genocide have gone unfettered in their efforts to exterminate certain groups of people,

Teaching About Genocide: Issues, Approaches, and Resources, 275–298
Copyright © 2004 by Information Age Publishing
275

while attempts to prevent or halt genocide have been, for the most part, unsuccessful. *The terrible fact is, in more cases than not, when the world community did act, it was after a particular genocide had already begun; and as a result of that, the murdered and the maimed already numbered in the tens of thousands, if not more.*

Indeed, far too little has been done to intervene in ethnic strife and other types of conflict *prior* to the outbreak of genocide. Granted, it is not a simple task to ascertain whether, for example, a civil war is going to creep toward or explode into a genocidal situation. Furthermore, in light of the plethora of violent conflict in the world, it is no small task to keep a hand on the pulse of such conflicts in order to ascertain the likelihood or possibility that any one of them is going to "go genocidal." Furthermore, not all extrajudicial killings or massacres of innocent victims culminate in genocide.

Many (scholars and activists, alike) have argued that in the information age it should be fairly easy to detect a potentially explosive/genocidal situation in the making. Such individuals have pointed out that the 24-hour news coverage by such outlets as CNN, along with the daily coverage of world news by major newspapers and the Internet, provide windows of opportunity into ascertaining genocidal situations early on. Furthermore, they argue, the hundreds of nongovernmental organizations—those whose focus is human rights, the plight of refugees, famine relief, and so forth—that operate across the globe constitute an unofficial and quasi early warning system. Finally, as some have pointed out, there also exists sophisticated government surveillance (e.g., space-based satellites) systems that provide around-the-clock information. Many have argued that with such an abundance of information sources, organizations, and mechanisms, there should be little problem in detecting potential genocidal situations. If only that were the case.

First, it must be noted that none of the aforementioned news sources have the mandate, let alone the means, to monitor, month-in and month-out, the day-to-day statements, decisions, actions, and reactions of human rights violators in order to ascertain the evolution of—let alone the intentions behind—such actions. Indeed, time and again, the media—due in part to the way information is gathered—have misinterpreted and misjudged the nature of mass slaughter. In fact, conflicting reports of the same situation by different media sources, humanitarian groups and/or government officials have often confused the issue. The 1994 Rwanda genocide presents a classic example of the aforementioned situation. In *The Limits of Humanitarian Intervention: Genocide in Rwanda*, Kuperman (2001) reports that

Starting on April 11, [1994,] just four days into the violence, news reports indicated that fighting in Rwanda had "diminished in intensity." Three days later it was reported that "a strange calm reigns in downtown" Kigali.... The commander of Belgian peacekeeping in Rwanda confirmed: "The fighting has died down somewhat, one could say that it has all but stopped." As late as April 17, UNAMIR commander [Romeo] Dallaire told the BBC that except for an isolated pocket in the north, "the rest of the line is essentially quite quiet."

Only on April 18 did a Belgian radio station question this consensus, explaining that the decline in reports of violence was because "most foreigners have left, including journalists." The exodus of reporters was so extreme that it virtually halted Western press coverage.... For example, France's *Le Monde* went silent for four days and Britain's *Guardian* for seven. Ironically, this was just as the killing peaked and spread to Rwanda's final two prefectures.

Three days into the killing, on April 10, the *New York Times* quoted an estimate of 8,000 dead in Kigali by the French humanitarian group *Médecins San Frontières* and "tens of thousands" by the International Committee of the Red Cross (ICRC). Three days later, the RPF [Rwandan Patriotic Front] offered its own estimate that "more than 20, 000" were dead in Kigali. However, during the second week, media estimates did not rise at all and so failed to approach levels that commonly would be considered "genocidal" for a country of 8 million people that included 650,000 Tutsi....

Death tolls of 20,000 are not uncommon in civil wars and generally are not considered genocidal. In order to make a determination of genocide in a specific case, one has to examine the details of the violence including whether the victims were noncombatants, were killed deliberately (rather than in crossfire), were members of a single group defined by ascriptive characteristics, and were being targeted exclusively because of their identity rather than for suspected actions such as supporting government rebels. In the absence of such details, only a high death estimate would suggest the possibility of genocide. Accordingly, given the early confusion about the nature of the violence in Rwanda, a death toll of 20,000 during the first week did not seem to indicate the occurrence of genocide. (pp. 26-27)

Variations of the above situation have been "played out" time and again throughout the last century when massacres have been perpetrated. Ascertaining what is and what is not "genocidal" is, obviously, not an exact science. At least not yet. Be that as it may, there are certain actions that can and should be implemented in order to better judge whether there is a possibility that a situation is evolving toward genocide or not. These shall be discussed shortly.

Second, when a major crisis hits a region, television news teams flock to the area to cover the events, but as soon as the events are no longer "newsworthy" or a crisis hits another part of the world, such teams are often enroute elsewhere. As a result, television news networks are not likely to be able to gauge whether a crisis-oriented situation is moving toward genocide. Newspaper reportage of major conflicts is not as "flighty" as television coverage, but even major newspapers' intensive coverage of an area has proved inadequate in detecting early signals (e.g., ever-increasing propaganda that criticizes, ostracizes, and ultimately calls for the exclusion, if not violence, against targeted groups; repressive policies and actions that discriminate against, ostracize, and isolate certain groups from the rest of society; repressive measures that become increasingly punitive and violent; extreme nationalism that results in repressive and oppressive policies) that genocide is possibly on the horizon.

Third, while surveillance operations conducted by independent nations are capable of detecting mass movements of people, piles of bodies, or huge gaps in the earth that may constitute mass burial grounds, what they detect is often taking place during the course of a genocide or in the aftermath—*again, not prior to it, when the most people could be saved from extermination*. Furthermore, if a neutral nation is not inclined to intervene, it is easy for a government bureaucrat or higher echelon official to ignore and/or not act on that which is detected in reconnaissance photos.

Fourth, although the field personnel of many nongovernmental organizations (NGOs) are often privy, if not outright witnesses, to major human rights infractions, it is not their job or within the realm of their expertise to track human right violations or other types of conflict for the purpose of ascertaining whether a potential genocide is on the horizon. Nongovernmental organization personnel dealing with famines, the massive movements of refugees, and other major crises, already have their hands full, often working in far-flung areas under chaotic and dangerous conditions. Thus, while such individuals and groups can, when possible, provide critical information regarding major human rights infractions, their efforts are often bound to be, at best, sporadic and fragmentary.

At one and the same time, it must be noted that in certain cases where the press and various NGOs only managed to provide fragmented information about incidents of cataclysmic violence that eventuated in genocide, the information would have proved invaluable *had* the reports been conscientiously followed up and fully investigated by the United Nations and/or independent governments. The reality, though, is that many individuals in the United Nations and officials of independent governments have cavalierly disregarded or outright dismissed such reports. Here again, Rwanda provides a classic example. In her hard-hitting book, *A People Betrayed: The Role of the West in Rwanda's Genocide*, Linda Melvern (2000), an investigative journalist, reports the following:

> In the years immediately before the 1994 genocide ... human rights groups [were] gathering information and becoming increasingly active in Rwanda.

They reported extensively on the Bagogwe massacres in January 1991 and the February 1992 massacre at Bugesera, describing the involvement of military and local government officials. In some communities, Tutsi had been repeatedly attacked and the military had distributed arms to civilians who supported [President] Habyarimana.

So bad did the situation in Rwanda become that, in January 1993, a group of international human rights experts from ten countries collected testimony from hundreds of people, interviewed witnesses and the families of victims, and reviewed numerous official documents. In March 1993, a report was published revealing that in the previous two years those who held power in Rwanda had organized the killing of a total of 2,000 of its people, all Tutsi.

The word genocide was not used in the report, being considered too highly charged by some of the group's authors, but a press release distributed with the report carried the headline: "Genocide and War Crimes in Rwanda."

There was little international concern when the human rights report came out [italics added]. Only the Belgian government recalled its ambassador from Kigali for consultations, and the Rwandese ambassador in Brussels was told that Belgium would reconsider its economic and military aid unless steps were taken to rectify the situation. The French ambassador dismissed the massacres as rumors. (pp. 55-56)

Similarly, Kuperman (2001) notes that prior to the 1994 genocide in Rwanda

death lists containing hundreds of names came to light in the succeeding two years. In spring 1992, the Belgian embassy in Kigali reported an anonymous allegation of a "secret headquarters charged with exterminating the Tutsi of Rwanda to resolve definitively ... the ethnic problem in Rwanda and to crush the domestic Hutu opposition." In March 1993 an international human rights panel reported that most victims of sporadic massacres in Rwanda were killed because they were Tutsi. During the first week of December 1993 several ominous indications arose: a Belgian cable reported the Presidential Guard conducting paramilitary training of youths; a Rwandan journal reported weapons being distributed to militias; and UNAMIR received an anonymous but credible letter, purportedly from leading moderate Rwandan army officers, warning of a "Machiavellian plan" to conduct massacres throughout the country starting in areas of high Tutsi concentration and also targeting leading opposition politicians. On December 17, the Rwandan press exposed details of a planned "final solution," including militia coordination, transportation arrangements, French military aid, and "identification committees" to compile death lists. These dangers signs were usually conveyed to Brussels within days via cable. (p. 102)

A complicating factor in some, if not many, pregenocidal situations, is that pieces of information and/or warnings may come from unknown parties, and thus it is difficult to ascertain their validity. One such case occurred during the period of the Holocaust. As Pauline Jelinek (2001) writes in her article "File: Allies Tipped Off to Genocide in Early '42," it is now believed that "The West may have been informed about Nazi Germany's plans for the Holocaust months earlier than previously thought. 'It has been decided to eradicate all the Jews,' says a newly declassified document believed to have been obtained by British and American intelligence by March 1942. Previously, historians have judged that the West didn't learn until August 1942 that the Holocaust was happening" (p. 4). As to why the British and U.S. governments did nothing in response to such news, Holocaust historian Richard Breitman answered with a question of his own: "Why would any British or American official pay particular attention to the views of an unknown Chilean diplomat in Prague?" (quoted in Jelinek, 2001, p. 4)

It is also true that when warning after warning of a pending genocide is reported over a period of months (and even years), those on the receiving end often begin to dismiss them. Once again, the 1994 Rwandan genocide offers an instructive example. A U.S. government official, Colonel Tony Marley, the U.S. State Department's political-military advisor in the region, asserted that "'We had heard them cry wolf so many times' that the new warnings fell on deaf ears" (quoted in Kuperman, 2001, p. 105).

Finally, intelligence gathering is not as strong as it could or should be when racial, religious, national, political, and ethnic strife that has the potential to become genocidal is taking place. As a result, early warning signals are either missed entirely or not pieced together in order to demonstrate that there is a pattern indicating the likelihood of genocide. There is also the problem in which one intelligence body neglects, for whatever reason, to share its intelligence with other parties (including the United Nations).

Ultimately, what is needed is an efficient, effective, and highly regarded system whose express purpose is the prediction of potential genocides—and then the will by the international community (or at least members of it) to act immediately and effectively.

DETECTING AN ACTUAL CASE OF GENOCIDE EARLY-ON

Barbara Harff (1993), a political scientist at the U.S. Naval Academy, has argued that

> The UN is in principle structurally capable of dealing with human rights abuses. One problem lies in the lack of data that allows for accurate predictions that would give adequate time to deal with impending crises. The

establishment of an early warning system similar to forecasting is indispensable to both prevention and readiness to respond in a manner that would
prevent escalation. Databanks must include background information on
pre-conflict situations. Early warning models need to be tested against real
world conflict scenarios. Although early warning of escalation may not lead
necessarily to effective responses, it makes planning relief and peacekeeping
efforts feasible. (p. 4)

Such a system must be capable of both detecting and tracking those situations that have the potential to turn genocidal. To be recognized as a
legitimate source of information, such a system will have to have the
imprimatur of the international community (e.g., United Nations or the
International Criminal Court [ICC]). That said, while it needs to be supported by the international community, the system, ideally, should be
operated by an independent agency that is not constrained by the whims
of the United Nations—and particularly the U.N. Security Council. The
system must not only be equipped with the latest and best monitoring,
computing, and communication technology, but be operated by non-
political, objective personnel with an expertise in the operation of early
warning systems and early forecasting of conflict escalation. Ideally, members of the team should, collectively, have an expertise in the areas of
international law, human rights, and genocide. It is a given, or at least
should be, that such a system must be based on the latest and best knowledge regarding the preconditions of genocide, risk assessment factors and
data-risk banks (Gurr, 1998), early warning indicators, and "accelerators"
(e.g., "accelerating events resemble the last stage of a crises before open
conflict occurs") (Harff, 1998, p. 71). Ultimately, such a system needs to
be capable of providing fairly definitive statements regarding situations
that have the likelihood of degenerating into genocide.

The significance and value of early warning cannot and should not be
underestimated. As Davies and Gurr (1998) state:

The goal of early warning … *is proactive engagement* in the earlier stages of
potential conflicts or crises, to prevent *or at least alleviate their more destructive
expressions* [italics added]. The chances for successful, cost effective, preventive conflict peace-building and pre-emptive peacemaking initiatives … are
usually more cost-effective strategies than late attempts at crisis avoidance,
containment (peacekeeping) and/or humanitarian assistance (as in Bosnia
or Congo/Zaire). The former requires longer-term risk assessments, but
even reliable short-term early warnings allow for more effective late-stage
crisis management, facilitating advance planning and deployment of supplies and personnel. (p. 2)

Among some of the many situations an early warning system would
need to monitor are: massive human rights violations; extreme national-

ism resulting in the scape-goating, discrimination, and ostracism of certain groups; "the exclusion of the victim from the universe of obligation" (Fein, 1990, p. 5); "pervasive racialistic ideologies and propaganda [endemic] in the nation-state's society" (Porter, 1982, pp. 17-18); extreme political strife; internal strife/civil war; state failure; ethnopolitical conflict; food crises (Davies & Gurr, 1998, p. 3); mass exodus of refugees; mass internal displacement of citizens; "the possibility of retaliation for genocidal acts by kin of the victims of an earlier genocide" (Porter, 1982, pp. 17-18); and a total breaking off of communication/interaction with other nations, and/or the total closing of borders by a nation (e.g., such as the Khmer Rouge's actions in Cambodia in 1975). The "exclusion of the victim from the universe of obligation" might include, for example, any of the following: governmental actions that dehumanize and/or isolate groups of people from the body politic; threats or actions taken to commit "ethnic cleansing" (as in Bosnia-Herzegovina in the 1990s); and television, radio, or newspaper commentary in a country that systematically disparages, maligns, or attempts to ostracize a particular group (e.g., the Nazis' scurrilous comments on the radio and newspapers about Jews and others) and/or calls for the removal, harm, or destruction of a particular group of individuals (e.g., broadcasts calling for the destruction of the Tutsis on the Rwandan government-controlled radio prior to and during the 1994 Rwandan genocide; and the television commentary in the former Yugoslavia during the 1990s in which both Slobodan Milosevic and Franjo Tudjman took over the media and used them as propaganda tools to incite their followers against their declared "enemies").

TENTATIVENESS TO INTERVENE

As noted earlier, over the past century or so most nations have been extremely tentative to undertake or even support humanitarian efforts to attempt to prevent genocide. Among the key factors impacting such tentativeness are: the primacy of sovereignty, the concept of internal affairs, and realpolitik. Integrally tied to the latter three factors is the lack of political will.

The Issue of Sovereignty (and "Internal Affairs"): A Sticking Point That Is Undergoing a Metamorphosis of Sorts

Since the Treaty of Westphalia in 1648, state sovereignty has not only been a "defining principle of interstate relations and a foundation of world order" (Weiss & Hubert, 2001, p. 5), but a notion that most world leaders have—at least in regard to their own states—perceived as being almost sacrosanct. It is also true that the concept "lies at the heart of both customary international law and the United Nations Charter and remains both an essential component of the maintenance of international peace and security and a defense of weak states against the strong" (Weiss & Hubert, 2001, p. 5). That is the positive side. The negative side is that over the centuries—through today—many leaders have interpreted sovereignty to mean that they are allowed to treat their citizens in any way they wish, and no one—certainly not another nation nor any intergovernmental bodies—have the right to interfere with their nation's "domestic" or "internal" affairs. As a result, such leaders have tortured, maimed, starved, enslaved, murdered, and committed genocide at will with the "understanding" that no one but no one would interfere with their actions. Furthermore, since the establishment of the United Nations in 1948, whenever a nation or group dared to criticize or threaten to intervene in a situation involving massive human rights violations (or "internal affairs"), the targeted state, more often than not, cited Article 2(7) in the United Nations Charter that provides that "nothing contained in the present Charter shall authorize the United Nations to intervene in matters that are essentially within the domestic jurisdiction of any State or shall require the Members to submit such matters to settlement under the present Charter."

Be that as it may, a whole host of factors have resulted in a weakening of the "sanctity" of sovereignty—or at least it appears that way today. Among such factors are the following: the new international human rights regime (e.g., the U.N. Declaration of Human Rights, the U.N. Convention on the Prevention and Punishment of Genocide, as well as others), the end of the Cold War, and the information age. As Weiss and Hubert (2001) note: "Not only have technology and communications made borders permeable but the political dimensions of internal disorder and suffering have also often resulted in greater international disorder. Consequently, perspectives on the range and role of state sovereignty have, particularly over the past decade [or so—1990-present], evolved quickly and substantially" (p. 5). Part and parcel of the reasons for this evolution is the way the international community—or at least significant segments of it—currently perceive and interpret the issue of "domestic" or "internal" affairs. More specifically, the latter note that Article 1(2) of the U.N. Charter asserts that "All Members, in order to ensure to all of them the rights and benefits resulting from membership, shall fulfill in good faith the obligations assumed by them in accordance with the present Charter." They further note that such obligations require member states to "promote and encourage respect for human rights and fundamental freedoms *for all without distinction as to race, sex, language, or religion*" [italics added]. As Weiss and Hubert (2001) further note: "The Charter elevates the solution of economic, social, cultural, and humanitarian problems as well as human rights, to the international sphere. By definition, these matters cannot be said to be exclusively domestic, and solutions cannot be located exclusively within the sovereignty of states. Sovereignty therefore carries with it primary responsibility for states to protect persons and property and to discharge the functions of government adequately within their territories" (p. 8). What this means, then, is that those nations that mistreat their citizens—be it through starvation, torture, massacres, genocide or in other ways that contravene the U.N. Declaration of Human Rights—have, for the time being, in a sense, abdicated their "right" to sovereignty.

On a related note, another major and new challenge to the principle of sovereignty is

> the broadening interpretation of [that which constitutes a] threat to international peace and security, [which is] the Charter-enshrined license to override the principle of intervention.... Collective efforts by the UN to deal with the *internal* [italics added] problems of peace and security, and gross violations of human rights, including genocide, have therefore run against the grain of the claim to sovereignty status as set out in the Charter.... [This broadening of the interpretation] actually began during the Cold War with the [UN] Security Council's coercive decision in the form of economic sanctions and arms and oil embargoes against apartheid in Southern Rhodesia and South Africa.... An affront to civilization was packaged as a threat to international peace and security in order to permit action." (Weiss & Hubert, 2001, p. 9)

Thus, the precedent was set for conducting interventions in the face of potential and/or actual cases of genocide.

Certainly three other events that point to the shifting perception of sovereignty are the relatively recent establishment of the International Criminal Tribunal for Rwanda (ICTR), the International Criminal Tribunal for the former Yugoslavia (ICTY), and the International Criminal Court (ICC). All three bodies, which were established by the United Nations, have the mandate to arrest and try individuals suspected of crimes against humanity, war crimes, and genocide.[1]

GARNERING THE POLITICAL WILL TO ACT IN ORDER TO *EFFECTIVELY* HALT A GENOCIDE

Without the willingness of an international body, a group of nations, or, at the very least, a single nation, to act to prevent a genocide or intervene early on, *not even the most sophisticated and most efficiently operated genocide early warning system will be of much, if any, use.* As Donald Krumm (1998), Humanitarian Affairs Consultant with the U.S. Committee for Refugees, asserts, what is of crucial importance is "urgently engaged early action by governments and international organizations based on having received warning. Many would argue that, had warning been matched with decisive action, the disintegration of the former Yugoslavia and the genocide in Rwanda could have been averted.... [W]hat is missing is the vision and sense of mission to redesign the international response system to move quickly to prevent or limit the damage resulting from conflict" (p. 248). At the heart of such indecision is the critical component of political will (or, rather, the lack of political will) by intergovernmental bodies such as the United Nations and/or individual nations—including, of course, the United States, the sole superpower today—to abide by the United Nations Convention on the Prevention and Punishment of Genocide to conscientiously and consistently work to prevent genocide from becoming a reality.

Speaking of the lack of political will and the disastrous results of such, genocide scholar Helen Fein (1992) has argued that

a monitoring scheme and an "early warning" system alike depend both on political will to pay the costs of sanctions, deterrence or intervention, and reliable reports on life integrity violations and patterns of discrimination (p. 50); state terror and genocide have been repeated in part because of the inviolability of perpetrators in the international system, *and* the lack of censure or sanctions from other states, regional and international organizations, from trading partners, patrons and allies. *Thus, we must consider not only how to target or move dangerous states to change but also how to move the movers to act* [italics added]. (pp. 50-51).

Discussing various factors that complicate the issue of "political will," Bruce Jentleson (1998), a professor of public policy and political science at Duke University, argues that:

Even if early warning is achievable, there remains the problem of mustering the political will necessary to act. Although the essence of the strategic logic of preventive diplomacy is to act early, before the problem becomes a crisis, it is often the same lack of a sense of crisis that makes it more difficult to build the political support necessary for *taking early action. In a traditional realist calculus, such situations have had difficulty passing muster both because the immediate tends to take priority over the potential, and because many contemporary ethnic conflicts involve areas which, as the U.S. ambassador to Somalia candidly put it, are "not a critical piece of real estate for anybody in the post-Cold War world."* [Italics added]

The latter comment parallels the cavalier assertion by then U.S. Secretary of State James Baker regarding the Bush Administration's (1988-

1992) position vis-à-vis the conflict brewing in the former Yugoslavia: "We don't have a dog in that fight" (quoted in Ronayne, 2001, p. 115).

Continuing, Jentleson (1998) states that

It is only partially parochial to say that this problem is particularly bad in post-Cold War American politics. But it is true in two senses. First, although the United States cannot and should not be expected to act unilaterally or always shoulder the major share of responsibility, U.S. leadership continues to be crucial to concentrated multilateral action. In Bosnia, although there was plenty of responsibility to go around, the corollary to the impact that the United States had, once it finally became serious about playing a lead role, is the debilitating effect of its earlier halting policies. In Rwanda, despite the historically-based lead roles of Belgium and France, Adelman and Suhrke (1996) argue that "by acts of omission, the United States ensured that neither an effective national response nor a collective UN effort to mitigate the genocide materialized" (p. 73).

There is also a problem of political will for international institutions. Although the United Nations has institutional weaknesses that are its own fault and responsibility, it does also get unfairly blamed for inaction and indecisiveness when the lack of will really resides with its members. The UN authorities in charge of UNPROFOR [the UN Protection Force] deserve much of the criticism they received for how they managed the Croatia- Bosnia peacekeeping mission, but the exceedingly limited mandate UNPRO-FOR was given from the start was the doing largely of the key permanent members of the Security Council: the United States, Britain, France, and Russia. Similarly, in Rwanda, one of UNAMIR's [the U.N.'s Assistance Mission to Rwanda] problems all along was the refusal by the major powers to provide sufficient financing or mandate. And when the early warning was sounded about another crisis brewing in Burundi in late 1994 and early 1995, including a charge by former [U.S.] President Jimmy Carter that the willingness to send troops to Bosnia but to keep doing very little in Burundi was racist, the most the Security Council mustered was a resolution for more contingency plans. Nor is this only true of the Security Council (pp. 306-308).

In the Executive Summary of the Organization of African Unity's (OAU, 2001) "International Panel of Eminent Personalities: Report on the 1994 Genocide in Rwanda and Surrounding Events," which was formed to investigate the Rwandan genocide and to contribute to the prevention of further conflicts in the region, it is reported that

The Panel endorsed the finding of the earlier Carlsson Inquiry report that "the U.N.'s Rwandan failure was systemic and *due to a lack of political will*" [italics added] (p. 140). The Panel found that "[j]ust about every mistake that could be made was made" (p. 140). The Panel found that the U.N. did not perceive the U.N. Assistance Mission to Rwanda (UNAMIR) as a particularly difficult mission, and so did not provide UNAMIR with an adequate force or mandate. The Panel also argued that the U.N. had compromised its integrity by maintaining "insistent and utterly wrong-headed neutrality regarding the genocidaires." (p. 140)

The Panel suggested that the U.N. Security Council and Secretariat had paid too much attention to cease-fire negotiations rather than ending the massacres.... The Panel found clear evidence that "a small number of major actors," including Belgium, France, and the United States, could have directly "prevented, halted, or reduced the slaughter." (p. 140)

The Panel called for a substantial re-examination of the 1948 Genocide Convention with attention to, *interalia*: 1. the definition of genocide; 2. a mechanism to prevent genocide; and 3. the legal obligation of states when genocide is declared. The Panel also proposed the institution of a special Rapporteur for the Genocide Convention, within the office of the U.N. High Commissioner for Human Rights, to provide the U.N. Secretary General and Security Council with pertinent information concerning situations that are at risk for genocide." (p. 140)

The question that remains is: How is pressure to be applied to the United Nations and independent nations to act expediently when a potential genocide is on the horizon? To make any headway at all in addressing the intervention and prevention of genocide, it is imperative that scholars, policymakers, and activists focus their attention on this most complex and contentious of issues. Among some of the many issues worthy of serious examination are: (1) way(s) to alter the tentativeness of the political will of the United Nations and/or independent nations to act, including the application of pressure by the constituents of such bodies—in the case of the latter, it would be petitions, letters, and rallies directed at the nation's leaders by their citizens; (2) ways in which the general public can be quickly and thoroughly informed of a pending genocide, and efficiently polled regarding its opinion about whether to intervene or not; and (3) methods for applying the most effective pressure on the United Nations and the governments of independent nations to act, sooner rather than later, to prevent genocide from becoming a reality. Finally, the few situations where the international community mustered the political will to act in order to prevent a potential genocide should be analyzed to assess what can be learned and applied to future pregenocidal situations. Among two of the most notable situations are those of the Bahá'í community in Iran in the 1980s, and that of East Timor at the close of the twentieth century. (Both are discussed later herein.)

A STRONG MANDATE, A WELL-TRAINED AND
WELL-RESOURCED FORCE, AND A SPEEDY RESPONSE

Assuming that early warning indicators are acted on in an expedient fashion and there is the political will to act (and that is a huge assumption),
there is still the critical need for the intervening bodies to implement an
effective effort to halt the genocide. Anything less than a strong mandate
with a well-trained, well-equipped, and adequately sized contingent of
personnel working in a timely manner is bound to be insufficient—if not
an abysmal failure.

Time and again, though, in the recent past (e.g., the 1990s), weak
mandates with under-resourced, and/or sorely under-manned, and slowly
implemented operations were sent to Rwanda, Bosnia, and Kosovo—and
in every case, the results were disastrous. As Jentleson (2000) points out,
"While the first decade of the post-Cold War era did have some preventive successes, it was more marked (or marred) by missed opportunities.
Even in such 'success' cases as Kosovo (1999) and East Timor (1999),
whatever may have been achieved was achieved only *after* mass killings,
only *after* scores of villages were ravaged, only *after* hundreds of thousands
were left as refugees. Yes, these conflicts were stopped from getting
worse—but they were already humanitarian tragedies" (p. 5).

As for the 1994 Rwanda genocide, it is now a commonly known fact
that Canadian Major General Romeo Dallaire, the former commander of
UNAMIR force, "argued forcefully and consistently that the United
Nations could have halted the genocide had it been willing to commit
more troops with authority to take decisive, preventive action" (Ronayne,
2001, p. 176). Testifying at a trial conducted by the U.N. International
Criminal Tribunal for Rwanda (ICTR) in February 1998, Dallaire
reported how he had repeatedly requested that the United Nations provide him with additional troops and enlarge his mandate—and how these
requests were made before Rwandan President Habyarimana's plane was
shot down—but the requests fell on deaf ears. When he was asked during
his testimony if he and his troops could have possibly halted the genocide, Dallaire, stated: "Yes, absolutely. We had a time frame of two weeks
[the first two weeks of the crisis] easily where we could have made the task
of killing much more difficult for these people" (quoted in Ronayne,
2001, p. 176). Continuing, Dallaire stated that

> one obvious tactic would have involved preventing Hutus from establishing
> their network of roadblocks and barricades that kept Tutsis from fleeing and
> aided in rounding them up for massacres. Instead, according to the general,
> his force was *too small, ill-equipped,* and inadequately trained and without

legal mandate to *intervene* [italics added]. Dallaire said that an effective force could and should have been assembled and deployed in April 1994, but the United Nations lacked the political will to do so. (p. 176)

Unlike Rwanda, the situation in Bosnia-Herzegovina did not result in an absolute failure. While the horror in Bosnia-Herzegovina was great and many innocent people were killed, the international community expended a great deal of effort, time, and resources in an attempt to temper the conflict and prevent the ongoing "ethnic cleansing" and bloodshed. And, in fact, great numbers of lives were saved. However, it is also true that the mission in Bosnia was badly flawed. Roy Gutman (1996) argues as much in an article entitled "Bosnia: Negotiation and Retreat":

The mission in Bosnia was a humiliating exercise in "how not to proceed," according to one U.N. ground commander. Even as massive war crimes were being committed, the superpowers of the U.N. Security Council looked the other way, refusing to take a consistent position, distorting historical precedent when convenient. Peace-keeping is not feasible in a raging war, nor is diplomacy, unless it is backed my military force. While U.N. personnel saved many lives, they also may have prolonged the conflict. (p. 187)

Elsewhere in his article, Gutman (1996) discusses the debacle of Srebrenica, one of the so-called U.N. "safe areas." In 1993,

after Srebrenica came under renewed attack ..., the U.N. Security Council, under pressure by non-aligned states, responded in mid-April by proclaiming the town a "U.N. safe area." Three weeks later, the Security Council extended the concept to five other mostly Muslim cities: Tuzla, Gorazde, Zepa in the east, Bihac in the northwest, and Sarajevo. In June 1993, it authorized "necessary measures, including the use of force" to deter bombardment or armed incursion and to ensure the movement of humanitarian aid. This was classic "mission creep," with a vague statement of goals and no agreement on means. Secretary-General Boutros Boutros-Ghali said 34,000 troops would be needed to police the safe areas, but in the event, U.N. members promised 7,000 and only 3,400 were finally deployed, according to Human Rights Watch, quoting U.N. observers in the field. When foreign ministers met in Washington in late May, with diplomacy failed and the U.S. government backing off from its "lift and strike option," they endorsed the plan to protect "safe areas" as the alternative.

But two years later, when Bosnia Serbs conquered Srebrenica, the "safe area" policy collapsed. All the flaws of the U.N. Mission combined into a single disaster: the refusal to confront the Bosnian Serbs blocking food deliveries; the absence of military intelligence; the emphasis on diplomatic process and partition over the protection of the population; and the failure to implement the Geneva Conventions, or even to report on violations. (p. 201)

The so-called safe area at Srebrenica was overrun on July 11, 1995, as the outnumbered Dutch troops watched the Serb forces divide the people within the safe area into two groups, forcing the oldest people, women and children onto trucks and buses and removing the thousands of men and boys to secret locations. Ultimately, approximately 7,000 Muslims were slaughtered, resulting in the single, largest massacre in Europe since the Holocaust.

In many cases, even when an intervention was agreed on, extreme efforts were made by various nations to avoid sending in ground troops. Such a limited intervention can and has had tragic consequences. For example, in March 1999, a decision was made by the North Atlantic Treaty Organization (NATO) to use air power—versus ground troops— aimed at Serb troops accused of violence, repression and "ethnic cleansing" against the ethnic Albanian majority in Kosovo. The controversial NATO air attack lasted for 78 days. As Alvarez (2001) notes in *Governments, Citizens, and Genocide: A Comparative and Interdisciplinary Approach*, ironically and tragically, "the attacks incited the Serb leadership to speed up the policies of ethnic cleansing of Kosovar Albanians, resulting in wide-spread massacres and the uprooting and dislocation of the majority of the population" (pp. 139-140). Had troops been on the ground, they may have been able to prevent the Serbs from carrying out their destructive aims.

Various scholars, nongovernmental personnel, and policymakers have raised the issue of establishing a rapid action force for the express purpose of staving off and/or intervening to halt a genocide. Such a force, many claim, should be voluntary in order to avoid, as much as possible, the evasive and inadequate commitment of troops by individual nations as a result of realpolitik and a lack of political will. Some have strongly argued that it should be a volunteer force under the auspices of the United Nations but not the Security Council, thus removing it from the whims of the U.N. Security Council. Still others have suggested that it be established under a special unit within the U.N.—a unit whose specialty is the prevention of genocide and which has both a genocide early warning system at hand and the mandate to act when a genocide is on the horizon. All have noted that such a force must be well-equipped and well-trained. As one can imagine, this is a controversial issue, but one that deserves ample consideration by all concerned parties.

THE CRITICAL NEED FOR A SYNERGY AMONGST SCHOLARS IN DIFFERENT FIELDS

Over the past 25 years or so, genocide scholars (as opposed to, for example, those in such fields as humanitarian intervention, peace studies, conflict resolution, and international law) have basically focused on nine major concerns vis-à-vis the intervention and prevention of genocide:

1. defining what genocide is and is not;
2. examining and delineating the processes of genocide;
3. analyzing specific genocidal events, including their causes, the individuals, and groups involved (e.g., perpetrators, collaborators, victims, bystanders, rescuers), the ways in which the genocides are perpetrated, the horrific results, and the aftermath;
4. analyzing data from genocidal incidents in order to develop potential early warning signals;
5. developing risk databases;
6. undertaking incipient work on the development of genocide early warning systems;
7. delineating and analyzing the adverse impact of the denial of past genocides;
8. arguing in favor of trying and punishing perpetrators of genocide, and analyzing the adverse impact of impunity; and
9. developing and implementing educational efforts at different levels of schooling (secondary, college and university) and/or within governmental agencies (including legislative, judicial, and executive bodies, as well as the military).[2]

As important as the aforementioned developments are, they only begin to touch the proverbial tip of the iceberg in regard to what needs to be done in order to develop the most effective methods possible to intervene and/or prevent genocides from being perpetrated.

The fact is, an eclectic group of organizations and individuals outside the field of genocide studies are working on various issues related to the intervention and prevention of genocide. Unsurprisingly, certain individuals, as will be noted below, are working on some of the same issues as genocide scholars. Tellingly, some research efforts and mechanisms developed outside of genocide studies are not only more sophisticated in certain ways but further along in addressing certain factors that result in genocide and/or that mitigate against early and effective actions to address potential genocidal situations. Among some of the many issues that individuals and groups outside of genocide studies are considering, analyzing, developing, and/or implementing are: international law (accords, covenants, conventions, treaties); information gathering and analysis; intelligence sharing; early warning signals; data risk bases; early warning systems; "confidence building measures"; preventive diplomacy, Track I diplomacy, and/or Track II diplomacy; conflict resolution; conflict prevention; conflict management; peacekeeping; diplomatic peacekeeping; peacemaking; peace enforcement; sanctions (and in certain cases, "smart sanctions" or combinations of sanctions, as well as the use of "carrots" or inducements); partitioning; temporary protection measures for refugees fleeing internal and other types of conflict; humanitarian intervention; policing efforts (such as regional police forces, constabulary forces, private security forces); military intervention; and institution building. If there is to be any hope whatsoever of developing effective means to intervene and prevent genocide, genocide scholars and others must undertake a joint effort to a create a synergism between and among the aforementioned efforts. Not to do so will almost inevitably result in an ongoing, fragmented, hit and miss approach to this life and death issue.

THE CRITICAL NEED TO ADDRESS SYSTEMIC ISSUES THAT CONTRIBUTE TO CONFLICT AND, IN CERTAIN CASES, GENOCIDE

If there is to be any hope of *preventing* genocide (versus intervening after a genocide has already begun), then it is crucial for intergovernmental organizations, nongovernmental organizations, individual governments, policymakers, and scholars to address the underlying root causes of human rights infractions that often serve as catalysts to mass violence—and, in certain cases, genocide. In other words, sincere efforts and real progress must be made in undertaking key structural changes underlying violent conflict. Not to do so is bound to result in situations where the international community is "condemned" to reacting to grave crises already under way in which hundreds, thousands and tens of thousands or more may have already been killed.

Among some of the many issues that need to be addressed—and the earlier the better—are: racism; rabid or racial antisemitism; other types of extreme prejudice and/or discrimination that "mark" a particular group of people as "other"; international and/or national economic policies that exacerbate social tensions and fuel conflict (including severe economic inequalities); and scarcity of resources (land, food, water, energy, etc.) and/or over-population, which, in turn, exacerbates the former.

Each and every one of these problems is complex, and will require in-depth study in order to address them in an effective manner. To neglect them entirely, to address them in a piecemeal fashion, or to attempt to "patch" them with a band-aid approach is almost tantamount to inviting genocide to rear its ugly face yet again.

PROGRESS AND HOPE ON THE HORIZON

Despite all of the bad news, there is good news in that several real advances have been made in the battle against the perpetration of genocide. Six of the most notable shall be mentioned herein. First, numerous institutes and think tanks have been established in various parts of the world whose goals are, in part, conducting research germane to the issues of intervention and prevention of genocide. Such organizations are based in such diverse locales as London; Copenhagan, Denmark; Jerusalem, Israel; New York City; College Park, Maryland; and Washington, D.C.

Second, as Kuper (1985) noted—and as previously mentioned above— one of the most successful cases of genocide prevention may have involved the effort to protect the Bahá'í community in Iran in the 1980s. More specifically, "in this case, there was an immediate protective response by the United Nations. The explanation for this radical departure from normal U.N. practice is to be found partly in the somewhat pariah status of Iran ..., [b]ut a more important factor was the role of the Bahá'í International Community in conducting a skillful campaign, in the nature of an international alert, sharply focused on the United Nations" (Kuper, 1985, p. 163). This case, along with the 1999 intervention in East Timor, need to be carefully analyzed in order to ascertain what can be learned from them and applied to future situations that have the potential to degenerate into genocide.

Third, in 1993 and 1994 two ad hoc international courts were established by the United Nations Security Council to try those indicted for genocide, crimes against humanity, war crimes, and violations of the Geneva Conventions, as they pertained, respectively, to the violence in the former Yugoslavia and Rwanda. Both the International Criminal Tribunal on Yugoslavia (ICTY) and the International Criminal Tribunal on Rwanda (ICTR) have been relatively successful (if lumbering) in bringing successful prosecutions. In September 1998, for example, the ICTR found Jean-Paul Akayesu, a former town mayor, guilty of the crime of genocide. This was a historical occasion in that it was the first time an international court had issued such a verdict for the specific crime of genocide.

Fourth, in June of 1998, a treaty (the Rome Statute of the International Criminal Court) established the first International Criminal Court (ICC), "a permanent body on call to deal with rouge leaders in a systematic way so that a mastermind of death like the late Pol Pot would not pose a jurisdictional problem if caught" (Crossette, 1998, p. 3wk). The Statute gives the Court, which was established by the United Nations under the aegis of the Security Council, jurisdiction for the crime of genocide, crimes against humanity, and war crimes. The Court became operative on July 1, 2002, after a minimum of 60 UN countries had ratified it. Notable among those refusing to ratify was the United States—though the establishment of the ICC proceeded without the United States—which promptly sought, and achieved, an agreement within the United Nations that would place U.S. citizens serving in foreign postings outside the Court's jurisdiction.

Fifth, as previously mentioned, there was the notable international intervention in East Timor in 1999. In 1975 Indonesia invaded and annexed East Timor after Portugal, which had ruled the territory for close to four centuries, suddenly withdrew. An estimated 200,000 people were murdered during Indonesia's brutal 24-year rule. In 1999, the territory voted overwhelmingly for independence in a U.N.-sponsored referendum. Immediately following the election, the Indonesian military and pro-Indonesian militia gangs went on a rampage that resulted in an estimated 1,500 people being killed and the mass destruction of the territory (e.g., about 70% of East Timor's buildings were destroyed, telephone exchanges were wrecked, electrical lines ripped out, and farms burned to the ground). An estimated 800,000 people were forced into military-controlled camps in Indonesian-controlled West Timor. The terror and killings drew international attention, and pressure was placed on the United Nations to act by various foreign ministries as well as human rights organizations such as Amnesty International (the latter of which issued an urgent alert in September 1999). The killing and destruction only came to an end once international peacekeepers (who were authorized by the U.N. Security Council to use "all necessary measures" against violent militias) intervened to halt the violence and place East Timor under United Nations rule on September 20, 1999, some three weeks after the election. Ultimately, and notably, Indonesia voluntarily relinquished its claim to the territory. This came as a relief to the countries taking part in the intervention for though the declared mission was to rescue civilians, "the unspoken premise was potentially explosive: the international community was effectively coming to the aid of a separatist movement in a sovereign nation" (King, 1999, p. A9). It almost goes without saying that had the international community not intervened when it did, many more innocent victims would have been slaughtered—possibly in the tens of thousands or more.

Sixth, slowly but surely the notion of "sanctity" sovereignty is changing in favor of intervention in the face of massive human rights violations—and particularly genocide. In and of itself, this "movement" has the power to radically alter the tentativeness of individual nations, as well as the United Nations, to intervene in the domestic affairs of individual states and to possibly, at least in certain ways, temper the realpolitik and lack of political will that has so often been evident as individual nations have done all they could to avoid both recognizing genocide for what it is and from preventing it from engulfing tens and hundreds of thousands in its maw.

CONCLUSION

As Rudolph Rummel (2002), a political scientist and a scholar of genocide and democide studies, has trenchantly noted, "Instead of 'Never Again,' the fact of the matter is that genocide reappeared in the last half of the twentieth century again, and again, and again" (p. 173). He knows of what he speaks. Among some of the many cases of genocide perpetrated in the post-World War II years are as follows: the Indonesian massacres of suspected communists (1965-1966); the Bangladesh genocide (1971); the Burundi genocide (1972); the so-called autogenocide in Cambodia (1975-1979); the Iraqi gassing of the Kurds (late 1980s); and the aforementioned genocides in Rwanda genocide (1994) and the former Yugoslavia (1990s).

Almost every time a genocide is perpetrated, politicians, Holocaust and genocide scholars, human rights activists, and various members of the public decry the horrific situation, and assert that it must not happen again. *But it always does, somewhere else, time and again.*

Up to this point in time it seems as if the international community has either not cared enough to put in the will, time, effort, and resources to devise effective ways to prevent genocide or it has perceived the task as next to impossible. Though it will be extremely complex, difficult, time-consuming, and costly, it is not impossible. If humanity truly cares about its fellow human beings, then the means must be devised, and sooner rather than later, to prevent genocide. It is a given that the longer it takes to accomplish this task an ever greater number of victims will fall prey to genocide—again, and again and again.

NOTES

1. For an in-depth and outstanding discussion of these and related matters, see *The Responsibility to Protect* by the International Commission on Intervention and State Sovereignty (2001) (and its supplementary volume that is subtitled "Research, Bibliography, Background").

2. Among some of the earliest ideas put forth in regard to the intervention and prevention of genocide are the following: the establishment of a Geno-

cide Bureau (or genocide early warning system) that would monitor "hot spots" around the globe that have the potential to explode into genocidal acts (Knight, 1982); a Committee on Genocide that would periodically report on situations that have the potential to degenerate into genocide and/or actual genocidal actions (Knight, 1982; Whitaker, 1985. In 1988 Charny recommended that such a Committee be "empowered to indict a State against which charges of genocide were raised" [p. 25]); the convening of mass media professionals to examine and develop more effective ways of disseminating information about genocidal acts (Charny, 1982); a specially organized and systematic effort to collect first-person accounts of targeted groups by relief workers and journalists in areas where a potential genocide was brewing (Totten, 1991); and the development of a World Genocidal Tribunal that would have the authority to try individuals as well as governments that have committed genocide (Kutner & Katin, 1984).

Although some of the ideas were spawned as early as 1982, none have been implemented. It is, of course, one thing to conjure up such ideas/concepts, and something altogether different to implement them. The lack of implementation does no one any good, especially potential victims of genocide.

REFERENCES

Alvarez, A. (2001). *Governments, citizens, and genocide: A comparative and interdisciplinary approach*. Bloomington: Indiana University Press.

Charny, I. W. (1982). *How can we commit the unthinkable: Genocide; The human cancer*. Boulder, CO: Westview Press.

Charny, I. W. (Ed.). (1988). Intervention and prevention of genocide. In *Genocide: A critical bibliographic review* (pp. 20-38). New York: Facts on File.

Crossette, B. (1998, November 29). Dictators (and some lawyers) tremble. *The New York Times*, pp. 1wk, 3wk.

Davies, J. L., & Gurr, T. R. (Eds.). (1998). Preventive measures: An overview. In *Preventive measures: Building risk assessment and crisis early warning systems* (pp. 1-14). Lanham, MD: Rowman & Littlefield.

Fein, H. (1992). Dangerous states and endangered peoples: Implications of life integrity violations analysis. In K. Rupesinghe & M. Kuroda (Eds.), *Early warning and conflict resolution* (pp. 40-55). New York: St. Martin's Press.

Fein, H. (1990, Spring). Genocide: A sociological perspective. *Current Sociology, 38*(1).

Gurr, T. R. (1998). A risk assessment model of ethnopolitical rebellion. In J. L. Davies & T. R. Gurr (Eds.), *Preventive measures: Building risk assessment and crisis early warning systems* (pp. 15-26). Lanham, MD: Rowman & Littlefield.

Gutman, R. (1996). Bosnia: Negotiation and retreat. In B. Benton (Ed.), *Soldiers for peace: Fifty years of United States peacekeeping* (pp. 187-208). New York: Facts on File.

Harff, B. (1993, Fall). An early warning system is needed. *The ISG* [Institute for the Study of Genocide] *Newsletter, 11*, 3-5, 12.

Harff, B. (1998). Early warning of humanitarian crises: Sequential models and the role of accelerators. In J. L. Davies & T. R. Gurr (Eds.), *Preventive measures: Building risk assessment and crisis early warning systems* (pp. 70-78). Lanham, MD: Rowman & Littlefield.

International Commission on Intervention and State Sovereignty. (2001). *The responsibility to protect.* Ottawa, ON: International Development Research Centre.

Jelinek, P. (2001, July 3). File: Allies tipped off to genocide in Early '42. *Arkansas Democrat Gazette*, p. 4.

Jentleson, B. (1998). Preventive diplomacy and ethnic conflict: Possible, Difficult, necessary. In D. A. Lake & D. Rothchild (Eds.), *The international spread of ethnic conflict: Fear, diffusion, and escalation* (pp. 293-316). Princeton, NJ: Princeton University Press.

Jentleson, B. (2000). *Coercive prevention: Normative, political, and policy dilemmas.* Washington, DC: United States Institute of Peace.

King, L. (1999, September 31). Indonesian pullout marks end of struggle, dawn of independence. *Northwest Arkansas Times*, p. A9.

Knight, G. (1982, March 20). *A genocide bureau.* Text of talk delivered at the Symposium of Genocide, London.

Krumm, D. (1998). Early warning: An action agenda. In J. L. Davies & T. R. Gurr (Eds.), *Preventive measures: Building risk assessment and crisis early warning systems* (pp. 248-254). Lanham, MD: Rowman & Littlefield.

Kuper, L. (1985). *The prevention of genocide.* New Haven, CT: Yale University Press.

Kuperman, A. J. (2001). *The limits of humanitarian intervention: Genocide in Rwanda.* Washington, DC: Brookings Institution Press.

Kutner, L., & Katin, E. (1984). World genocide tribunal: A proposal for planetary preventive measures measure supplementing a genocide early warning system. In I. W. Charny (Ed.), *Toward the understanding and prevention of genocide: Proceedings of the International Conference on the Holocaust and Genocide* (pp. 330-346). Boulder, CO: Westview Press.

Melvern, L. R. (2000). *A people betrayed: The role of the West in Rwanda's genocide.* London: Zed Books.

Organization of African Unity. (2001). International Panel of Eminent Personalities: Report on the 1994 Genocide in Rwanda and Surrounding Events (Selected Sections). *International Legal Materials, 40,* 140-235.

Porter, J. N. (Ed.). (1982). Introduction. In *Genocide and human rights: A global anthology* (pp. 2-33). Washington, DC: University Press of America.

Ronayne, P. (2001). *Never again? The United States and the prevention and punishment of genocide since the Holocaust.* Lanham, MD: Rowan & Littlefield.

Rummel, R. J. (2002). From the study of war and revolution to democide—Power kills. In S. Totten & S. L. Jacobs (Eds.), *Pioneers of genocide studies* (pp. 153-177). New Brunswick, NJ: Transaction.

Totten, S. (Ed.). (1991). Introduction. In *First-person accounts of genocidal acts committed in the twentieth century: An annotated bibliography* (pp. xi-lxxv). Westport, CT: Greenwood Press.

Weiss, T. G., & Hubert, D. (2001). *The responsibility to protect: Research, bibliography, background: Supplementary volume to the report of the International Commission on*

Intervention and State Sovereignty. Ottawa, ON: International Development Research Centre.
Whitaker, B. (1985). *Revised and updated report on the question of the prevention and punishment of the crime of genocide.* (E/CN.4/Sub.2/1985/6, 2 July 1985).

CHAPTER 11

SELECT ANNOTATED BIBLIOGRAPHY

Teaching About Genocide[1]

Samuel Totten

JOURNALS

Holocaust and Genocide Studies (Oxford University Press, 2001 Evans Rd. Cary, NC 27513. Published in association with the United States Holocaust Memorial Museum, 100 Raoul Wallenberg Place, SW Washington, D.C. 20024-2126).

Published three times a year, this is an outstanding scholarly journal that regularly includes essays and book reviews on a wide variety of issues related, primarily, to the Holocaust.

Journal of Genocide Research (published four times a year by Carfax Publishing, Taylor & Francis Ltd., Customers Services Department, 325 Chestnut Street, 8th Floor, Philadelphia, PA 19106).

The first scholarly journal dedicated to the field of genocide studies (theory and research) and various genocidal events. Its founder and editor is Professor Henry R. Huttenbach.

NEWSLETTERS

The Genocide Forum (A publication of The Center for The Study of Eth-
nonationalism, c/o The History Department, The City College of The
City University of New York, Convent Ave. at 138th Street, New York,
NY 10031).

   The Genocide Forum, which appears bimonthly, is intended to serve as a
   convenient vehicle of exchange to discuss critical issues of common
   interest to students of Holocaust and genocide studies. The founder
   and editor of *The Genocide Forum* is Professor Henry R. Huttenbach.

The ISG Newsletter (A publication of the Institute for the Study of Genocide, c/o John Jay College of Criminal Justice, 899 Tenth Avenue, New
York, NY 10019).

   The Institute for the Study of Genocide exists to promote and dissem-
   inate scholarship and policy analyses on the causes, consequences,
   and prevention of genocide. *The ISG Newsletter* is sent, semi-annually,
   to all members of the ISG. *The Newsletter* covers news on key genocide
   issues, publishes short position papers/statements on issues germane
   to various facets of genocide, includes notes about conferences, and
   generally contains book reviews. Dr. Helen Fein, the executive director of IGS, is the editor of *The ISG Newsletter.*

BIBLIOGRAPHIES

Charny, I. W. (Ed.). (1988). *Genocide: A critical bibliographic review.* New
York and London: Facts on File and Mansell Publishing Limited,
respectively. 273 pp.

   This initial volume in the widely acclaimed "Genocide: A Critical Bibliographic Series" was the first bibliography to extend beyond individual occurrences of genocide, and to encompass both the totality of
   genocide as a process and the efforts that were being made to understand and combat it. Chapters herein address the study of genocide,
   the history and sociology of genocidal killings, the Holocaust, the
   Armenian genocide, the genocide in the U.S.S.R., the Cambodian
   genocide, other selected cases of genocide and genocidal massacres,
   understanding the psychology of genocidal destructiveness, and the
   literature, art and film of genocide.

Charny, I. W. (Ed.). (1991). *Genocide: A critical bibliographic review, Vol 2.*
London: Mansell. 432 pp.

   This volume is comprised of essays and accompanying annotations on
   such topics as the psychology of denial of known genocides, Holo-

caust denial, denial of the Armenian genocide, documentation of the Armenian genocide in Turkish sources, the status of basic genocide law, humanitarian intervention in genocidal situations, educating about the Holocaust and genocide, total war and nuclear omnicide, the professions and genocide, first-person accounts of genocidal acts, righteous people in the Holocaust, and the language of extermination in genocide.

Charny, I. W. (Ed.). (1994). *The widening circle of genocide: Genocide: A critical bibliographic review, Vol. 3.* New Brunswick, NJ: Transaction. 375 pp.

This volume includes essays and annotations on such issues as democracy and the prevention of genocide, religion, and genocide, documentation of the Armenian genocide in German and Austrian sources, genocide in Afghanistan, genocide of the Kurds in northern Iraq, the East Timor genocide, the fate of the Gypsies in the Holocaust, horizontal nuclear proliferation and its genocidal implications, and nongovernmental organizations working on the issue of genocide.

Hovannisian, R. G. (Ed.). (1980). *The Armenian holocaust: A bibliography relating to the deportations, massacres, and dispersion of the Armenian people, 1915-1923.* Cambridge, MA: Armenian Heritage Press. 43 pp.

This bibliography, by a noted historian at the University of California at Los Angeles, includes more than 400 citations on memoirs, accounts, collections of documents, and studies on the subject of the Armenian genocide.

Totten, S. (Ed.). (1991). *First-person accounts of genocidal acts committed in the twentieth century: Annotated bibliography.* Westport, CT: Greenwood Publishers. 351 pp.

The first bibliography to focus exclusively on first person accounts of a wide range of genocidal acts committed in the twentieth century, it is comprised of 1,275 annotations that address the following genocidal acts: the German extermination of the Hereros in Southwest Africa, the Armenian genocide, the Soviet manmade famine in the Ukraine, the Soviet Deportation of Whole Nations, the Holocaust, the fate of the Gypsies during the Holocaust years, the Indonesian genocide of communists and suspected communists, genocide in Uganda, genocide in Bangladesh, genocide of the Hutus in Burundi, the Indonesian-perpetrated genocide in East Timor, the Khmer Rouge-perpetrated in Cambodia, the threatened genocide of the Baha'is, and the genocide of various indigenous peoples.

Totten, S. (in press). *Genocide at the millennium*.

This is the fifth volume in the "Genocide: A Critical Bibliographical Series" inaugurated by Professor Israel W. Charny, Director of the Institute on the Holocaust and Genocide (Jerusalem, Israel). It includes essays and select annotated bibliographies on international law and genocide; the 1994 genocide in Rwanda; the genocide in the former Yugoslavia; the establishment of the International Criminal Tribunals in the former Yugoslavia and Rwanda, respectively; the establishment of the International Criminal Court (ICC); the role of nongovernmental organizations in addressing genocide; and the role of the United Nations as it applies to the intervention and prevention of genocide.

Totten, Samuel (in press). *The intervention and prevention of genocide: An annotated bibliography*. Westport, CT: Greenwood.

A major bibliography on a wide array of topics and issues (e.g., conflict, conflict resolution, early warning signals, early warning systems, international law, intervention, peace enforcement, peacekeeping, peacemaking, political will, prevention, sanctions, sovereignty, systemic issues, volunteer rapid action force, United Nations) germane to the issues of the intervention and prevention of genocide.

Totten, Samuel (in press). *The United Nations and the intervention and prevention of genocide: An annotated bibliography*. Westport, CT: Greenwood.

This major bibliography explores numerous issues germane to the United Nations and its role in the prevention and intervention of genocide (e.g., the U.N. Convention on the Prevention and Punishment of Genocide (UNCG), failures and successes, early warning programs, peace enforcement, peacekeeping, peacemaking).

Trynauer, G. (Ed.). (1989). *Gypsies and the Holocaust: A bibliography and introductory essay*. Montreal: Interuniversity Centre for European Studies & Montreal Institute for Genocide Studies. 51 pp.

This bibliography, one of the first (if not the first) bibliographies on the fate of the Gypsies, includes citations in English, French, and German on various facets of the Gypsies' subjugation, persecution and murder at the hands of the Nazis.

ENCYCLOPEDIAS

Charny, I. W. (Ed.). (1999). *Encyclopedia of genocide*. 2 Volumes. 718 pp. Santa Barbara, CA: ABC-CLIO.

The first encyclopedia on the topic of genocide, this two volume set includes entries by many of the most noted specialists in the field on a wide-range of topics and issues, including: key terminology, the psychology and ideology of genocide, typologies of genocide, the history of genocide, individual genocidal acts, victim groups, perpetrators, international law as it applies to genocide, the denial of genocide, and concepts and practices vis-à-vis the intervention and prevention of genocide.

Gutman, I. (Ed.). (1995). *Encyclopedia of the Holocaust*. New York Macmillan. Two Volumes. 1,905 pp.
This massive and major work contains entries on a wide array of critical issues by many of the most noted scholars of the Holocaust.

Laqueur, W. (Ed.). (2001). *The Holocaust encyclopedia*. New Haven, CT: Yale University Press. 765 pp.
This single-volume reference work provides both an overview of the subject of the Holocaust and detailed entries on major issues, events, policy, decisions, cities, and individuals. Scholars from 11 countries contributed the entries, drawing from a wide variety of sources in order to provide in-depth commentary on the political, social, religious, and moral issues of the Holocaust. Among the contributors are such noted scholars as Michael Berenbaum, Richard Breitman, Christopher Browning, David Cesarani, Saul Friedländer, Israel Gutman, Raul Hilberg, Michael Marrus, and Nechama Tec

Ramsbotham, O., & Woodhouse, T. (1999). *Encyclopedia of international peacekeeping*. Santa Barbara, CA: ABC-CLIO. 356 pp.
This reference book provides an overview of international peacekeeping operations on a wide array of issues, including but not limited to the following: the causes of and parties to conflict; the key differences between peacekeeping, peacemaking, peacebuilding; the ways in which air-power can be used in peace operations; key controversies over peacekeeping; military doctrines behind peacekeeping; historical summaries of major peacekeeping missions; biographical sketches of major military, civilian and political figures; and the United Nations' responsibility for peacekeeping operations.

Rozett, R., & Spector, S. (2000). *Encyclopedia of the Holocaust*. New York: Facts on File. 528 pp.
This reference work is comprised of both essays and alphabetical entries. The essays (ranging in length from 4 to more than 15 pages) cover topics such as different aspects of European Jewry, the Nazis' rise to power, the Allies' response, and the aftermath of the Holocaust. The entries are written by scholars associated with Yad Vashem, the Holocaust Martyrs' and Heroes' Remembrance Authority in Israel. The more than 650 alphabetical entries (ranging in length

from one paragraph to several pages) cover key individuals, places, events, and concepts. The volume also includes a detailed chronology (1933 to 1945) and a bibliography arranged by subject.

DICTIONARIES

Edelheit, A. J., & Edelheit, H. (Eds.). (1994). *History of the Holocaust: A handbook and dictionary.* Boulder, CO: Westview Press. 524 pp.
    An outstanding and useful reference work that includes a several hundred page section entitled "Dictionary of Holocaust Terms" (pp. 160-460).

Epstein, E. J. & Rosen, P. (1997). *Dictionary of the Holocaust: Biography, geography, and terminology.* Westport, CT: Greenwood Press. 416 pp.
    This dictionary contains some 2,000 entries that provide information on key figures, concentration and death camps, cities and countries, and significant events. It includes terms translated from German, French, Polish, Yiddish, and 12 other languages.

Totten, S., Bartrop, P., Jacobs, S., & Huttenbach, H. (Eds.). (in press) *Dictionary of genocide.* Westport, CT: Greenwood Press.
    The first dictionary to be compiled and published on the broad subject of genocide, this work includes over 900 entries. Among the numerous topics addressed are: key terms and concepts related to a wide array of issues germane to genocide; key documents and conventions; various genocidal events; perpetrators; and terms and issues germane to the preconditions, causes, and the intervention and prevention of genocide.

THEORETICAL, HISTORICAL AND OTHER KEY WORKS

Alvarez, A. (2001). *Governments, citizens and genocide: A comparative and interdisciplinary approach.* Bloomington: Indiana University Press. 224 pp.
    An examination of the crime of genocide through a social science lens, with specific references to the ideas and concepts that have been developed to explain criminal behavior. Beginning with the macro level, Alvarez, a criminologist, reviews the role of the state in perpetrating genocide, with the rationale that "genocide is invariably created and carried out by governments and their agents." He then explores the organizational level, in which individuals carry out genocide in their capacity as governmental administrators. He concludes

by examining the ways in which ordinary citizens end up participating in genocide.

Ball, H. (1999). *Prosecuting war crimes and genocide: The twentieth century experience*. Lawrence: University Press of Kansas. 288 pp.

Ball, a professor of political science, examines a host of issues, including: the history of war crimes and genocide from 1899 through 1939; the Nuremberg Tribunals, the Far East Tribunal, the postwar developments in international law; the Cambodian genocide; "ethic cleansing" in the Balkans and the International Criminal Tribunal for the Former Yugoslavia; the 1994 genocide in Rwanda and the International Criminal Tribunal for Rwanda; the International Criminal Court (ICC); and current and potential problems and prospects.

Barnett, M. (2002). *Eyewitness to a genocide: The United Nations and Rwanda*. Ithaca, NY: Cornell University Press. 215 pp.

A well-researched, hard-hitting, highly informative, and thought-provoking book by a professor of political science at the University of Wisconsin, Madison. Possibly the most disturbing message of this book—and there are many—is Barnett's contention that the culture of the United Nation contributed to the passive reaction to the genocide in Rwanda—and even led many at the U.N. to believe they were "doing the right thing."

Campbell, K. J. (2001). *Genocide and the global village*. New York: Palgrave. 178 pp.

The author examines why the international community has failed, time and again, to prevent, suppress, and punish contemporary genocide (e.g., in Bosnia, Rwanda, and Kosovo). He concludes by recommending a series of steps that the international community can take to improve its response to future genocidal situations.

Chalk, F., & Jonassohn, C. (1990). *The history and sociology of genocide: Analyses and case studies*. New Haven, CT: Yale University Press. 461 pp.

This book presents over two dozen examples of genocide, from antiquity to the present.

Charny, I. W. (1982). *How can we commit the unthinkable? Genocide: The human cancer*. Boulder, CO: Westview Press. 430 pp.

This text is comprised of three main parts: 1. What Are the Origins of Human Destructiveness?; 2. When Does Man Commit Genocide?; and 3. Why Can There Still Be Hope? (which includes a chapter entitled "Toward a Genocide Early Warning System").

Dobkowski, M., & Wallimann, I. (Eds.). (1998). *The coming age of scarcity: Preventing mass death and genocide in the twenty-first century.* Syracuse, NY: Syracuse University Press. 350 pp.

   The contributors to this book examine many of the systemic issues (e.g., population growth, land resources, per capita consumption) that can and have contributed to the outbreak of intrastate conflict and, in certain cases, genocide.

Fein, H. (1990, Spring). Genocide: A sociological perspective. *Sociology, 38*(1), 1-126.

   An outstanding publication by one of the most noted researchers in the field of genocide studies, this booklet is comprised of six chapters: 1. "Social Recognition and Criminalization of Genocide"; 2. "Defining Genocide as a Sociological Concept"; 3. "Explanations of Genocide"; 4. "Contextual and Comparative Studies I: Ideological Genocides"; 5. "Contextual and Comparative Studies II: Other Genocides"; and 6. "Punishment and Prevention: Implications for Social Policy and Research."

Fein, H. (Ed.). (1992). *Genocide watch.* New Haven, CT: Yale University Press. 204 pp.

   This collection of essays is comprised of three parts: 1. Definitions of Genocide and Research Findings; 2. Recognition, Denial, and Labeling of Cases; and 3. Approaches to Prevention and Punishment. Among the contributors to this book are Barbara Harff, Leo Kuper, René Lemarchand, James E. Mace, and Ervin Staub.

Gurr, T. R., & Harff, B. (Eds.). (1994, July). Early warning of communal conflicts and humanitarian crises. *The Journal of Ethno-Development, 4*(1), 1-131.

   This volume constitutes the proceedings of a workshop held at the Center for International Development and Conflict Management at the University of Maryland. The various talks are included under four headings: I. Theories and Models of Communal Conflict, Genocide, Politicide, and Humanitarian Crises; II. Evaluating Early Warning Models; III. Policy Uses of Early Warning Models; and IV. Assessments and Critiques of Early Warning Research. Among the contributors to this volume are Ted Robert Gurr, Barbara Harff, Rodolof Stavenhagen, Helen Fein, Kumar Rupesinghe, Juergen Dedring, and Howard Adelman.

Hinton, A. (Ed.). (2002). *Annihilating difference: The anthropology of genocide.* Berkeley: University of California Press. 405 pp.

   Among the genocides discussed in the essays herein by anthropologists and others are those of various indigenous peoples across the globe, the Holocaust, the 1994 Rwandan genocide, the Khmer-Rouge perpetrated genocide in Cambodia, and the genocidal massacres in Guatemala in the 1980s.

Hirsch, H. (2003). *Anti-genocide: Building an American movement to prevent genocide*. Westport, CT: Praeger. 232 pp.

Hirsch, a professor of political science at Virginia Commonwealth University, proposes the need for and ways to build what he calls "a politics of prevention." Focusing on the United States, he asserts that a political movement must be built that supports the politics of prevention in the international realm. More specifically, he argues that long-term prevention is contingent on changing how humans view each other, and that what is needed is the creation of a new ethic of life-enhancing behavior based on the ideology of universal human rights that is passed on from generation to generation via the process of political socialization.

Horowitz, I. L. (1989). *Taking lives: Genocide and state power*. New Brunswick, NJ: Transaction. 224 pp.

Utilizing genocide as the focal issue, Horowitz, a noted sociologist, derives a new typology of social systems, which distinguishes eight types of societies within a framework of state power rather than cultural systems. Horowitz views genocide as a totalitarian technique for achieving national solidarity, ultimately resulting in a state where there is law without justice.

Kuper, L. (1981). *Genocide: Its political use in the twentieth century*. New Haven, CT: Yale University Press. 255 pp.

Herein, Kuper, considered one of the doyens of genocide studies, provides a systematic, comparative study of numerous cases of genocide. In doing so, he discusses, among other issues, theories of genocide, genocidal processes, the concept of the sovereign territorial state, and concludes with a chapter entitled "The Non-Genocidal Society."

Kuper, L. (1985). *The prevention of genocide*. New Haven, CT: Yale University Press. 286 pp.

Herein, Kuper provides case studies of genocide, analyzes the major obstacles to effective U.N. action (and ways to overcome such obstacles), and discusses how to prevent genocide (including the establishment of an early warning system that would monitor events around the world, reporting on situations threatening people's human rights, and alerting the international community of the need to exert pressure on the offending government).

Mills, N., & Brunner, K. (Eds.). (2002). *The new killing fields: Massacres and the politics of intervention*. New York: Basic Books. 276 pp.

The contributors to this book (including Samantha Power, David Rieff, Geoffrey Robinson, William Shawcross, and Michael Walzer) discuss what the United States and its international allies did and did not do in the face of state-sponsored genocide in the 1990s in Yugoslavia, Rwanda, and East Timor.

Neier, A. (1998). *War crimes: Brutality, genocide, terror, and the struggle for justice*. New York: Times Books. 286 pp.

In this hard-hitting book, Neier, a noted human rights activist, examines the brutality and genocide committed in the Former Yugoslavia in the 1990s and in Rwanda in 1994, and questions the world community's totally inadequate response to such atrocities. He also explores various instruments used to hold those accountable who commit such crimes (e.g., "truth" commissions, local trials, dismissal from public office), and calls for a global commitment in regard to the establishment of a permanent war crimes tribunal.

Power, S. (2002). *"A problem from hell": America and the age of genocide*. New York: Basic Books. 610 pp.

In this detailed, highly readable, and extremely disturbing book, Power examines the United States government's reactions to genocide in the twentieth century. It is a scathing indictment of realpolitik at its worst.

Reimer, N. (Ed.). (2000). *Protection against genocide: Mission impossible?* Westport, CT: Praeger. 193 pp.

In this thought-provoking text, various authors call for strengthening the key institutions of a global human rights regime, developing an effective policy of prudent prevention of genocide, working out a workable policy of targeted sanctions (political, economic, military, judicial), and adapting a guiding philosophy of humanitarian intervention. Among its contributors are such notable scholars as Helen Fein, Saul Mendlovitz, and David Wippman.

Rittner, C., Roth, J. K., & Smith, J. M. (Eds.). (2002). *Will genocide ever end?* St. Paul, MN: Paragon House. 254 pp.

The short essays in this book by specialists working on the issue of genocide address a range of issues germane to the intervention and prevention of genocide. In addition to the editors, among some of the many other contributors are: Michael J. Bazyler, Helen Fein, Richard Goldstone, Barbara Harff, Henry Huttenbach, Steven L. Jacobs, Ben Kiernan, Mark Levene, Eric Markusen, Robert Melson, Linda Melvern, Roger W. Smith, Ervin Staub, and Samuel Totten.

Ronayne, P. (2001). *Never again? The United States and the prevention and punishment of genocide since the Holocaust*. Lanham, MD: Rowman & Littlefield. 222 pp.

An informative and telling examination of the United States' stance and actions in regard to the international effort to halt genocide. The author provides insights into the United States' reservations about and intransigence in regard to ratifying the U.N. Convention on Genocide, and discusses how the United States' efforts to prevent genocide have fallen sorely short of stated goals and repeated promises. At one and the same time, Ronayne notes the significant, though excruciatingly slow, progress that has been made in the effort to counter the genocidal efforts of various perpetrators.

Schabas, W. A. (2000). *Genocide in international law: The crimes of crimes*. New York: Cambridge University Press. 624 pp.

Herein, Schabas, a professor of international law, provides an overview of international law as it applies to the issue of genocide. Among the many issues/topics he addresses are: the origins of the legal prohibition of genocide, groups protected by the United Nations Convention on the Prevention and Punishment of Genocide, the physical element or *actus reus* of genocide, the mental element or *mens rea* of genocide, prosecution of genocide by international and domestic tribunals, prevention of genocide, and treaty law questions and the Convention.

Smith, R. (Ed.). (1999). *Genocide: Essays toward understanding, early-warning, and prevention*. Williamsburg, VA: Association of Genocide Scholars. 240 pp.

The essays in this book are were drawn from two conferences of the Association of Genocide Scholars (1995 and 1998, respectively). It is comprised of four parts: I. Introduction; II. Genocide in Rwanda: Causes and Implications; III. Genocide and Contemporary History: Some Neglected Cases from the Middle East, The Caribbean, and Europe; IV. Preventing Genocide: Early-Warning and Political Will; and V. Preventing Genocide: Technology, Institutions and Law. Among the contributors are Rouben Adalian, Howard Adelman, René Lemarchand, and Mark Levene.

Staub, E. (1989). *Roots of evil: The origins of genocide and other group violence*. New York: Cambridge University Press. 336 pp.

Applying the data and insights of behavioral and social science research, Staub examines how certain social and cultural conditions engender widespread hatred of certain groups, and delineates the psychological processes by which "ordinary" individuals become mass murderers. In doing so, he examines the origins of genocide and mass killing, the effects of difficult life conditions, the psychology of perpetrators (individuals and groups), steps along a continuum of destruction, and the creation and evolution of a caring and nonaggressive society. Among the cases of genocide he explores are the Holocaust, the Turkish genocide of the Armenians, and the Cambodian genocide.

Totten, S., & Jacobs, S. (Eds.). (2002). *Pioneers of genocide studies*. New Brunswick, NJ: Transaction. 616 pp.

The contributors to this volume discuss the genesis and evolution of their thought and work in the field of genocide studies; and in doing so, they discuss, among other issues, the strengths and weaknesses of the U.N. Convention on the Prevention and Punishment of Genocide, theories and processes of genocide, and provide insights into the barriers, challenges, and potential ways to overcome the latter in regard to the effective intervention and prevention of genocide. The contributors to book are: Raphael Lemkin, Leo Kuper, Israel W. Charny, Irving Louis Horowitz, Helen Fein, Barbara Harff, Richard Hovannisian, Rouben Adalian, Henry Huttenbach, James Mace, Yves Ternon, Herbert Hirsch, Robert Melson, R.J. Rummel, Colin Tatz, Vahakn Dadrian, Eric Markusen, M. Cherif Bassiouni, Gregory Stanton, Ervin Staub, Steven Leonard Jacobs, David Hawk, and Samuel Totten.

Totten, S., Parsons, W. S., & Charny, I. W. (Eds.). (1997). *Century of genocide: Eyewitness accounts and critical views*. New York: Garland. 488 pp.

This volume includes scholarly essays on a wide range of genocides perpetrated in the twentieth century (from the 1904 genocide of the Hereros through the mid-1990s genocide in the former Yugoslavia) by some of the most noted specialists in the field of genocide and Holocaust studies (James Dunn, Ben Kiernan, James Mace, René Lemarchand, Sybil Milton). Accompanying each essay are a series of first-person accounts of the genocide under discussion. The revised and third edition of this book—which is entitled *Century of Genocide: Critical Essays and Eyewitness Testimony*—is currently in production and will be available in 2004.

Wallimann, I., & Dobkowski, M. (Eds.). (2000). *Genocide and the modern age: Etiology and case studies of mass death*. Syracuse, NY: Syracuse University Press. 317 pp.

This book is comprised of two parts: Part 1. Conceptualizing, Classifying, Defining, and Explaining Genocide: Some Macro Perspectives; and Part II. Understanding Occurrences of Genocide: Some Case

Studies and Investigations of Related Social Processes. Among the contributors to the book are Roger W. Smith, Barbara Harff, Irving Louis Horowitz, John K. Roth, and Eric Markusen.

CURRICULA, TEACHING GUIDELINES, AND PEDAGOGICAL ESSAYS

Drew, M. A. (1991, February). Merging history and literature in teaching about genocide. *Social Education*, 55(2), 128-129.

A short but informative and thought-provoking article that addresses two key issues: "Link Between Atrocity and the Individual" and "History as the Framework."

Fein, H., & Apsel, J. (Eds.). (2002). *Teaching about genocide: An interdisciplinary guidebook with syllabi for college and university teachers*. New York: American Sociological Association for the Institute of the Study of Genocide. 214 pp.

A new and revised edition of *Teaching About Genocide: A Guidebook for College and University Teachers*, this booklet is comprised of two introductory essays (one on the history and status of genocide education, and one that addresses pedagogical issues germane to teaching about genocide), and course syllabi and other resources (e.g., study questions, topic reports, research, exams, and bibliographies). The section containing the syllabi are broken down into the following sections: Armenian Genocide; Holocaust; Genocide and Holocaust; Genocide; and Genocide, Human Rights and International Affairs.

Fein, H., & Freedman-Apsel, J. (Eds.). (1992). *Teaching about genocide: A guidebook for college and university teachers: Critical essays, syllabi and assignments*. New York: The Institute for the Study of Genocide. 102 pp.

Includes syllabi by anthropologists, historians, sociologists, political scientists, and psychologists. Among the contributors are Helen Fein, Richard Hovannisian, Rhoda Howard, Hilda Kuper, Leo Kuper, R.J. Rummel, Roger Smith, and Ervin Staub.

Fernekes, W. R. (1991). Defining genocide. *Social Education*, 55(2): 130-131.

The strategy outlined in this article is an excellent one. That said, educators who use it or adapt it to their purposes, really need to use more definitions than the author uses herein. While he makes use of the definition in the U.N. Convention on the Prevention and Punishment of Genocide as well as those developed by Israel W. Charny and Irving Louis Horowitz, the lesson would have been much stronger had he included an even wider range of definitions—particularly those by such individuals and scholars, for example, as Leo Kuper, Helen Fein, and Ben Whitaker.

Friedlander, H. (1979). Toward a methodology of teaching about the Holocaust. *Teachers College Record, 80*(5), 519-542.

An early and thought-provoking essay in the field of Holocaust education by a Holocaust survivor and scholar.

Parsons, W. S., & Totten, S. (1993). *Guidelines for teaching about the Holocaust*. Washington, DC: United States Holocaust Memorial Museum.

Though these guidelines focus on teaching about the Holocaust, certain points are germane to teaching about other genocides.

Sproat, P. A. (2001). Researching, writing and teaching genocide: Sources on the Internet. *Journal of Genocide Research, 3*(3), 451-461.

The author highlights and discusses some of the sites on the Internet devoted to the topic of genocide.

Totten, S. (2001, September). Addressing the "null curriculum": Teaching about genocides other than the Holocaust." *Social Education, 65*(5), 309-313.

Discusses "barriers" to teaching about genocide and ways to overcome them.

Totten, S. (2002). *Holocaust education: Issues and approaches*. Boston, MA: Allyn & Bacon. 195 pp.

Holocaust Education: Issues and Approaches raises a host of critical issues and highlights effective teaching strategies related to teaching about the Holocaust. For example, the essays address the need to reflect deeply on both the purpose and process of teaching such complex issues; methods for assessing student's knowledge base prior to teaching Holocaust history; and the critical need to teach the history accurately. Among some of the many chapter titles are: "Common Misconceptions and Inaccuracies That Plague Teaching and Learning About the Holocaust"; "'Complicating' Students' Thinking Vis-à-Vis the History of the Holocaust; "Do the Jews Constitute a Race? An Issue Holocaust Educators Must Get Right"; and "Diminishing the Complexity and Horror of the Holocaust: Using Simulations in an Attempt to Convey Personal and Historical Experiences."

Totten, S. (1999). The scourge of genocide: Issues facing humanity today and tomorrow. *Social Education, 63*(2), 116-121.

Part of *Social Education's* "Human Rights Series," this article includes the following sections: Barriers to Ending Genocide, The Study of

Genocide, Early Warning Systems, and The Role of Teachers in Addressing Genocide.

Totten, S., & Feinberg, S. (2001). *Teaching and studying about the Holocaust.* Boston, MA: Allyn & Bacon. 342 pp.

This book is comprised of 13 chapters by some of the most noted Holocaust educators in the United States. In addition to chapters on establishing clear rationales for teaching about the Holocaust, the book includes individual chapters on incorporating primary documents, first-person accounts, film, literature, art, drama, music, and technology into a study of the Holocaust. It also includes a major chapter on Holocaust historiography. It concludes with an extensive annotated bibliography especially designed for educators.

Totten, S., & Parsons, W. S. (Eds). (1991, February). Teaching about genocide [Special issue]. *Social Education, 55*(2).

Among the 17 pieces included herein are essays on: rationales and methodology for teaching about genocide; a historical overview of genocide; brief discussions of various genocides (e.g., the Armenian genocide, the Soviet manmade famine in Ukraine, the Holocaust, the genocide in Burundi, the Cambodian genocide); an examination of the denial of genocide; a discussion concerning the intervention and prevention of genocide; and units for teaching about genocide. Among the contributors are Leo Kuper, Israel Charny, Rouben Adalian, Sybil Milton, René Lemarchand, Deborah Lipstadt, Henry Friedlander, Ben Kiernan, Geoffrey Hartman, William S. Parsons, and Samuel Totten.

RESEARCH/CRITIQUES OF HOLOCAUST AND GENOCIDE CURRICULA

Dawidowicz, L. S. (Ed.). (1992). How they teach the Holocaust." In *What is the use of Jewish history?* (pp. 65-83). New York: Schocken Books.

This is a hard-hitting essay that examines both state department of education developed curricula as well as other key curricula on the Holocaust. Dawidowicz, a historian of the Holocaust, issues a scathing critique in which she asserts, and provides evidence, that most of the curricula are plagued with gaps in the history, rife with historical errors, and include learning activities that verge on the gimmicky and ahistorical.

Haynes, S. (1998). Holocaust education at American colleges and universities: A report on the current situation. *Holocaust and Genocide Studies, 12*(2), 283-307.

> Among the topics and issues examined in this essay are: the evolution of Holocaust education in North America; the techniques, activities and resources used in such pedagogical practices; a discussion about whether there is anything like a normative approach to teaching the Holocaust; and an outline of challenges for the future.

Riley, K., & Totten, S. (2002, Fall). Understanding matters: The Holocaust and social studies classrooms. *Theory and Research in Social Education, 30*(4), 541-562.

> Using the approaches of historical methodology and "historical empathy" as well as guidelines (*Guidelines for Teaching about the Holocaust*) developed by the United States Holocaust Memorial Museum, the authors critique the historical accuracy and pedagogy of four widely used Holocaust curricula and guides (one each from California, Connecticut, Virginia, and Florida). They find that all four resources are rife with historical inaccuracies and more prone to a frontal approach to teaching than a constructivist approach.

Totten, S., & Parsons, W. S. (1992, Spring). State developed teacher guides and curricula on genocide and/or the Holocaust. *Inquiry in Social Studies: Curriculum, Research, and Instruction: The Journal of the North Carolina Council for the Social Studies, 28*(1), 27-47.

> An early critique of nine state developed resources (curriculum and teacher guides) on the Holocaust and genocide. The critique notes the strengths and weaknesses of each in regard to their presentation of history and their suggested pedagogy.

KEY FILMS/VIDEOTAPES

"A problem from hell": America and the age of genocide by Samantha Power. (60 minutes, videocassette. Available from The C-Span Archives, PO Box 66809, Indianapolis, IN 46266-6809).

> A fascinating interview with Samantha Power about the focus and writing of her book, *"A Problem from Hell": America and the Age of Genocide.*

The Armenian case (45 min., color, videocassette. Available from Atlantis Productions, 1252 La Granada Drive, Thousand Oaks, CA 91362).

> In this film, survivors of the Armenian genocide and European and American eyewitnesses recall the historical events that were to shape

the destiny of the Armenian people. It includes documentary sequences on World War I, and the establishment of the Republic of Armenia and Soviet Armenia.

The forgotten [Armenian] genocide (29 min., color, videocassette. Available from Atlantis Productions, 1252 La Granada Drive, Thousand Oaks, CA 91362).

This film presents the story of the genocide of the Armenian people committed by the Ottoman Turks between 1915 and 1918.

Forsaken cries: The story of Rwanda (35 min., color, videocassette. Available from Amnesty International-USA, 322 Eighth Avenue, New York, NY 10001).

This documentary, which examines Rwanda as a case study of the human rights challenge of the twenty-first century, incorporates historical footage, interviews, and analysis of the genocide that resulted in the deaths of up to one million people in 1994.

From Yugoslavia to Bosnia (50 min., color, videocassette. Available from Landmark Media, 3450 Slade Run Drive, Falls Church, VA 22042).

This video relates the story of the war in the former Yugoslavia from its inception up to the events that took place in 1994. The narrator discusses the atrocities/genocide and "ethnic cleaning" as well as European and U.S. political evasion and apathy to the crisis. The film features President Bill Clinton, Prime Minister Margaret Thatcher and two historians who contradict accepted opinion on the causes of the situation.

Gacaca: Living together again in Rwanda (55 minutes, videocassette. Available from First Run/Icarus Films, 32 Court Street, 21st Floor Brooklyn, NY 11201).

This documentary film features a Rwanda-based grassroots justice system called the Gacaca Tribunals. Gacaca is an attempt to unify a nation that in 1994 suffered a genocide in which more than 800,000 of the country's Tutsi minority, and many Hutu moderates, perished. Citizens will act as judges in an attempt to democratize the justice system. By exploring the process of building and sustaining this new justice system, the film portrays the Tutsi and Hutu peoples' struggles to rebuild their lives and communities by dealing with the emotional trauma of their past and reconciling their deep differences. Includes first-person accounts by survivors and prisoners, and a bird's-eye view of the first of a series of open-air "Pre-Gacaca" hearings.

Genocide, 1941-1945 (World at War Series) (50 minutes, c, videocassette. Available from Arts and Entertainment, 800-423-1212, or write A&E Home Video, P.O. Box 2284, South Burlington, VT 05407).

This documentary presents the story of the destruction of European Jewry through the use of archival footage and testimonies of victims, perpetrators, and bystanders. An excellent film for presenting an overview of many key issues. Ideal for use with middle school, high school, and adult audiences.

A good man in hell: General Roméo Dallaire and the Rwanda genocide (60 minutes, videocassette. Available from The C-Span Archives, PO Box 66809, Indianapolis, IN 46266-6809).

An astonishingly honest and remarkably thought-provoking interview of General Roméo Dallaire, the head of the U.N. Forces in Rwanda prior to and during the 1994 genocide, conducted by television commentator Ted Koppel. This is a must resource/interview (conducted on June 12, 2002) for any educator planning to teach about the 1994 Rwanda genocide. The transcript of the interview is available, free of charge, on the web page of the United States Holocaust Memorial Museum.

Guatemala: Personal testimonies (20 min., color, videocassette. Available from Icarus Films, 200 Park Avenue South, Suite 1319, New York, NY 10003).

This series of testimonies from Guatemalan Indians clearly bears witness to the widespread abuse of human rights and genocidal actions committed by the government of General Rios Montt.

Harvest of despair: The unknown holocaust (55 min., color with black and white sequences, videocassette. Available from the Ukrainian Research Centre, St. Vladimir Institute, 620 Spadina Ave., Toronto, Ontario, Canada M58 2HY).

A documentary of the Soviet manmade "terror famine" of 1932-1933. It includes interviews with survivors and scholars about various aspects of the genocide, and rare photographic evidence.

The killing fields (137 min., color, videocassette. Available from Swank Film Programmer, 6767 Forest Lawn Dr., Hollywood, CA 90068).

This moving feature film tells the horrific and true story of one man's (Dith Pran, an interpreter for *New York Times* reporter Sydney Schanberg) traumatic experiences in Cambodia during the mid- to late-1970s during which the Khmer Rouge killed over 3 million out of its 7 million fellow citizens in a purported attempt to create a "new society."

No man's land (98 Minutes. Available from Amazon.com.)

Considered one of the best foreign language films of 2001, *No Man's Land* tells the story of two wounded soldiers, a Bosnian and a Serb, trapped in a trench between enemy lines during the 1993 Bosnian war. Neither the Bosnian Ciki nor the Serbian Nino trust one another. Constantly arguing about who was responsible for starting the war, they not only accuse each other but threaten one another with a weapon. The focal point, though, of this powerful anti-war film is a third man, Cera, Ciki's comrade, who was shot and ostensibly killed. Under his body, a Serbian soldier has placed a spring-loaded American-made bomb specifically manufactured to blow up when the corpse is recovered. Thus, when Cera comes to, Ciki and Nino realize that if Cera moves, they will be killed along with him. Ultimately they come to the realization that their only hope is to attract the attention of the United Nations humanitarian force which patrols the area. Rated "R."

The Pinochet case: Remembering the other September 11th (109 minutes, with English subtitles. Available from International Media Resource Exchange, 124 Washington Place, New York, NY 10014).

This documentary by Patricio Guzman, director of the classic documentary trilogy *The Battle of Chile*, presents the story of what took place on Sepetmber 11, 1973, when General Augusto Pinochet carried out a miliary coup that overthrew the democratically elected Chilean socialist government of Salvador Allende. Responsible for the torture and death of thousands of his opponents, Pinochet ruled Chile until 1990, and retired from politics in 1998 with legal immunity granted by his self-designated position as "senator for life." This documentary presents an examination of the the 1998 arrest of Pinochet while on a shopping spree in London after a Spanish judge, Baltasar Garzón, issued a warrant charging him with human rights violations. Guzmán details the precedent-setting legal efforts to make the tyrant answer for his crimes. Equally important is the platform the film gives many of the survivors who recount egregious human rights abuses that have gone unpunished for nearly three decades.

Saddam's killing fields (52 min., color, videocassette. Available from Landmark Media, 3450 Slade Run Drive, Falls Church, VA 22042).

In this film, historian Michael Wood presents evidence of the Saddam Hussein regime's sustained and murderous destruction of the Shia Marsh Arabs, whose way of life went back 5,000 years.

Shadow on the cross (52 minutes, c, videocassette. Available from Landmark Films, Inc., 3450 Slade Run Drive, Falls Church, VA 22042).

This documentary examines the strained and tragic story of Jewish-Christian relations over the past 2,000 years; and in doing so, discusses the influences of Christian antisemitism through the ages on the Third Reich. Comprised of two parts, part one provides a summary of the history of religious antisemitism. In part two, noted theologians discuss the implications that the Holocaust has for Jewish-Christian relations in today's world. Ideal for use with high school or college/university students.

Shoah (9 hours, color, video cassette. Available in most video centers and libraries. It can be purchased from the Simon Wiesenthal Center, 9760 West Pico Blvd., Yeshiva University of Los Angeles, Los Angeles, CA 90035).

This extremely powerful documentary includes interviews with perpetrators, bystanders, and victims, and is set in various locations, including former camps, railway areas, and towns with significance to the Holocaust. The nine hours can be divided into 5 to 10 days of viewing time. Recommended for high school and college/university audiences.

Srebrenica: A cry from the grave (PBS, Thirteen/WNET, 450 West 33rd Street, New York, NY 10001; 212-560-2888).

This powerful documentary is packed with first-person accounts by many of the participants, though not the Serbs, of the massacre of some 7,000 Muslim boys and men by Serb forces in 1995. The accounts are interspersed with contemporary news footage taken of heartrending aspects of the massacre. The film relates the story of the 10 days during which the so-called "safe-area" fell to the Bosnian Serbs when the Dutch U.N. peacekeeping force stood aside and basically allowed the ethnic cleansing and mass murder to take place.

Year zero (53 min., color, videocassette. Available from American Friends Service Committee, Pacific Southwest, Region, 980 N. Fair Oaks Ave., Pasadena, CA 91103).

Produced in Cambodia in September 1979 for Britain's Associated Television, this film presents gruesome evidence of the genocide committed between 1975 and 1979 when over an estimated one to two million people were murdered, worked, and starved to death by the Khmer Rouge.

KEY WEBSITES[2]

www.amic.org

The official website of the American Non-Governmental Organizations Coalition for the International Criminal Court (ICC), it includes information about the ICC, U.S. public opinion regarding the establishment of the ICC, U.S. official opposition to the ICC, and grassroots activity supporting the establishment of the ICC. In mid-2003, the ICC began investigating cases of suspected genocide, war crimes, and crimes against humanity.

www.genocidewatch.org

The official website of Genocide Watch, it includes information about the following: Genocide Watch's international campaign against genocide, a detailed definition of genocide, predictors of genocide, copies of "Prevent Genocide International News" (2002-2003), resources on genocide, and updates on genocide alerts.

www.hrw.org

The home page of Human Rights Watch, a major international human rights organization, this website includes numerous publications that contain the findings of field-based research conducted by HRW researchers into various genocidal incidents, including those in Iraq (1988), the former Yugoslavia (the 1990s), and Rwanda (1994). Also available are Iraqi government documents delineating aspects of the genocide it perpetrated against its Kurdish population in northern Iraq in the late 1980s.

www.icc.org

This is the official web site of the recently established International Criminal Court (ICC), a court that is mandated to try crimes against humanity, war crimes, genocide and contraventions of the Geneva Conventions.

www.ictr.org

This is the official web site of the International Criminal Tribunal for Rwanda (ICTR), the tribunal that is trying those suspected of planning and perpetrating the 1994 genocide in Rwanda in which some 800,000 people—mainly Tutsis but also many moderate Hutus—were slain by radical Hutus. In addition to providing an overview of the purpose and procedures of the ICTR, it includes the full transcripts of each of the trial sessions as well as the judgments rendered.

www.icty.org

This is the official web site of the International Criminal Tribunal for Yugoslavia, the tribunal that is trying those suspected of contravening the Geneva Conventions, and/or committing crimes against humanity, war crimes, and/or genocide between 1991 and 1999 in the former Yugoslavia. In addition to providing an overview of the purpose and procedures of the ICTY, it includes the full transcripts of each of the trial sessions as well as the judgments rendered.

www.thro@uc.edu

Teaching Human Rights Online (THRO) provides free exercises and interactive study guides for use as course modules in political science, international relations, philosophy, history and women's studies, as well as professional classes in law, education, and business. Students who work the problems online or on CD-ROM receive immediate, individualized feedback to forced choice and short answer self-assessment questions. Each peer-reviewed case can also be printed from a single file for use in hard copy. Cases may be used for class discussion, simulation, and transnational Internet conferences arranged by THRO.

Among previous cases have been the following: "The International Court of Justice Considers Genocide," and "Rape and Genocide in Rwanda: The ICTR's Akayesu Verdict."

www.ushmm.org.conscience

The official website of the United States Holocaust Memorial Museum's Committee on Conscience, this site includes the following information: the focus of the Committee on Conscience, genocide warnings, and the status of the various international tribunals trying suspected perpetrators of genocide.

www.yale.edu/cgp

In addition to a wealth of information on various aspects of the 1975-1979 Cambodian genocide (including first-person accounts by survivors of the genocide), this web page of Yale University's Cambodian Genocide Program (CGP) includes the following: news on the tribunal trying the perpetrators of the Cambodian genocide, key databases, photographs of the Cambodian genocide, maps of Cambodia, and information about the Genocide Studies Program at Yale.

NOTE

1. Each Case Study in Chapter 6 ("Genocides in the Twentieth Century: An Overview of Individual Cases of Genocide") concludes with a short list of

recommended readings, thus works on specific genocides are not listed herein.

2.   For additional information about other websites that address the topic of genocide, see Sproat (2001).

REFERENCE

Sproat, P. A. (2001). Researching, writing, and teaching genocide: Sources on the internet. *Journal of Genocide Research, 3*(3), 451-461.

ABOUT THE CONTRIBUTORS

Paul R. Bartrop is a Fellow in the Faculty of Arts at Deakin University, Melbourne, Victoria, Australia, and a member of the teaching staff at Bialik College, Melbourne, where he teaches history, international studies, and comparative genocide studies. His published works include: "The Holocaust, the Aborigines, and Bureaucracy of Destruction: An Australian Dimension of Genocide" in the *Journal of Genocide Research*; "The Relationship Between War and Genocide in the Twentieth Century: A Consideration" in the *Journal of Genocide Research*; and *Surviving the Camps: Unity in Adversity During the Holocaust* and *Australia and the Holocaust, 1933-45*.

Michael Berenbaum is the incoming Director of the Sigi Ziering Institute for the Study of the Holocaust and Ethics at the University of Judaism. He is the former President and CEO of the Survivors of the Shoah Visual History Foundation and former Project Director of the creation of the United States Holocaust Memorial Museum and later the director of its research institute. The author and editor of 12 books on the Holocaust and modern theology, he has also been a consultant, historian, and executive producer to several Holocaust documentaries and dramas.

Robert Cribb is Senior Fellow in Pacific and Asian History, Research School of Pacific and Asian Studies, Australian National University. He is the editor of *The Indonesian Killings of 1965-1966: Studies from Java and Bali*.

Craig Etcheson is a Visiting Scholar at Johns Hopkins University School of Advanced International Studies in Washington, and an Advisor to the

323

Documentation Center of Cambodia in Phnom Penh. From 1980 through 1990, he was a lecturer and an adjunct faculty member at the University of Southern California's School of International Relations, specializing in extreme forms of violence. He was Executive Director of the Cambodia Campaign in Washington, D.C., from 1992 through 1994, helping to bring about passage of the 1994 Cambodian Genocide Justice Act. In 1995, Etcheson cofounded the Documentation Center of Cambodia, serving as its director in 1995 and 1996. From 1994 through 1998, he was a principal of Yale University's Cambodian Genocide Program, serving as program manager and acting director. In 1998 and 1999, Etcheson was program director for the International Monitor Institute. Since then, he has continued his efforts to help achieve accountability and reconciliation in Cambodia. Etcheson is the author of *The Rise and Demise of Democratic Kampuchea*; *The Number: Quantifying Crimes Against Humanity in Cambodia*; and *Crimes of the Khmer Rouge: The Search for Peace and Justice in Cambodia*. He is currently preparing a manuscript dealing with problems of retribution and reconciliation in Cambodia.

William R. Fernekes is supervisor of social studies at Hunterdon Central Regional High School in Flemington, New Jersey. He has published widely in professional journals on Holocaust and genocide education, human rights education, and children's rights. Among his publications are: *The Oryx Holocaust Sourcebook*; "The Babi Yar Massacre: Seeking Understanding Using a Multimedia Approach" in *Teaching Holocaust Literature*; and "Education for Social Responsibility: The Holocaust, Human Rights and Classroom Practice" in *The Holocaust's Ghost: Writing on Art, Politics, Law and Education*.

Richard G. Hovannisian is professor of Armenian and Near Eastern History and holder of an endowed Chair in Modern Armenian History at the University of California, Los Angeles. Among his numerous publications are *Armenia on the Road to Independence* and the four-volume *The Republic of Armenia*. He has edited and contributed to *The Armenian Genocide in Perspective; The Armenian Genocide: History, Politics, Ethics; Remembrance and Denial: The Case of the Armenian Genocide;* and *Looking Backward, Moving Forward: Confronting the Armenian Genocide*.

Henry Huttenbach is professor of East European history at The City College of the City University of New York. He is also the founder and editor of *The Genocide Forum* and the *Journal of Genocide Research*. Among his many publications in the field of genocide studies are: "The Philosophical and Practical Limitations of Genocide Prediction and Prevention" (Yale University Genocide Studies Program, No. 14); "Locating the Holocaust on the Genocide Spectrum: Toward a Methodology of Categorization" in *Holocaust and Genocide Studies: An International Journal*; and "Anticipating Genocide" in *Will Genocide Ever End?*

Rounaq Jahan is currently affiliated with the Southern Asian Institute at Columbia University as a Senior Research Scholar. She was a Professor of Political Science at Dhaka University, Bangladesh, from 1970-1993. Jahan received her PhD in Political Science from Harvard University in 1970 and did postdoctoral research at Columbia, Chicago, Harvard, Boston, and Bergen. Her publications include *Pakistan: Failure in National Integration* and *Bangladesh Politics: Problems and Issues*. Jahan also worked for the United Nations for many years. She was the coordinator of the Women's Program at the United Nations Asia-Pacific Development Centre in Kuala Lumpur, Malaysia (1982-1984) and Head of the Rural Women's Employment Program at the International Labour Organization in Geneva, Switzerland (1985-1989).

Michiel Leezenberg teaches in the Department of Philosophy of the Faculty of Humanities at the University of Amsterdam. He has conducted extensive research on the Anfal operations and subsequent developments in Iraqi Kurdistan, and made several field trips to the region. Among his main publications relevant to the Anfal operations are: "Between Assimilation and Deportation: History of the Shabak and the Kakais in Northern Iraq" in *Syncretistic Religious Communities in the Near East*; "De Anfal-operaties in Iraaks Koerdistan: tien jaar straffeloosheid" (The Anfal Preparations in Iraqi Kurdistan: Ten years of Impunity") in *Soera*; "Politischer Islam bei den Kurden" in *Kurdische Studien*; and *Genocide and Responsibility: International Reactions to Iraq's Chemical Attack Against Halabja*.

René Lemarchand is Professor Emeritus of Political Science at the University of Florida (Gainesville). He has written extensively on Rwanda, Burundi and the Congo. His book, *Rwanda and Burundi* received the Melville Herskovits Award of the African Studies Association in 1970. He served as Regional Advisor on Governance and Democracy with USAID from 1993 to 1998, first in Abidjan and then in Accra. He has been Visiting Professor at Smith College, Brown University, the University of Copenhagen, the University of Bordeaux, and the University of California at Berkeley.

James Mace is Professor of Political Science, Kiev-Mohyla Academy National University, Kyïv, Ukraine. Previously he was a research associate at the Harvard Ukrainian Research Institute (1981-1986), Executive Director of the United States Commission on the Ukraine Famine (1986-

1990), Senior Research Fellow at the Harriman Institute of Columbia University (1990-1991), Research Fellow in the Ukrainian Research Program of the University of Illinois at Urbana-Champaign (1992-1993), and a Supervising Research Fellow in the Institute of Ethnic and Political Studies of the National Academy of Sciences of Ukraine in Kyïv (throughout the 1990s).

Eric Markusen is a professor of social work and sociology at Southwest State University in Marshall, Minnesota. He is also the Director of Research with the Danish Institute for International Studies, Department for Holocaust and Genocide Studies in Copenhagen. He is the coauthor of *The Genocidal Mentality* and *The Holocaust and Strategic Bombing: Genocide and Total War in the Twentieth Century*. He also served as an associate editor of the *Encyclopedia of Genocide*.

Martin Mennecke, who has a LLM from the University of Edinburgh, is a doctoral candidate in international law at the University of Kiel, Germany. His thesis concerns the most recent evolution of the legal definition of genocide and is supported by the Friedrich-Ebert-Stiftung and Herbert-Quandt-Stiftung. Currently affiliated with the Danish Institute for International Studies, Department for Holocaust and Genocide Studies, he is mainly working on issues of international criminal law. Among his publications are "The International Criminal Tribunal for the Former Yugoslavia and the Crime of Genocide" (with Eric Markusen) in *Genocide: Cases, Comparisons and Contemporary Debates*.

Carol Rittner, R.S.M., a member of the Religious Sisters of Mercy, is Distinguished Professor of Holocaust and Genocide Studies at The Richard Stockton College of New Jersey (United States), where she has taught since 1994. She is the author, editor, or coeditor of numerous publications, including *The Courage to Care, Different Voices: Women and the Holocaust, The Holocaust and the Christian World, Beyond Hate*, and *Will Genocide Ever End?*

Samuel Totten is a member of the Council of the Institute on the Holocaust and Genocide (Jerusalem, Israel). From 1991 to 1993 he was an educational consultant to the United States Holocaust Memorial Council and from 1992 to 1994 he was a consultant to the United States Holocaust Memorial Museum (USHMM). As part of his duties at the USHMM, he developed the Museum's first series of talks on genocide theory and genocide in the twentieth century. Among the books he has edited and coedited on genocide are: *First-Person Accounts of the Genocide Committed in the Twentieth Century: An Annotated Bibliography*; *Genocide in the Twentieth*

Century: Critical Essays and Eyewitness Accounts; *Teaching and Studying the Holocaust*; *Pioneers of Genocide Studies*; *The Intervention and Prevention of Genocide: An Annotated Bibliography*; *The United Nations and Genocide: An Annotated Bibliography*; and *Century of Genocide: Critical Essays and Eyewitness Accounts*. He served as one of the associate editors of the *Encyclopedia of Genocide*, and is currently the book review editor for *The Journal of Genocide Research*.

INDEX

287